STANDING ROOM ONLY

Previous Publications

Standing Room Only: Strategies for Marketing the Performing Arts, coauthored with Philip Kotler, 1997.

Arts Marketing Insights: The Dynamics of Building and Retaining Performing Arts Audiences, 2007.

Standing Room Only

Marketing Insights for Engaging Performing Arts Audiences

Second Edition

Joanne Scheff Bernstein

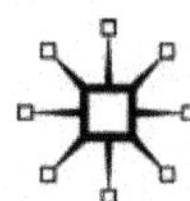

STANDING ROOM ONLY

First published in 1997 by Harvard Business Review Press, Boston, MA.

First published in 2014 by
PALGRAVE MACMILLAN®
in the United States—a division of St. Martin's Press LLC,
175 Fifth Avenue, New York, NY 10010.

Where this book is distributed in the UK, Europe and the rest of the world, this is by Palgrave Macmillan, a division of Macmillan Publishers Limited, registered in England, company number 785998, of Houndmills, Basingstoke, Hampshire RG21 6XS.

Palgrave Macmillan is the global academic imprint of the above companies and has companies and representatives throughout the world.

ISBN: 978–1–137–28293–4

Library of Congress Cataloging-in-Publication Data

Bernstein, Joanne Scheff, 1945–
Standing room only : marketing insights for engaging performing arts audiences / Joanne Scheff Bernstein. — Second edition.
pages cm
ISBN 978–1–137–28293–4 (hardback)
1. Performing arts—Marketing. I. Kotler, Philip. Standing room only. II. Title.

PN1590.M27B47 2007
791.06'98—dc23 2013041296

A catalogue record of the book is available from the British Library.

Design by Newgen Knowledge Works (P) Ltd., Chennai, India.

First edition: May 2014

10 9 8 7 6 5 4 3 2 1

For Jacob
May the arts offer you a lifetime of delights and inspiration

This book is also dedicated to the many artists of generations past and present whose work inspires, challenges, and sustains me

CONTENTS

Tables and Exhibits

Tables

Exhibits

Foreword

When Joanne and I published the first edition of SRO in 1997, we recognized an intense need among performing arts managers and marketers to increase their understanding and practice of the principles, strategies, and tactics that would help them to improve their effectiveness and efficiency in developing and retaining audiences. Since that time, there have been significant changes in the lifestyles, behavior, and preferences of current and potential audiences that have made it increasingly difficult to fill seats for performances.

- People want to select exactly which programs to attend and, as a result, subscriptions have declined dramatically.
- People are more spontaneous and more commonly make their purchase decisions within a short time before the performance, making it harder for managers to plan.
- Pricing strategies have become complex. With the advent of dynamic pricing in the arts world, and trends to deeply discount ticket prices, a strong emphasis on price has obfuscated the true value of attending performing arts events. It is crucial that arts marketers fully understand pricing principles and options and how best to employ them.
- Digital marketing has become central to the marketer's communication efforts. Arts marketers must know how best to leverage the power of the Internet, email, social media, and big data.
- People have come to expect personalized offers and great customer service. Performing arts organizations need to train their staff in a thoroughgoing customer service philosophy.

In the light of these developments, many of the audience development strategies that worked in the past are no longer as viable as they once were. A mark of sound management of any organization is to review the new market conditions, the competition—both inside and outside of the arts world—and the mind-set of consumers, revise old assumptions, and develop new marketing initiatives for breathing fresh life into the performing arts business.

We know that a significant proportion of the population truly enjoys attending arts events. They look forward to their visits to symphonies, theaters, operas, and dance companies. They need reliable information describing the art and performers in order to select carefully among the many offerings. Today's art goers are price and time conscious and everything must be done to give them a picture of what to expect with any performance.

Performing art organizations face tight budgets. They need to set their prices carefully in relation to the value of the experience. They need to improve their fund-raising by satisfying their audiences and other potential donors. Marketing is the art of creating genuine customer value. The marketer's watchwords are quality, service, and value.

Joanne Scheff Bernstein has not only taught arts management and marketing at Northwestern University, but has also consulted with many arts organizations throughout the United States and in many other countries. She also ran an orchestra for two years and improved its income and capital, and importantly, greatly increased its audience size. She is one of the most effective consultants and educators on the performing arts. She is the best person I know to guide you to the new market factors and the new strategies, based on sound marketing principles. Her advice in the book will improve your audience development results now and into the future. She will help you to understand the changing customer mind-set and to leverage the many new opportunities created by the digital revolution and its manifestations in web-marketing, email, and the use of social media. I encourage every arts manager and marketer, student and teacher of arts management, board members of arts organizations, and other stakeholders in the performing arts to read this book and discuss every chapter with their peers. New ideas are bound to emerge to help your arts organization build your brand and deeply satisfy your target customers.

Philip Kotler
S. C. Johnson & Son Distinguished Professor of International Marketing
Kellogg School of Management
Northwestern University

Introduction

In March 2011, one of the worst disasters in history occurred in Tōhoku, Japan. The most powerful earthquake to have ever hit Japan (one of the five most powerful in the world) triggered huge tsunami waves reaching heights of 133 feet, which traveled up to six miles inland and caused atomic reactor meltdowns and explosions. Thousands of people died, hundreds of thousands of buildings collapsed or were severely damaged, and all areas within 50 miles of the Fukushima Daiichi Nuclear Power Plant were evacuated. The entire region suffered from lack of electrical power, closed roads and rail lines, poor access to food and medical supplies, and the massive fear that accompanied the aftershocks.

When the earthquake hit, the members of the Sendai Philharmonic had been rehearsing, but had to evacuate their badly damaged hall. Fortunately, the orchestra members and their instruments were unharmed. The musicians quickly established the Center for Recovery through the Power of Music with the slogan "Connecting hearts, connecting lives." They presented what they called "Reviving Concerts" at various locations, including street corners and evacuation shelters. In some of the shelters, they got people to participate, not just listen. A mere 15 days after the earthquake, the Sendai Philharmonic Orchestra performed at the Kenzuiji Temple, near Sendai Station, playing Barber's *Adagio for Strings*, which was also performed at the funeral of President John Kennedy. The concert was as much about hope as about mourning the dead, and it ended with the audience joining in the song *Furusato* (Homeland). In addition to the live performances, the orchestra recorded the school song of each school that had been closed, for the memorial value to the students and their families.

The Sendai Philharmonic's mission in these efforts, according to its website, is "to aid in the reconstruction after the disaster and to enhance the resilience of the devastated area, even though it may take a long time. We sincerely believe that music has the power of healing and recovery." Said one orchestra musician,

> For the average Japanese where I live, the point was to reassure local children who were so traumatized they were afraid to walk to school in case walls collapsed on them. For teachers it was to carry on with children's

> education while helping children stay calm and come to terms with the situation. People's responsibilities were simply to keep the country running, to keep themselves stable so that as much help as possible could be offered in the right way.[1]

This music making had a therapeutic effect on the players as well as their audiences.

The late conductor James DePreist saw music as presenting "a mirror of our better selves—as something to which we can relate on a very personal and emotional level. We can admire it architecturally. We can admire it in terms of its harmonic structure. But fundamentally there is a visceral reaction, and that visceral reaction is sheer magic."[2]

In Littleton, Colorado, in 1999, two teenagers devastated a community and shocked the country when they walked into Columbine, their suburban high school, and killed 12 students and one teacher. In 2013, the American Theater Company in Chicago presented *columbinus*, an overwhelmingly powerful play, written by Artistic Director PJ Paparelli and Stephen Karam, that confronts this horrendous event. Based on interviews with the survivors and community members in Littleton, and with teenagers across the country, this poignant drama weaves together the testimonies with excerpts from diaries and home video footage. The shaken audiences grapple with the ultimate questions: Why? How could this happen? How well do we really know our children?

Theater possesses the unique conditions for laying out the complexity of issues, giving people the distance to talk about difficult topics, and for opening up discussions that need to happen. One Columbine student, who hid under a table in the school library on that fateful day, saw and heard and felt his friend being killed just inches away from him. For 13 years after the tragedy, he was unable to discuss his terrifying and tragic experience with anyone. Then he traveled to Chicago to see *columbinus*, which opened a floodgate of emotions for him. The play served to give him the catharsis he needed to begin to heal and move forward in his life. It allowed him to begin to talk about his experiences; to express his feelings; to pick up his guitar, which had remained untouched in a corner of his room all those years; and to relate better with his family and friends.

Performances by the Sendai Philharmonic in the earthquake disaster area and the play *columbinus* are vivid examples of how arts attendance can be a truly transformational experience.

Every evening, worldwide, there are thousands of performances of symphonic and chamber music, operas, dances, plays, and musicals that people attend for their emotional resonance, aesthetic appeal, intellectual stimulation, educational value, for social bonding, for an escape from reality, and/or purely for personal pleasure. In the arts, intrinsic value varies from production to production and from patron to patron.[3] Furthermore, people's interests and needs transform over the course of their lives. To further complicate matters, in our dynamic society, the behavior of current and prospective arts patrons is changing dramatically.

Thus, the central importance of marketing. The role of marketing is to bring an audience to the art and the art to the audience in ways that are relevant, meaningful, and compelling.

In recent years, since the publication of the first edition of this book, the responsibilities and tasks of arts marketers have become more complex. Ever-increasing costs and fluctuating contributed income have placed a great deal of stress on arts managers of organizations large and small. Since the beginning of this century, performing arts audiences have been shrinking. Subscriptions have been declining steadily and significantly, as many people want to be spontaneous in their ticket purchases and choose exactly which shows to attend. This situation puts increased pressure on marketers to attract more people on an individual ticket basis and determine new ways to develop engagement and loyalty. The rapid advancements in high technology mean that arts organizations must keep up with new trends and consumer expectations, while maintaining and improving effective traditional marketing methods and leveraging limited financial and human resources. Marketers are increasingly challenged to attract the widest possible audiences and concurrently maximize their ticket revenue.

Some arts managers have been at the forefront of designing and implementing marketing programs that respond to evolving customer needs and preferences. Many other marketers are eager to change their practices but do not know where to turn for direction on viable approaches.

Arts organizations everywhere must learn new ways to attract the resources they need to sustain their mission and quality. Managers and marketers must improve their skills for increasing and broadening their audience base, improving accessibility to various art forms, and learning how to better meet the needs and interests of specific audience segments.

This book addresses a wide range of questions and concerns common to performing arts managers and marketers, such as:

- How can an arts organization attract and engage new audience members?
- How can an arts organization increase the frequency of attendance of its current audiences and develop loyalty among its patrons?
- How can an arts organization develop a better understanding of its customers: their interests, attitudes, and motivations?
- How can an arts organization create offerings, services, and messages to which its target audiences will respond enthusiastically?
- How can arts organizations make their performances and services an integral part of people's lives, woven into the fabric of the community?

- In today's complex and rapidly changing environment, how can arts organizations develop long-term strategic plans?
- How can an organization determine how well it is satisfying the interests of various constituents?
- How can an arts organization learn to be fiscally responsible, to live within its means?
- What role should the marketing function have in an arts organization?

Standing Room Only will significantly help performing arts organizations focus on strategies and techniques to improve their impact and practices while ensuring that they remain true to their artistic and public missions. This book combines theory, strategy, tactics, and innovative examples, all with the objective of improving the ability of arts managers to better meet the needs of audience segments and thereby increase audience size and engagement. This book explains not only the *what*, but also the *why*—why some approaches that have been ingrained in the performing arts industry for decades no longer resonate with many current and potential audiences; why new ways of thinking and new strategies are essential for success.

Some readers may expect that on these pages they will discover "quick fix" solutions. Many of the ideas suggested in this book can be and have been implemented with rewarding results. However, specific tactics must be developed for each organization in the context of its own goals and circumstances. This means that better planning is needed as well as better, more thorough and continuous implementation, evaluation, and modification of the plans. Therefore, this book consists of both new approaches to audience engagement and many of the commonly accepted best practices in marketing theory that support them.

From cultivating an organization-wide marketing mind-set, developing a strategic marketing plan, building a brand identity, conducting market research, and understanding your target market to communicating effectively with current and potential customers; designing attractive offerings for various market segments; leveraging the Internet, email marketing, and social media; and delivering great customer service, this book covers everything you need to know to put marketing programs in place, manage them, and adapt them for the future.

Standing Room Only offers dozens of examples of innovative and effective marketing strategies from performing arts organizations all over the world—strategies to help the performing arts develop a more diverse audience base and prosper in the midst of an evolving economic and technological landscape. The examples derive from organizations large and small, from the United States, from Italy, Japan, and other countries. I am most fortunate to live in the Chicago area, where there is an abundance of fine performing arts organizations, many of which have provided me with not only endless hours of enjoyment and enrichment, but also rich information to demonstrate many of the principles and strategies I present in this book.

Who Should Read This Book

Standing Room Only is an indispensable tool for arts managers, marketers, fundraisers, board members, arts management educators, and students. It is also valuable for others who work closely with arts organizations and desire a deeper understanding of their issues such as foundation directors, corporate sponsors, consultants, managers of arts service organizations, and the artists themselves.

Marketing is a mind-set for the entire organization, not simply a function of the marketing department. Therefore, it is critical for upper level management and board members to understand the principles of customer-centered marketing. This book serves as a comprehensive text for those relatively new to the subject and as an inspiration and challenge to those more experienced in the sector to become more relevant to their current and potential audiences.

It is useful for people in organizations of all sizes. Large institutions often continue to employ the strategies that made them successful in the past, even as they experience new challenges. At the other end of the spectrum, the smallest companies with negligible marketing budgets should consider that their offer must be strategically priced, packaged, and described to best appeal to target audiences. All these efforts require up-to-date strategic marketing practices.

The Organization of This Book

Chapter 1 sets the stage. It presents a brief history of the performing arts industry; an overview of the current state of theaters, operas, dance companies, and symphony orchestras; and describes the issues that face modern arts marketers and managers. It also suggests the trends that affect people's arts attendance and ticket purchasing behavior. Chapter 2 addresses the evolution of business orientations to the marketing function over the decades and the core principles of modern marketing.

Marketing success is a function of how well the organization understands its current and potential patrons in order to be relevant to them. Chapter 3 presents theories about how consumers think, how people make decisions, what benefits people seek in their performing arts attendance experience, and what barriers keep them away. Effective marketing communications appeal to each target customer's core values, lifestyles, and interests. Chapter 4 addresses age and life cycle status, gender, and ethnicity as well as personal, cultural, social, and psychological factors and the ways these characteristics affect various audience segments' attitudes, behaviors, expectations, and preferences.

The strategic marketing process is the subject of chapter 5. It offers the foundation for developing the organization's mission, objectives, goals, and plans; approaches for implementing, analyzing, and controlling the strategic plans; and principles for budgeting. The basic methodology for market planning is segmenting the total available market, targeting key segments with specific offers, and positioning the offers to these targeted segments. Chapter 6 describes segmentation variables and characteristics of various segments that are useful for marketers, and principles and strategies for targeting and positioning. If the strategic marketing process is the foundation of marketing planning, then market research can be considered its structural support. Research plays a critical role in understanding customer attitudes and behavior and in planning marketing strategy. Chapter 7 offers an overview of some common research approaches and methodologies for their application.

The key elements of marketing planning are known as the four Ps of marketing: product, place, price, and promotion. Chapters 8, 9, and 10 discuss, respectively, defining and designing the offer, choosing performance venues and ticket outlet options, and developing pricing strategies—a topic that has become ever more complex in recent years. These elements are discussed in light of the focus that marketers should give to customer value rather than on factors internal to the organization. Branding has become a marketing buzzword in recent years, but what is a brand? Why is it important to have a strong brand identity? How can an organization build a brand image that resonates with its publics? Chapter 11 offers responses to these questions.

Marketing is fundamentally a matter of influencing behavior, which is largely based in communication. Chapter 12 addresses the principles of formulating communication strategies, while chapter 13 delves into various approaches for communicating: advertising, personal selling, sales promotion, public relations, and crisis management. The Internet, email, and social media have irrevocably changed the daily lives of consumers and have increased the responsibilities and challenges marketers face. Furthermore, digital marketing evolves constantly, so marketers need to know both the basic principles of high technology marketing and how to adapt them in our dynamic environment. Chapter 14 discusses how to harness and leverage the power of digital marketing.

Chapter 15 deals with a major issue facing arts marketers in the twenty-first century: building commitment and loyalty. From the perspective of many organizations, the subscriber is the ideal audience member, but, since the mid-1990s, increasingly more audience segments find subscribing unattractive. Not only are arts marketers less successful in attracting new subscribers, but each year, fewer current subscribers are likely to renew. This chapter presents the pros and cons of subscriptions from both the organization's and the customer's perspectives, suggests a new mind-set on the part of the arts marketer as to the meaning of a valuable customer, and recommends ways to build the subscriber base and alternatives to subscriptions for audience development. Many people *like* being single ticket buyers. This changing preference is not short term; rather it is part of a larger societal trend that will affect arts organizations into the foreseeable future. This chapter also focuses on how to attract single ticket buyers and build their frequency of attendance, while showing that infrequent buyers can be reached more efficiently and effectively than ever before.

Another major societal trend is heightened expectations for excellent customer service. It is crucial that arts marketers listen to their customers, learn their needs and preferences, and provide attentive, high-quality customer service. This is the topic of chapter 16. Chapter 17 offers some final insights: insights for understanding and engaging audiences, for creating relevance and accessibility; insights to help arts organizations professionalize their management and marketing and to approach their tasks strategically in light of a continually changing environment.

The purpose of this book is not only to offer insights on new theories and processes that improve the effectiveness and efficiency of the marketing function, but also to help arts managers and marketers develop their own insights in the face of a changing environment and changing customer values, so that arts organizations will survive and prosper for now and into the future.

ACKNOWLEDGMENTS

Piccolo Teatro di Milano is an outstanding role model in the performing arts industry in terms of its superb artistry, innovative audience development practices, and excellent general management. I am especially grateful to Sergio Escobar, general manager; Lanfranco Li Cauli, head of marketing; and Giovanni Soresi, strategic marketing consultant, for graciously and generously sharing the rich information about Piccolo Teatro included in this book, for hosting me, and for being tirelessly responsive to my questions.

I would also like to thank the following arts professionals who have contributed their personal time to share their insights and experiences with me, which have significantly helped to enrich the material content and focus of this book: Eugene Carr, Scott Harrison, Miho Ito, Chris Jones, Lori Kleinerman, Gerri Morris, Andrew McIntyre, Tom O'Connor, PJ Paparelli, and Jim Royce.

The wisdom of Philip Kotler is always an essential guide and inspiration to me in my work. My editor, Laurie Harting, has offered perceptive suggestions. Bob Bernstein, my husband, is a superb sounding board for ideas and perpetually encourages and supports me in my work.

CHAPTER 1

THE PERFORMING ARTS: HISTORY AND ISSUES— AN ONGOING CRISIS? A GROWING CRISIS?

THE NONPROFIT PERFORMING ARTS INDUSTRY IN AMERICA AND MANY PERFORMING arts organizations around the world are facing crises on a variety of fronts. In April 2011 the Philadelphia Orchestra, one of the largest and most well-respected orchestras in the United States, filed for bankruptcy, facing a structural deficit of $14.5 million and other economic woes. The orchestra has since exited from bankruptcy protection, but continues to face financial challenges. In 2010, the Detroit Symphony found it necessary to make drastic cuts in musicians' salaries to help rectify their shortfall of $8.8 million. Following a six-month strike, the musicians settled for a 23 percent salary cut. These situations are not unique.

In a study of the participation rates of the arts, the National Endowment for the Arts (NEA) found that during the period 1982–2008, classical music audiences declined by 29 percent and opera audiences by 30 percent. From 1998 to 2008, attendance at nonnonmusical plays declined a massive 33 percent. In a study for the five-year period 2008–2012, the NEA observed that attendance declined 9 percent at musicals and declined 12 percent at nonmusical plays. During those five years, classical music attendance declined 5.4 percent.[1] These dramatic signs of decline are not exclusive to the United States. Heloisa Fischer, founder and director of Viva Musica in Rio de Janeiro, says, "In Brazil, classical music plays only a small part in the life of most people." Asks Fischer, "Is someone working in this segment rowing against the tide? Is classical music an art form which is able to be consumed in contemporary societies?"[2]

The Theatre Communications Group (TCG), in its annual publication *Theatre Facts*, reports that from 2003 to 2012, there was a 5.2 percent decline in ticket sales at US theaters. Most significantly, between 2003 and 2012, subscriptions, which are considered to be the lifeblood of performing arts organizations, declined 23 percent in theaters. Since 2000, single ticket income has exceeded subscription income for the 108 theaters surveyed annually by TCG, but marketing single ticket sales is far more expensive and labor intensive than marketing subscriptions. Furthermore, subscribers are the primary donors to arts organizations, and as the number of subscribers declines, so does the potential source for all-important contributions.

In a 2011 survey of 350 US theaters, TCG learned that the top five concerns and priorities of theater managers were: audience development (66 percent), cultivation of donors for individual gifts (64 percent), board development (45 percent), corporate giving (33 percent), and strategic planning (28 percent).[3] Clearly, audience development and cultivation of individual donors go hand in hand.

According to the study on arts participation conducted by the NEA in 2008, educated audiences are the most likely to attend or participate in the arts, and college-educated audiences (including those with advanced degrees and certifications) have curbed their attendance in nearly all art forms. Ballet attendance has declined at the sharpest rate—down 43 percent since 1982. Less-educated adults have significantly reduced their already low levels of attendance.

Many important organizations have been forced to eliminate programs; others have closed altogether. Lifestyles and interests are changing among younger generations of potential attenders. The arts face growing competition from other, more accessible and less-expensive forms of entertainment as leisure time options proliferate and the costs of attending performing arts programs increase dramatically. In the United States, cutbacks in arts education in the schools are affecting younger generations of potential audiences. The arts have also been hard hit by declining contributed support. Cuts in government funding have become severe. Corporate arts sponsorships have declined since September 11, 2001, and rose only slightly in the middle of the first decade of this century between economic crises. Furthermore, many private foundations have been earmarking grants for specific programs, leaving less money for general operating budgets. Some other foundations have discontinued their contributions to the arts to support what they consider to be more critical services.

Arts organizations in many other countries worldwide have traditionally been protected from the vagaries of the marketplace and dependence on private donors because their governments provide extensive financial support for the arts. However, government funding cutbacks over the past decade, especially in Europe, have been devastating. Government funding is being slashed in countries like Italy, the Netherlands, Great Britain, and Greece. In 2012, Portugal completely abolished its Ministry of Culture, and the world-famous La Scala Opera House in Milan, Italy, faced a $9 million shortfall because of reductions in subsidies from the Italian government. The following spring, La Scala announced a dramatic reduction in the number of productions it would present the next season. Despite the projected shortfall, La Scala and similar large institutions are in a better position to fend for themselves than smaller companies, especially those engaged in experimental and avant-garde efforts. European arts organizations are newly in the position of having to justify what they do artistically and economically and compete for limited funds.[4]

In the UK, during the 2012 -2013 season, attendance at orchestra concerts was 16 percent higher that three years earlier but earned income fell 11 percent due to lower ticket prices. During the same time period, public funding declined 14 percent.[5] In England, arts organizations are reaching out to individuals for contributions for the first time in their history and are increasing their emphasis on corporate sponsorships, which currently account for about 10 percent of British arts organizations' revenue. Arts organizations in eastern European and former Soviet countries are seeking new sources for contributed support because, under the new regimes, their governments no longer fully sustain them financially.

Arts organizations everywhere face a constant upward spiral of operating costs. Their debts are rising and they are finding it increasingly difficult to locate the resources they need to sustain their mission and quality.

History: The Arts Expand over the Decades

In his famous ordering of national priorities, statesman John Adams said that he had to study politics and war so that his sons could study mathematics and philosophy, in order to give their children a right to study painting, poetry, music, and architecture. Although he highly valued the fine arts, Adams feared that a strong focus on the arts might be linked with an excess of luxury, a trend he had observed in France during his diplomatic service at the court of Versailles. Therefore, Adams reasoned, too great an interest in the arts might corrupt his fellow citizens, who were trying to forge a new nation in a vast wilderness and were preoccupied with more "practical" things. It was nearly another two hundred years before the US government committed itself to a program of sustained, direct financial support for the arts.[6]

Before the Civil War, formal systems for presenting high culture did not exist in the United States. Serious art and popular works intermingled in public venues, where sponsoring organizations might present a relatively austere program of classical music one week and a popular extravaganza (such as "Mr. Mutie, his African monkey, and several Chinese Dogs") the next. After the Civil War, nonprofit enterprise in the arts as we know it today began to develop as the emerging urban upper classes attempted to define and legitimate a body of art that they could call their own and that would serve as a source of honor and prestige among their peers. Says sociologist Paul DiMaggio, "The institutionalization of high culture... can best be understood as an antimarket social movement, aimed at defining a corpus of sacred art beyond the reach of profane commercial concerns."[7] Cultural entrepreneurs of the late nineteenth century utilized the nonprofit form to bring performers under their direct employment, protecting art from government intervention and from the whims of the masses. Says DiMaggio,

> Only the power to hire and fire performers, to demand their exclusive services, and to place them under the authority of a conductor hired by the nonprofit entrepreneur or trustees enabled the orchestras of the twentieth century to develop a musical canon and modern performance standards. In purely market terms, the nonprofit form was less efficient than the conventional combination of proprietary band and proprietary sponsor; but it attained ends that were unrealizable through market exchange alone.[8]

The nation's first permanent, independent, and disciplined orchestra, the Boston Symphony Orchestra, was founded in 1881 by philanthropist Henry Lee Higginson. Higginson once asserted, "I alone am responsible for the concerts of the Symphony Orchestra." He defined his role and purpose as "to pay the bills, to be satisfied with nothing short of perfection, and always to remember that we were seeking high art and not money: art came first, then the good of the public, and the money must be an after consideration." It was Higginson who gave the money to build the symphony's hall, to manage operations, and to single-handedly cover a deficit that by 1914 had accumulated to $900,000. It was also Higginson who chose the conductors, who had ultimate authority to hire and fire musicians, and who insisted that the musicians devote themselves exclusively to the orchestra and avoid such vulgar activities as performing for dances. He also persisted in segregating the musical fare so that "light" music was increasingly relegated to summer performances by the Boston Pops Orchestra while the Symphony Orchestra "purified" its programs, presenting music that Higginson and his conductors considered worthy, whether or not it was popular with the audiences.[9]

By the 1930s, courses in art and music were being taught in all major universities, and radio brought music to small towns and rural areas; so by the post–World War II period, art became more popularized. Most orchestras, however, continued to be administered by their founders and supported by wealthy individuals.

In the 30 years from the mid-1950s to the mid-1980s, the arts sustained rapid and persistent growth. A promotion boom made arts attendance more accessible and compelling to greater numbers of people, and expanding attendance levels stimulated the growth of new and larger performing arts organizations. From 1965 to 2010, the number of professional orchestras in the United States increased from 58 to more than 1,200 (including community orchestras); opera companies from 27 to more than 120; dance companies from 37 to 250; and professional resident theater companies from 12 to more than 1,500. In addition, there are literally thousands of nonprofit presenting organizations that engage artists and groups to perform in their halls: youth symphonies, university theaters, dance groups, and more.

In Europe, symphonic music and symphony orchestras emerged during the seventeenth century. Courts and churches financed the earliest orchestras from the oppressive taxes they inflicted on the lower classes. By the early eighteenth century, musical activity shifted from courts and churches, which had become unable to cover the increasing costs of musicians and composers, to cities and towns, where the tastes of the public at large became more important and middle-class patrons became the primary source of financial support. Since World War II, European arts organizations have relied on government subsidies to cover the great majority of their operating budgets, while US arts organizations have relied primarily on private donations to fund the 50 percent or more of the budget that is not covered by ticket sales.

Economist Robert Flanagan, in his analysis of orchestral revenue systems of many countries, has found that in Australia, Continental Europe, and Scandinavian countries, subsidies from national, regional, and local government agencies make up about 60–90 percent of total revenues for most orchestras. In Canada and the United Kingdom, government sources provide 30–40 percent of orchestras' revenues while in the United States, direct public subsidies comprise only about 5 percent of revenues at orchestras.[10] (The US government, by offering tax deductions for donations to nonprofits, gives significant indirect subsidies for gifts from private individuals and corporations. These gifts are market-driven in that the donor chooses what to support.) Private philanthropy and investment income, so substantial in the United States and so essential for US arts organizations, barely exist in other countries. Although US arts managers have long been envious of the substantial direct support their foreign colleagues receive from their governments, the US system has an inherent flexibility that the others do not. When one major private donor or foundation ceases making contributions, the organization has many other funding resources to explore and court. However, in most other countries, when the governments slash their cultural budgets, arts managers have nowhere else to turn.

ENTER THE CRISES

In recent years the performing arts industry worldwide has had to confront the realities of stagnant or declining audience sizes, aging audiences, declining subscriptions, spiraling expenses, dwindling work for musicians, deficits of crisis proportions, and reductions in donations from some formerly dependable sources.

EXPENSES AND REVENUES

The systems put in place by US arts organizations during their period of seemingly unstoppable growth in the 1960s through the 1980s are a root cause of many of the problems the industry has faced in recent years and with which it will grapple for years to come. During those years, many arts organizations displayed their newfound affluence in more elaborate productions, larger management staffs, and new performance facilities with more seats to fill. Midsize orchestras gave their musicians full-year contracts instead of fee-for-service agreements, thereby providing them with a welcome measure of financial security formerly enjoyed only by musicians employed by large symphony orchestras in major cities. These changes reflected optimism about continued growth in both audiences and contributions. Yet, in recent years, growth has ground to a halt.

The US orchestra industry's annual deficit grew from $2.8 million in 1971 to $23.2 million in 1991. Expenses of symphony orchestras had increased dramatically from $87.5 million in 1971 to $698.9 million in 1991 (an increase of 137.5 percent after adjusting for inflation). Most important, both the amount and the pace of increase were greater for expenses than for revenues.[11] In 2011, more than two-thirds of orchestras operated at an average annual deficit of $700,000.[12]

Across the United States, in 2012, between 42 percent and 76 percent of theaters—depending on budget size—were operating with deficits, meaning that the average theater was borrowing funds to meet day-to-day cash needs and current obligations. The average working capital at all theaters was a negative $1.3 million. From 2003 to 2012, overall expense growth exceeded inflation by 14 percent and income growth by about 2 percent, despite the fact that nearly one-third of the theaters have significantly reduced their operating budgets. In 2012, earned income supported 3.2 percent less of total expenses than it did in 2008.[13]

Disappearing Work for Musicians and Other Artists

In 2006, the two hundred and fiftieth anniversary year of Mozart's birth, small orchestras everywhere took advantage of the fact that the great master's symphonies require only about 35 musicians, approximately half the number for which most symphonies are scored. Large orchestras contract their musicians for a full season of work, meaning that their musicians are paid whether or not they perform. But small and some mid-sized orchestras typically hire their musicians on a fee-for-service basis, which means that they can dramatically reduce their artistic fees when celebrating Mozart's special birthday and programming other works less grand in scale than the likes of Beethoven and Mahler.

Arts presenters and many Broadway producers contend they have no choice but to cut musicians amid shrinking budgets. John Tomlinson, executive director of Paul Taylor Dance Company, said it would have cost the company an extra $450,000–500,000 to use live musicians for its three-week season at Lincoln Center and he wasn't able to raise that much money in a difficult economic climate. With smaller orchestras scaling back and musicals and dance productions using fewer players or none at all, professional musicians are facing an increasingly tough time. They are being forced to piece together bits of freelance work, take on heavy teaching schedules, or leave the business altogether. Over the first decade of the current century, the memberships of the Associated Musicians of Greater New York Local 802 shrank from about 15,000 to 8,500.[14] Musicians are not the only artists who find it increasingly difficult to find steady work. Some dance companies are reducing the number of dancers in their companies and theaters are ever more frequently programming plays that require smaller casts.

The New York City Opera, after a year of acrimonious negotiations with the unions that represent their singers, choristers, stage manager, assistant directors, and orchestra members, reached agreements that allowed them to move ahead with a sharply reduced budget. Facing persistent deficits and a shrinking endowment, in 2011 the company left its home at Lincoln Center—which it could no longer afford—and slashed its budget from $31 million to $13 million. The new agreement ended the salaries that made sense in the previous decade when the opera featured as many as 100 performances per year. In hindsight, managers found that this former level of activity was overly optimistic and financially unsound. Under the new budget, City Opera initiated a system of pay per rehearsal and performance, saying that it could not afford more given that the reduced season would consist of 16 opera performances. The unions resisted, saying that their compensation would be cut by about 90 percent, but it became clear to all parties that the old systems could no longer work. Although this was a drastic move for the company to undertake, "these contracts ensure our financial solvency," said George Steel, the company's general manager and artistic director.[15] Yet, sadly, in October 2013, after 70 years of performances, the organization closed its doors.

Contributed Income Challenges

"In order to maintain its ideal form, theater needs to be subsidized," says Robert Brustein, founding former director of American Repertory Theater (A.R.T.) and the Yale Repertory Theatre. US government subsidies began in the 1930s, when a small portion of the New Deal Works Progress Administration budget was earmarked to support the Federal Theatre Project (FTP). These funds were instrumental in helping the American stage culture flourish until 1939, when Congress canceled funding in reaction to the left-wing character of many FTP productions. Arthur Miller, Elia Kazan, and John Houseman were among those who launched their careers under FTP.

In the 1950s, cultural visionary W. McNeil Lowry, director of the Ford Foundation's Office of Humanities and the Arts, began infusing millions of dollars of capital financing into the infrastructure of arts organizations—donating $80 million to orchestras alone. Lincoln Kirsten, a founder of the New York

City Ballet, called Lowry "the single most influential patron of the performing arts that the American democratic system has produced."[16] Other foundations quickly followed the Ford Foundation model. In 1955 the arts received $15 million in contributions from foundations and corporations; by 1990 that amount had increased to $500 million.

This cultural emphasis was given official life in 1965 when Lyndon Johnson signed the law that created the NEA, declaring, "Art is a nation's most precious heritage, for it is in our works of art that we reveal to ourselves, and to others, the inner vision which guides us as a nation."[17] NEA support was accompanied by a surge in grants from the Ford, Rockefeller, Mellon, and Shubert Foundations, thereby energizing about four hundred regional theaters across the United States. In 1966, Robert Brustein started the Yale Repertory Theatre with $300,000 from the Rockefeller Foundation.[18]

By the late 1980s, the entire nonprofit performing arts industry was faced with declining contributed support due to an economic recession, changing philanthropic priorities among foundations and corporations, severe funding cuts from government sources, and increased competition for contributions resulting from the significant growth in the sheer number of arts organizations competing for funds. In the mid-1990s, the very existence of the NEA and the National Endowment for the Humanities was at stake. What began with isolated cases of criticism of some NEA-funded arts projects by certain conservative members of Congress had become a major debate over government funding of the arts. Now, state and local funding sources are at risk, and many have been subject to deep cuts that have severely affected the size and number of their grants to arts organizations.

Increasingly, funders—especially government agencies and foundations—are designating their grants for specific purposes, and less funding is available for general operating support. Corporate support has become more commercial than philanthropic and is often conditioned on arts organizations becoming leaner, more business-oriented, and able to meet the corporations' own marketing objectives. These factors are affecting the quality and quantity of programs and services the arts organizations can undertake, and in many cases, funding cutbacks and restrictions are threatening organizational survival.

By the 1980s, in conjunction with the great boom in audience size, wide adoption of subscriptions among loyal attenders, and sophisticated efforts among development directors, gifts from individuals became the primary source of contributions to US arts organizations. To broadly generalize, about half the revenue in US performing arts organizations comes from ticket sales and other earned sources such as hall rentals, concessions, and the like. Half or more of contributed revenue comes from individuals; the rest comes from foundations, corporate sponsorships, and although it is much discussed and debated, only about 5 percent of revenue is provided by a combination of federal, state, and local governmental agencies.

Changing Audience Demographics, Habits, and Lifestyles

Subscribers and Single Ticket Buyers

The biggest change performing arts organizations have faced in recent years in terms of earned revenue sources is the transition from subscriptions to single ticket purchase. From the late 1960s, when full-season subscriptions were first offered, through the 1980s, subscription sales increased dramatically for many organizations. Subscriptions were relatively stable through the 1990s, but since the turn of this century, people have become more spontaneous in choosing their entertainment options and are less likely to commit a year in advance to specific dates or to an entire series of performances. This situation has strained arts marketers, who must work much harder to maintain and grow ticket sales levels by offering smaller packages and by promoting subscriptions and single tickets in compelling ways.

Shrinking Audience Size

The classical music industry is facing a dramatic change in audience size and composition. Overall classical music participation rates consistently declined between 1982 and 2008, from 12.9 percent to 9.3 percent of the population. Paid attendance declined by 8 percent between 2002 and 2007.[19] Between 2002 and 2008, there was a decline of 2.9 million people attending at least one classical music concert annually.[20]

Most importantly, losses occurred in nearly every generational cohort with participation rates declining between subsequent generations, as well as within each generation as they age. This finding challenges the widely accepted belief that people attend more classical music concerts as they enter the "core audience" segment of people aged 45 and older. If recent participation trends remain unaddressed, the audience for live classical music could decline by an additional 2.7 million people, or 14 percent, by 2018. Furthermore, it is projected that the overall participation rate will decline at a faster rate than in the past. This is due to the aging out of core cohorts (e.g., Silents and Baby Boomers) and the slower replacement rate among Gen Yers, who are participating at substantially lower rates than preceding generations.[21]

Aging Audiences, Declining Audiences

Arts managers everywhere are deeply concerned about the graying of their audiences. Between 1982 and 2008, the portion of the live classical music audience over the age of 45 increased from 40 percent to 59 percent, while the proportion of the US adult population in that age group has increased from 42 percent to 51 percent.[22] Since 1982, attendance at classical music concerts among people aged 18–24 has dropped 37 percent, according to a 2009 study by the NEA. Attendance among people aged 45–54 has dropped 33 percent.[23]

Furthermore, there is growing competition from less expensive and more convenient forms of entertainment, especially a variety of movie streaming options and an extensive array of music listening resources, while the cost—both in dollars and convenience factors—of attending performing arts programs has been increasing dramatically. Since cultural performances can be seen in the movies or on television, and music is commonly listened to on iPods, compact discs, and radio, technology limits prices through competition with mass media. There is a theory that as people are exposed more and more to technological reproductions, they will increasingly crave and appreciate live performances, but this theory has not played out in reality as live attendance has been shrinking, especially among the younger generations.

The advent of opera in high definition (HD), initiated by the Metropolitan Opera, has been heartily embraced by many current opera goers, many of whom are substituting this art form for the live stage. Several symphony orchestras and theaters are also offering their productions in HD at local movie theaters for a much lower price than a live performance, and with special features such as backstage interviews during intermission. The film *Pina* in 3-D, which is a magnificent homage to the work of choreographer Pina Bausch, greatly enlivens what used to be a lesser two-dimensional offering of dance on film. While these HD and 3-D productions bring fine-quality performances to people who would not be able to access them otherwise, such high-tech offerings are likely to become an increasingly significant challenge to audience development for live performances.

ETHNICITY

The country's ethnic demographics are changing dramatically. In 1990, one out of six workers belonged to an ethnic minority; by 2000, the proportion was one in three. The Census Bureau reports that during the first decade of the twenty-first century, the US Hispanic population surged 43 percent, rising to 50.5 million in 2010 from 35.3 million in 2000. Nearly 92 percent of the nation's population growth over that decade—25.1 million people—came from minorities of all types, including those who identified themselves as mixed race. Children make up about one-third of the entire Hispanic population while they only constitute about one-fifth of whites. In the 2010 census, 16.3 percent of people identified themselves as Hispanic or Latino of any race; 63.7 percent identified as white; 12.2 percent identified as black; 4.7 percent as Asian; and 0.7 percent as American Indians or Alaska Natives. Other races made up the rest. The Census Bureau has estimated that the non-Hispanic white population will drop to 50.8 percent of the total population by 2040, then drop to 46.3 percent by 2050.[24] This change has dramatic implications for many sectors of society. For the arts, which have primarily appealed to middle-aged and older white people, intensive efforts must be made to become relevant to various ethnic groups and to attract ethnically diverse people to become members of their audiences, boards of directors, donors, and volunteers.

ECONOMIC CHALLENGES

OVERGROWTH OF THE INDUSTRY

During the 1960s through the 1980s, a period of rapid growth, arts organizations used their surplus income to invest in larger performance halls, higher-quality productions, larger staffs, more musicians, and longer seasons. During the years 1977–1987, overall ticket sales increased 50 percent in inflation-adjusted terms, but during the same period, employment was boosted by 161 percent in nonprofit theaters and by 83 percent in orchestras and opera companies.[25] The decisions made in the years of surplus and growth continue to haunt arts organizations as it is much more difficult to scale down than to sustain or increase spending on staff, programming, quality of the art, and other fixed expenses.

Many orchestras, which scheduled performances based on current demand in the growth years, responded to pressure from their musicians for year-round employment. "Now," says one symphony marketing director, "my main problem is that I have 30 percent too many concerts to sell." If managers had been more conservative and anticipated the possible slackening of demand and shrinking funding, which typically follows a period of high growth, they could have used their surpluses—or at least part of them—to create endowments for future security instead of locking in such high fixed expenses. Managers and boards of directors today are saddled with organizations and budgets too big to manage and they are struggling to manage change.

PRODUCTIVITY IN THE ARTS

The problem of spiraling expenses is persistent in the arts industry because performing arts organizations do not benefit from the productivity gains realized by other sectors of society. For most of the twentieth century, increasing efficiency in our technology-oriented, for-profit economy has been continuous and cumulative. Output per man-hour has doubled approximately every 29 years. Suppose that, due to technological advances and other factors, an automobile worker's productivity increases 4 percent per year. If, each year, the worker's wages increase by 4 percent, the ratio between total labor cost and total output remain virtually unchanged. Productivity and wages rise in tandem.

In contrast, productivity in the arts has actually decreased relative to the rest of the economy. A live performance of a 45-minute Schubert quartet will take the same three man-hours to produce as it did a century ago, and always will.

But musicians' wages rise over time, even if their productivity does not. If a string player provides just as many performances as she did the previous year, but her wage is 4 percent higher, the cost per performance has risen correspondingly, and it will continue to increase in proportion to the performer's income. The other costs of managing organizations and mounting performances—managerial salaries, materials, rents, advertising, and so on—also keep increasing. And while Broadway theaters run productions for as long as they can attract an audience, nonprofit theaters mount several new productions every season, each requiring massive efforts in rehearsing, directing, development of sets, lighting, costumes, and promotional activities.

Nonprofit performing arts organizations, no matter how successful artistically, typically suffer from what Baumol and Bowen call an inevitable "cost disease" of growing financial pressures and an ever-widening gap between income and expenses.[26] In the classical music industry, performance revenues averaged 60 percent of performance expenses in the mid-twentieth century. By the 2005–2006 season, performance revenues averaged just 41 percent of expenses. And few in the industry believe that this income gap can be substantially reduced through increasing ticket prices, which are already as high as the market will bear in most organizations.

It is largely because of the cost disease that performing arts organizations have needed to shift part of the financial burden back to their performers and managers. It is also why the recent wave of demands by musicians' labor unions has put many symphony orchestras at risk. Interestingly, since the late 1980s the wages of symphony musicians have increased more rapidly than the wages of most other workers. These pay increases were not strongly correlated with the financial performance of orchestras; rather, musicians' wages were strongly correlated with private donations to orchestras. Says Flanagan, "The large number of orchestra bankruptcies over the past 20 years demonstrates that a wage policy that ignores measures of an organization's economic strength will have serious consequences for both musicians and music lovers."[27]

Productivity increases can be realized through the use of longer seasons, more performances of each production, and larger theaters that enable organizations to serve larger audiences with a near-constant expenditure of effort. Of course, benefiting from increased scale requires that the organization attract consistently large and growing audiences. Overall, the arts can never hope to match the remarkable productivity growth achieved by the economy as a whole.[28]

Other Financial Issues

While a constant upward spiral of operating costs is likely to remain characteristic of the industry, other economic factors affect the financial health of arts organizations as well. Arts organizations are revenue intensive, meaning that they rely heavily on current income and advance ticket sales to support current expenses. They often carry sizable debt in accounts payable, which tends to accumulate slowly over a period of years. Differing audience response to various programs leads to earned-income swings. Most organizations have little or no endowment, no significant cash reserves, and limited or no lines of credit. Many organizations have been using what endowment funds they do have to meet their rising operating expenses. The desire to fulfill the organization's artistic mission often leads directors to spend all available money on short-term artistic pursuits.

SOCIAL AND POLITICAL QUESTIONS

ART VERSUS ENTERTAINMENT

"The whole notion of entertainment is confused and diffused," says actor John Lithgow. "Half of television now is 'reality television,' where you have regular people forcing themselves into the limelight and everybody watching happy amateurs failing before their very eyes. Being voted off an island or being fired by Donald Trump is the new drama: that's where dramatic tension is being generated, rather than from the minds of writers, actors, and filmmakers."[29] Furthermore, entertainment delivered cheaply to a home-based large screen TV, a laptop, or handheld device beats theater and concerts on price and convenience.

A century and a half ago, Alexis de Tocqueville prophesied the difficulty, if not the impossibility, of supporting a serious culture in a democratic society.[30] Contemporary writer Fran Liebowitz concurs, saying that our society should be democratic, but democracy in our culture only serves to lower quality and standards of our art.

Although much art deals with universal topics such as life, death, fear, joy, love, war, and peace, there is a perception that the fine arts reflect the taste of a very small cultural elite. Some sociologists recognize a cultural hierarchy that tends to polarize into a conflict between traditional high culture (fine art) and popular culture. To high culture devotees, the product of popular culture is "kitsch": sentimental, manipulative, predictable, vulgar, unsophisticated, and superficial. Critics of popular culture see it as a threat to Western culture itself. They maintain that popular culture is undesirable because it is mass produced by entrepreneurs for profit, because it debases high culture, produces spurious gratification, and is emotionally harmful to its audience. To popular culture fans, the product of high culture is overly intellectual, effeminate, snobbish, and superficial. Those who enjoy popular fare believe they have just as much right to their taste as those who enjoy Beethoven, theater of the absurd, and opera.[31]

People tend to make a sharp distinction between art and entertainment and have a strong, even exclusive, preference for one or the other. Composer Charles Wuorinen says, "I think there's a very simple distinction, and it doesn't diminish entertainment in any way, because we all want it and we all enjoy it. Entertainment is that which you receive without effort. Arts is something where you must make some kind of effort, and you get more than you had before."[32] This effort is partly a matter of repeated exposure and partly a function of arts education.

The sharp distinction between the "nobility" of art and the "vulgarity" of mere entertainment is due in part to the systems under which they operate. The performing arts are predominantly distributed by nonprofit organizations, managed by artistic professionals, governed by prosperous and influential trustees, and supported in a large part by funders. Popular entertainment, on the other hand, is sponsored by profit-seeking entrepreneurs and distributed via the market.[33] However, the differences between high art and popular culture are often greatest in the minds of their enthusiasts, especially in terms of the social status conferred by participation or nonparticipation, rather than in the intrinsic nature of the art itself.[34] Consider the fact that Mozart's *Magic Flute*, which today is performed in the world's finest opera houses, was originally commissioned by a music hall to entertain its populist audience. And, in response to criticism from the classical avant garde for writing music they considered too accessible, Pulitzer Prize–winning composer Morton Gould quipped, "I'm sorry I wrote something a lot of people like. I'll try never to do it again."[35]

To some high culture elitists, what is considered "popular" or "entertaining" is whatever draws a huge crowd. Says the director of an ethnic music festival, "There's an idea out there that if something is 'popular,' it's not very good, that if we're selling out the house we must be doing something wrong."[36]An attitude of "art versus entertainment" on the part of arts managers and artistic decision-makers can only work to the detriment of arts organizations. At its best, art is highly entertaining and, likewise, entertainment is highly artistic. It is counterproductive for arts managers, marketers, and artistic personnel to consider them to be mutually exclusive. Artist Leopold Segedin reminds us that "not only great artists and connoisseurs, but all persons have the potential to achieve satisfaction and fulfillment through creative and aesthetic experiences."[37]

ART-CENTERED VERSUS MARKET-CENTERED PRODUCT CHOICE.

In most sectors of society, marketing calls for satisfying the customer. But, in the arts, is satisfaction the goal? If the purpose of art is to broaden human experience, it may be necessary for the artist to take audiences through an uncomfortable period of "unfreezing" before they come to terms with the new possibilities. However, when audience members clamor for the traditional and familiar, it can be counterproductive for an organization to present too much contemporary and avant-garde programming. Henry Fogel, former president of the Chicago Symphony Orchestra, reflects that if he were to receive significant negative feedback from the audience on such a matter, it would certainly affect the amount of contemporary music he would program for the next season. In order for a performing arts organization to survive, it must both meet the current needs of its audiences and assist in the developmental process that will cause audiences to seek and respond to a product that is closer to the director's artistic vision.

One of marketing's major theorists, Theodore Levitt of Harvard, argues that any business must try to satisfy its customers. According to Levitt, the basic premise of the marketing concept is that a company should determine what consumers need and want, and try to satisfy those needs and wants, provided that doing so is consistent with the company's strategy and that the expected rate of return meets the company's objectives.[38]

However, this purely market-centered philosophy is inconsistent with what the concept of art is all about. Should high customer satisfaction even be the objective of a performing arts organization? If patrons were all satisfied, artistic directors wouldn't be living up to their responsibility to challenge and provoke. Said a Rockefeller report on the problems facing the performing arts, "Entertainment which makes no demand upon the mind or the body offers neither permanent enrichment of the spirit nor a full measure of delight."[39]

Furthermore, say Morison and Dalgleish,

> When the goal is creating a love affair between people and a certain artist's vision of art, then changing the product does not help to accomplish that end; it betrays it. It is totally inappropriate for those who govern, manage or market the arts to suggest that the product be changed to make it sell better. In theory, they have accepted the vision of the artistic leadership and their responsibility is to support that vision and to communicate clearly and effectively its values, meaning and benefits to those who might find enjoyment and enrichment in participating as audience. If that is done and no one is interested, then eventually the maintenance of the organization will no longer be justified. Or, as an alternative to institutional euthanasia, a board of trustees may choose to select artistic leadership whose tastes and values are closer to those of a larger public. In that case, however, it would appear that institutional survival rather than artistic purpose is what is being pursued.[40]

The artist's perspective transcends audience input. Art is visionary, and, when successful, it leads an audience on a journey that for many is a previously unimagined experience.

On the other hand, the essence of the performing art experience resides in the communication between the performers onstage and the members of their audience. Unless the performers are speaking, singing, playing, or dancing in a language that audience members can appreciate, relate to, or be moved by, there is an element of futility in the very act of performing. Some people want to fully engage and learn something every time they go out, while others prefer a more passive, disconnected experience. A small segment of the arts-going public seeks to be challenged by unfamiliar art, while many more arts goers prefer the comfort of revisiting familiar works. Besides, there are always people experiencing *Swan Lake* or Beethoven's *Fifth* for the first time. As the range of sophistication levels in the audience widens over time, so do expectations for fulfillment. Furthermore, "orchestras have an imperative to innovate," says Michael Tilson Thomas, music director of the San Francisco Symphony and founder and artistic director of Miami's New World Symphony. "We cannot rely on audiences simply appearing," he says. "We need to have a role in creating new audiences."[41] As a result, it gets increasingly difficult for an arts organization to satisfy its various patron segments.

Along the continuum of art-centered and market-centered choices, the ultimate product is the one in which the two concepts become one.

Arts managers must understand that listeners and viewers are coproducers of their artistic product. Says Pulitzer Prize–winning playwright Paula Vogel, "I write the script, the directors and actors and designers write the production, and the audience writes the play. If there are 200 audience members, there are 200 plays."[42]

ART FOR ART'S SAKE OR ART FOR SOCIAL PURPOSE?

Since the 1960s, when government agencies, foundations, and corporations began to fund the arts substantially, art has taken on a social function, serving purposes previously confined to education, religion, and politics. Ronald Berman, a former chairman of the National Endowment for the Humanities, believes that with this transformation, "art becomes a political commodity. There is assumed a direct connection between the appropriation of funds and the resolution of one or more social problems."[43]

Over the decades, many claims have been made for the value of the arts in American society, largely by lay patrons and community leaders. The arguments advanced for the arts almost universally accepted the role of art as a means to some other end, rather than as an end in itself. However, say arts consultants and authors Bradley Morison and Julie Dalgleish,

> Art does not exist to serve practical purposes. It is misguided to try to justify its support on the basis of community prestige, economic impact, urban development, corporate image, enlightened self-interest or even the chamber of commerce quality-of-life. We must accept art as art, as an end in itself, and strive to make it part of the lives of all because, quite simply, it is the essence of civilization.[44]

Said Robert Brustein, "Given the limited resources available for both social and cultural programs, the humanitarian agencies that disburse grant money no doubt believe that a single dollar can fulfill a double purpose.... An unidentified contributions manager is bluntly quoted as saying: 'We no longer "support" the arts. We use the arts in innovative ways to support the social causes chosen by our company.'"[45]

Staggering under large-scale deficits, our nonprofit cultural institutions are being asked to validate themselves not through creative contributions, but on the basis of community services. As Alexis de Tocqueville once predicted: "Democratic nations...will habitually prefer the useful to the beautiful, and they will require that the beautiful be useful."[46]

Boston Globe critic Richard Dyer believes that

> pieces should never be programmed simply because they are by a woman, a gay, a representative of some ethnic or racial community, or indeed simply because they are new (or simply because they are old).... A concert can rise above political, careerist or even aesthetic concerns to reveal the life of the community through the efforts of individuals conveying their aspirations, fears, convictions and concerns. The value of that cannot be called into question.[47]

"[Art has] a noble and important mission, but it is a mission we may be in danger of losing if we try to replace it instead with politically correct and socially worthy goals," said Richard Christiansen, former chief theater critic of the *Chicago Tribune*.[48]

EDUCATION AND THE ARTS

At a performance of *Macbeth* during Chicago's International Theater Festival, England's Royal Shakespeare Company took three curtain calls and returned to the stage to applaud the audience. One commonly asks what makes a great performance. But the flip side of that question is, what makes a great audience? Says writer Fran Liebowitz, "A very discerning audience with a high level of connoisseurship is every bit as important as the artist."[49]

A study conducted by the NEA indicated that most Americans have not had any form of arts education. The study found that, in general, public school systems do not provide opportunities for most students to become culturally literate; they do not teach the arts sequentially in kindergarten through high school. The study also reported that many arts teachers are not prepared to teach art history and critical analysis. The study warns that "the artistic heritage that is ours and the opportunities to contribute significantly to its evolution are being lost to our young people." Says arts critic Samuel Lipman, "The direct result of this neglect is a downtrodden army of cultureless children marching into a barren and depleted adulthood and taking American civilization with them. And the sheer numbers of these future citizens confront the nation with prospects of a diminishing cultural future."[50]

A 2008 study on arts education in America, conducted on behalf of the NEA, reported that in 2008, only half of all 18-year-olds (49.5 percent, or 2.2 million) had received any arts education in childhood, a decline of 23 percent since 1982.[51] The relationship between arts education and adults' rates of arts participation has been consistently strong throughout the survey's history.

This situation poses a serious threat to the transmission of culture in this country and will have a long-term negative impact on the number of people interested in participating in the performing arts.[52] It creates special challenges for arts organizations both in attracting and satisfying audiences and in taking on the role of educating current and potential future audiences.

Yet, a Harris Poll conducted about the attitudes of Americans toward arts education revealed that 93 percent of Americans agree that the arts are vital to providing a well-rounded education for children. Additionally, 54 percent rated the importance of arts education a "ten" on a scale of one to ten. The survey reveals additional strong support among Americans for arts education: 86 percent of Americans agree that arts education encourages and assists in the improvement of a child's attitudes toward school; 83 percent of Americans believe that arts education helps teach children to communicate effectively with adults and peers; 79 percent of Americans agree that incorporating arts into education is the first step in adding back what's missing in public education today.[53]

The Complexities of Marketing the Arts

The art marketer's role is complex. In the commercial world the customer reigns. Goods and services are produced and distributed according to demand and profitability considerations. But the purpose of a nonprofit arts organization is to expose an artist and his or her message to the widest possible audience, rather than to produce the artist and the message that the largest audience demands. Businesses typically innovate only when there is economic justification to do so. Nonprofit arts organizations innovate and explore in the pursuit of social or aesthetic value, even when there is no assurance of economic success.

This situation creates three major marketing problems.[54] First, the organization must *find* a market for its offerings. Because it is presenting productions for which there may be little or no existing demand, the organization must create new needs in the marketplace, rather than just meet existing needs. Consider the contrast between a Broadway commercial theater that will run a production for as long as it can attract an audience ("*Cats*—now and forever!") and a nonprofit theater that will close a successful show in order to open the next, probably unfamiliar production, which may run at a financial loss. Second, the organization must *expand* its market. Any market is usually of a limited size. But since art is often ahead of its audience (consider contemporary classical music at any time in history), the organization limits itself from producing the works that would most quickly expand the audience. It takes a long time to develop and educate an audience, and it entails great financial risk. Third, an arts organization must *keep* its audience. While continually innovating and experimenting to promote the growth and exposure of artists, organizations drop successful trends and repeatedly make new demands on their audience members. So an artistic organization that tries to fulfill its mission and to innovate regardless of economic rationale must operate under severe handicaps in its efforts to find, expand, and keep an audience. Given these conditions, arts marketers must be aware of and sensitive to the different and perpetually changing interests and needs of a wide variety of audience segments.

Accessibility Issues

Various surveys conducted over the years show that many more people are interested in attending the performing arts than currently do so. Some arts organizations are finding that several factors act as barriers and keep away many people who are attracted to theater, dance, and music.

Some nonattenders claim to feel that the arts are not for them, that they would feel uncomfortable or out of place at performing arts events. In her book *Is Art Good for Us?*, Jolie Jensen says that this perspective derives, at least in part, from the attitude disseminated, consciously or not, by arts organizations, which say in effect that "the arts are good medicine, especially in today's 'sick' society. The mass media, in contrast, are bad medicine, poisoning a healthy society." Jensen continues, "For many social and cultural critics, the passive, unproven effects of Muzak (the music heard in fast food joints and shopping malls) are a form of social control, while the passive, unproven effects of a single Mozart sonata are examples of social benefit." The problem, says Jensen, is the elitist notion that high culture *does* something (it is good for us), rather than that it *is* something (good in and of itself). Arts supporters cannot persuade people to attend the arts by defining them as "cultural spinach."[55]

Accessibility problems often pertain to mundane issues such as time of the concert, price, ticket availability, and concern about where to park, what to wear, or when to applaud. Many arts organizations fail miserably in their marketing communications at answering questions prospective patrons have in their minds: Will I like this show? Will I understand it? Do I need to understand it to enjoy it? Will I feel comfortable there and "fit in"? What relevance does this performance have to my life?

Arts marketers must make a sincere effort to delve inside the heads and hearts of their publics. Managers who blame external forces for their decline, rather than modifying their strategies from what worked in the past, are likely to fail.

Organizations that implement marketing strategies relevant to their target audiences certainly are not overcoming all the issues they face, yet they are realizing significant success.

Ticket Pricing Issues

Although ticket price is commonly assumed by arts managers to be a primary barrier to attendance, research repeatedly shows that this factor ranks relatively low among those who have interest in attending. Most people who do not attend would not do so if the tickets were even half price. Low ticket price is an operative factor for those who could not otherwise afford to attend and among those who have, in recent years, become accustomed to discount pricing schemes that have "taught" people to wait until the last minute for cheaper seats.

Research demonstrates that people want to choose exactly which performances to attend and are typically willing to pay to have the best possible experience. This attitude has driven the development and implementation of dynamic pricing strategies, with which prices for popular shows are increased (or decreased) during the run of the show so the organizations can benefit financially from high (or low) demand.

The primary operative factor for patrons in the ticket price equation is *value.* There are many complexities to this topic, which will be covered in depth in a later chapter.

The High-Technology Age

As businesses have increasingly leveraged the power of high technology to personalize their product offerings, consumers have grown to expect—and to respond best—to customized experiences. Users of On Demand or streaming programming choose exactly which television shows and films to watch and when to watch them. Netflix and Amazon build a recommendation list for each patron, based on the customer's reviews and on what other people with similar taste have liked. When people listen to music in their daily lives, they upload just what they want to hear to their iPods and choose one kind of music while dining, another kind while exercising. Contrast this with arts organizations that offer preset programs at a fixed time in a single location and ask patrons to purchase a series of these programs many months in advance for a considerable price.

Arts organizations face significant competition from the Internet and other mass media. According to the NEA's 2008 Survey of Public Participation in the Arts, the Internet and mass media are reaching much larger audiences for the arts than arts organizations offering live performances. This research found that 30 percent of adults who use the Internet download, watch, or listen to music, theater, or dance performances online at least once a week. More Americans view or listen to broadcasts and recordings of arts events than attend them live (live theater being the sole exception). The good news about these statistics is that substantial numbers of people are interested in the arts and may be attracted to live performances with the right messages and incentives.

The greatest challenge for arts marketers is to keep up with people's evolving preferences regarding communication methods for different types of content. In today's society, websites, emails, YouTube, and social media sites such as Facebook and Twitter appear to have the most staying power and popularity, but in this rapidly changing environment, arts organizations must remain current with people's habits, and furthermore, must know how best to use each of these resources and integrate them into an overall strategy.

As people play video and computer games, and attend multisensory, highly stimulating events, they are moving farther and farther away from enjoying fixed, static experiences. Says arts researcher and consultant Alan Brown, "The sum of all these trends is higher demand for more intense and more pleasurable leisure and learning experiences. Arts managers who think that their organizations are immune to the shifting sands of demand are sadly mistaken.... We cannot invent a more viable future for our institutions and agencies without a deep understanding of how consumers fit art into their lives."[56]

MANAGERIAL ISSUES

THE CRISIS MODE OF MANAGEMENT

As managers and board members attempt to alleviate current problems, they often take certain decisions and actions that contribute to the crisis instead. Crisis management, which is the process of dealing with emergency situations, becomes the norm, preventing strategic planning both for the short and the long term. As their debt increases, organizations become driven by financial concerns—often at the expense of artistic considerations—further complicating their problems. As arts consultants Nello McDaniel and George Thorn point out, "Stress from accumulated deficits is debilitating. It affects the way they [arts managers] are working, the way they're thinking. It controls what they're able to do—and not do. Institutions become debt-driven rather than art-driven."[57] Cash shortages create the necessity for immediate and impassioned fund-raising, triggering emotional responses from both staff and board members and undercutting strategic fund-raising efforts. In the process, organizations become secretive about their true state of affairs in an effort to keep up appearances and to maintain funding and community support. Systems falter and organizations lose "memory" as staff and board membership turn over rapidly due to stress and burnout. This description is especially true of smaller organizations and of those that produce more challenging work, because they rarely have as strong a foundation of community and financial support on which to rely in difficult times as older, more established organizations.[58] Yet, the large number of troubled mature institutions confirms that a confluence of factors has developed during the last 10–15 years that can overwhelm even the best of arts managers and boards.

McDaniel and Thorn have observed growing patterns of stress in arts organizations, indicating that a fundamental problem exists throughout the field. They say that "the number of arts professionals and volunteers who are experiencing anger, frustration and feelings of failure is growing at an alarming rate."[59]

Arts organizations must learn to provide for intervention, diagnosis, and stabilization before serious crises occur.

Professionalizing Management

The first generation of arts managers exhibited an impresarial style, combining traditionalistic authority, charisma, and entrepreneurship. The impresario was a connoisseur and a gentleman, primarily engaged in wooing contributors and satisfying wealthy trustees, while dominating performers and employees by an autocratic imposition of his will. By the 1960s, arts organizations had grown much larger, in terms of both number of employees and size of budgets, causing more bureaucratic structures to emerge. Also, the large number of funding agencies and other interest groups have increasingly held the organizations and their managers formally accountable. "This demand for formal accountability," says sociologist Richard Peterson, "in turn puts a premium on employing arts managers adept at working in the administrative rather than the impresarial mode."[60]

Managing arts organizations has become much more complex as organizations have taken on auxiliary activities such as operating retail stores, dining facilities, educational programs, real estate ventures, music recording contracts, and membership tours. Today's arts administrators must negotiate formal written contracts with artists' managers and with musicians' and backstage technicians' labor unions; they must understand labor laws, workers' compensation laws, retirement plans, medical insurance benefits, workplace safety codes, and record and television performance royalties. Furthermore, new economic developments have led to a greater focus on the private sector and the marketplace, producing an increased emphasis on marketing, fund-raising, profit-making ventures, and the commercialization of artistic enterprises. In the process, marketing and fund-raising functions have become more sophisticated and competitive, requiring the employment of more highly educated managers. The days of continuous growth and expansion have given way to the need for austerity, consolidation, and careful planning. It is no longer enough to know basic operational skills such as accounting, database management, grant-writing, and how to produce a play. Subtler and more sophisticated communication and negotiating skills are critical for managing and marketing an arts organization.

A dramatic change in the functions and functioning of arts organizations is inevitable. This change can be most productive when it is the result of strategic planning, effective implementation, and ongoing evaluation. The ability to analyze and strategize requires fluency and comfort with current management theory and techniques. Arts managers have often been content to evaluate themselves based on the "goodness of their cause" and to substitute good intentions for results. Winston Churchill once claimed that "success consists of going from failure to failure with undiminished enthusiasm."[61] But such enthusiasm becomes increasingly difficult to muster on the part of the managers, the artists, the audiences, and the funders alike. Contemporary arts managers need the tools to build and sustain an organization that supports its artistic mission while facing the challenges of an uncertain and changing environment.

RESISTANCE TO MARKETING

Some arts professionals, especially artistic directors, refuse to join with administrators in marketing-oriented activities and to include marketers in decisions about programming. Sometimes they actively seek to frustrate efforts in which market response is given "undue consideration." Says Robert Kelly of such professionals, "They are an 'enemy within'; not because they consciously wish to place their arts organizations in jeopardy but because their blind resistance to marketing may, in the end, have that consequence."[62]

Some artistic directors are at the forefront of creating relevance for and engaging new audiences. Some are seemingly stuck in their preference for doing things as they always have. In some cases, says Kelly, "marketing is feared and even hated for what it might do to the arts; on the other [hand], there is an unquestioning conviction that marketing can work miracles for the arts. To compound matters, these contradictory beliefs are often held by the same persons." Kelly believes that arts professionals must become more, rather than less, heavily involved in the marketing decision-making process, since avoidance will lead to ineffectual marketing, not to an absence of marketing.[63]

THE MARKETING RESPONSE

Superb marketing cannot alone resolve the myriad issues arts organizations face in the current dynamic environment of changing interests, lifestyles, economics, and funding sources. But the outlook for the health and sustainability of arts organizations could be greatly improved if organizations work to accommodate broader and more diverse audiences by making better use of marketing procedures and concepts. As long as arts organizations continue to focus their efforts on serving only the people they have traditionally served, those that survive and thrive will be the exception rather than the rule. On the other hand, organizations looking elsewhere to build their audiences and contribution sources must be sure to give excellent attention to their current patrons, the resources most immediately at their disposal and most likely to remain fruitful for many years to come.

For effective marketing to take place, now and in the future, it is necessary to understand strategies, principles, and tactics for increasing and broadening the audience base; for increasing the accessibility of various art forms and of more "difficult" productions; and for supporting projects that meet the needs of specific audience segments. It is also necessary to look to new opportunities in the marketplace, to identify changes in consumer behavior and attitudes, and to develop innovative strategies that will keep the organization and its offerings relevant for its current and potential audiences over time.

The essence of art is in its communication with the audience member. Therefore, arts organizations must shift their focus from a pure product-centered focus to one that balances the artistic decision-making process with audience needs and preferences. The role of the arts organization is to serve as liaison, as facilitator, as a distribution channel between artists and audiences; therefore it must be sensitive to both. The artist is the visionary, the creator, the raison d'être for the organization; but it is the audience members who vote for the programs they want to experience each time they purchase a ticket—or not. And the "audience" is not a monolithic being; the term represents a multitude of individuals with varying experiences, tastes, preferences, interests, and constraints, all of whom must be taken into consideration when planning and marketing programs. The artistic director, managing director, and marketing director should collaborate to create experiences that are ultimately satisfying to both the artists and the audiences.

Sang the artist George Seurat in the second act of Stephen Sondheim's great musical *Sunday in the Park with George*, "Art isn't easy."[64]

CHAPTER 2

THE EVOLUTION AND PRINCIPLES OF MARKETING

ONE CAN BE A SUCCESSFUL MARKETER ONLY IF ONE HAS ADOPTED THE PROPER marketing mind-set. This means having a clear appreciation for what marketing comprises and what it can do for the organization. Marketing, as it relates to the arts, is not about intimidation or coercion or abandoning an artistic vision. It is not "hard selling" or deceptive advertising. It is a sound, effective technology for creating exchanges and influencing behavior that, when properly applied, *must be beneficial to both parties* involved in the exchange.

DEFINING MARKETING

Traditionally, marketing management has been defined as the analysis, planning, implementation, and control of programs designed to create, build, and maintain beneficial exchange relationships with target audiences. Marketing the arts is the process by which an organization relates creatively, productively, and profitably to the marketplace, with the goal of creating and satisfying customers *within the parameters of the organization's mission* for the purpose of achieving the marketer's objectives.

The key feature of this definition is that it *focuses on exchanges.* Marketers are in the profession of creating, building, and maintaining exchanges. Because exchanges take place only when a target audience member takes an action, the ultimate objective of marketing is to influence behavior. Also, this definition clearly allows for the centrality of the mission.

According to Philip Kotler, "The meaning of marketing in this new age is the marketing of meaning. Marketing should no longer be considered as only selling and using tools to generate demand. Marketing should now be considered as the major hope of a company to build and restore consumer trust."[1]

In 2008, the American Marketing Association modified its longstanding definition of marketing to read: "Marketing is the activity, set of institutions, and processes for creating, communicating, delivering, and exchanging offerings that have value for consumers, clients, partners, and society at large."[2] This definition includes the role marketing plays within society at large, and defines marketing as a science, educational process, and a philosophy—not just a management system. It also expands the previous scope of the term to incorporate the concept that one can market something to "do good."

The objectives of marketing are not, ultimately, either to educate or to change values or attitudes. It may seek to do so as a *means* of influencing behavior. If someone has a final goal of imparting information or knowledge, that person is in the profession of educating, not marketing. If someone has a final goal of changing attitudes or values, that person may be described as a propagandist, a lobbyist, or perhaps an artist, but not a marketer. Though marketing may use the tools of the educator or the propagandist, its critical distinguishing feature is that its ultimate goal is to influence behavior, either by changing it or by keeping it the same in the face of other pressures.

THE EVOLUTION OF THE MARKETING PHILOSOPHY

Over the decades, marketing has evolved. It started with a *product* focus and the belief that people will buy if the product meets or exceeds their needs. Later, marketing evolved to a transaction-oriented *sales* focus. Eventually, marketing became *relationship* oriented, focusing on how to keep a consumer coming back and buying more. More recently, marketing has shifted to a *collaborative* focus, encouraging consumers to participate in the company's development of products and communications.

Consumer-centricity is empowered by the information age. Information technology has made it possible to target specific groups of customers with information and offers relevant to them. But today's consumers are well informed, readily seek their own information, critical reviews, word-of-mouth commentaries, and can easily compare several similar offerings. So now, product value is defined by the consumer. Says Kotler, "Today's marketers try to touch the consumer's mind and heart. Unfortunately, the consumer-centric approach implicitly assumes the view that consumers are passive targets of marketing campaigns. This is the view of the customer-oriented era."[3]

Today we see marketing as transforming once again in response to the new dynamics in the environment. Companies are expanding their focus from products to consumers to humankind issues; they are shifting from consumer-centricity to human-centricity.[4] People are looking for not only functional and emotional fulfillment but also human spirit fulfillment in the products and services they choose. Now, instead of treating people simply as consumers, marketers need to approach them as whole human beings with minds, hearts, and spirits. The new marketing, which Kotler calls Marketing 3.0, complements emotional marketing with human spirit marketing.[5] This is a values-driven era and supplying meaning is an important value proposition in marketing. The arts have a great advantage in that they are perfect vehicles for helping people tap into strong values and to realize their spiritual side. Arts marketers do not need to show customers that their company is helping to support an organization of value as many businesses do; the arts themselves have inherent value.

Furthermore, when individual consumers experience the product, they personalize the experience according to their own unique needs and wants.

To understand modern marketing management, it is useful to trace the evolution of different business orientations toward marketing over the last one hundred years.

THE PRODUCT ORIENTATION

Marketing first emerged as a distinct managerial function around the turn of the last century, during an era that venerated industrial innovation in the design of new products. It was a period that saw the development of the radio, the automobile, and the electric light bulb. In this early period, marketing also was decidedly product oriented. The belief was that to be an effective marketer, you had simply to "build a better mousetrap," and, in effect, customers would beat a pathway to your door. As the strippers sang in the musical *Gypsy*, "You gotta' get a gimmick."[6]

A product orientation toward marketing holds that consumers will favor those products that offer the most quality, performance, and features. Managers in these product-oriented organizations focus their energy on making good products and improving them over time. These managers assume that buyers admire well-made products, can appraise product quality and performance, and are willing to pay more for product "extras." Many of these managers are caught up in a "love affair" with their product and fail to appreciate that the market may be less "turned on" and may even be moving in a different direction. Many organizations firmly resist modifying their product even if this would increase its appeal to others, and even if modification would have little or no impact on the organization's artistic integrity.

THE SALES ORIENTATION

Primary emphasis on selling is another approach that many firms take to the market. This approach assumes that consumers typically show buying inertia or resistance and have to be coaxed into buying more. The selling orientation led to significant expansion of the roles of advertising and personal selling in the marketing mix. As "salesmanship" became the byword of successful marketing in the middle of the twentieth century, playwright Arthur Miller created Willy Loman, who achieved a central role in American folklore. Said Willy in *Death of a Salesman*, "Oh, I'll knock 'em dead next week, I'll go to Hartford. I'm very well liked in Hartford."[7]

A sales orientation toward marketing holds that success will come to those organizations that best *persuade* customers to accept their offerings rather than competitors' or rather than no offering at all. But selling is only the tip of the marketing iceberg. Peter Drucker, a leading management theorist, explains: "There will always, one can assume, be need for some selling. *But the aim of marketing is to make selling superfluous.* The aim of marketing is to know and understand the customer so well that the product or service fits him and sells itself."[8]

The selling orientation is still pervasive today. Organizations often believe they can substantially increase the size of their market by increasing their selling effort. They increase the budget for advertising, personal selling, sales promotion, and other demand-stimulating activities. This is why the public often identifies marketing with advertising and hard selling. Sales-oriented steps will undoubtedly work to produce more customers in the short run. But their use in no way implies that the organization has adopted a marketing strategy that will generate higher sales in the long run. One of the biggest problems that orchestras face is that about 80 percent of people who attend a concert once never return to that organization. Researchers find it difficult to learn the reasons for this situation, known in the industry as "churn." But it is likely that some first-time attenders felt pressured or somehow misled by marketing efforts and the actual experience was very different (and inferior, from these people's perspective) from how it was promoted.

Thus, selling, to be effective, must be preceded by strategic marketing planning. If the marketer does a good job of identifying consumer needs; developing appropriate products; and pricing, distributing, and promoting them effectively, they will do a better job of satisfying their audiences and giving them good reason to return.

THE CUSTOMER ORIENTATION

The orientations that characterized the earliest stages in marketing's historical development had one thing in common. Marketing planning always began with the organization and what it wanted to offer. However, as consumers became more sophisticated and discerning, they also became more selective and more responsive to custom-tailored options, and were less willing to settle for just anything the market tried to persuade them to buy. And marketers began to realize that it is the consumer—not the marketer—who ultimately decides if and when transactions occur. They recognized that they had had the marketing equation turned backward. They had been trying to change consumers to fit what the organization had to offer. But since what the customer chooses to buy determines the organization's success, the customer is truly sovereign.

It follows, then, that marketing planning must begin with the consumer, not with the organization. Outside-inside marketing must replace inside-outside marketing. As Willy Loman's son Hap once told him, "The trouble with you in business was you never tried to please people."[9]

A customer-centered mind-set requires that the organization systematically study customers' needs and wants, perceptions and attitudes, preferences and satisfactions. Then the organization must act on this information to improve its offerings to better meet its customers' needs.

This does not mean that artistic directors must compromise their artistic integrity. Nor does it mean that an organization must cater to every consumer whim and fancy, as many managers fear. Jim McCarthy, CEO of Goldstar, says that "for those who can't distinguish between being audience-oriented and 'pandering,' I refer you to Matthew Bourne." About his new interpretation of *Sleeping Beauty,* Bourne says: "I'm thinking of the audience when I'm making work, always. I'm not just pleasing myself.... I naturally want to communicate, I want to give to audiences, I pick my dancers... to be generous with the audience." Continues McCarthy, "Being audience-oriented in many ways is the key to selling out. All the pricing and promotion in the world will not make a success of a show that an audience doesn't want."[10]

Even if an organization ought not, will not, or cannot change the selection of works it performs or presents, the highest volume of exchange will always be generated if the way the organization's offering is described, priced, packaged, enhanced, and delivered is fully responsive to the customer's needs, preferences, and interests. Marketing will help maximize exchanges with the *targeted* audience.

As an example, a university research project in the Buffalo, New York, area revealed that many consumers who indicated that they thought they might like to attend a concert did not do so because they expected the occasion to be very formal. As these potential consumers put it, "We can't go because we don't have the proper clothes. We would feel really uncomfortable around all those fancy-dressed people." The orchestra itself was seen as distant, formal, and forbidding. Once the Buffalo Philharmonic understood the barriers preventing potential customers from attending, the organization took great pains to humanize the orchestra and the concert-going experience. Orchestra members began playing shirt-sleeved chamber music programs at neighborhood art fairs and other local outdoor events. The orchestra performed a half-time show at a Buffalo Bills football game. The conductor, Michael Tilson Thomas, began appearing on local television and giving brief, informal talks to audiences at specific concerts. Formality was no longer a barrier keeping potential patrons away from concerts, and rising attendance figures clearly reflected this new customer-centered orientation.

A customer-centered organization is one that makes every effort to sense, serve, and satisfy the needs and wants of its clients and publics *within the constraints of its mission and budget.* A common result of a customer-centered orientation is that the people who come in contact with such organizations report high personal satisfaction. Since word of mouth is known to be the best advertising for a performing arts organization (or for any product), satisfied consumers become the most effective advertisement for these institutions.

Detecting an Organization-Centered Marketing Orientation

Many managers wish to be customer-centered, and, in fact, wrongly believe that they are. There are several clues that reveal an organization's organization-centered philosophy. Among them are the following.

The organization considers its offerings to be inherently desirable. Arts administrators often find it hard to believe that any right-thinking person would not wish to attend their productions. Waving the banner of the artistic "imperative," arts managers often place themselves above the marketplace, attributing lack of organizational success to customer ignorance, absence of motivation, or both. Some managers admit that they simply haven't yet found the right way to communicate the benefits of their offerings or created the right incentives to overcome inertia among target consumers. However, a significant number of nonprofit managers actually view the customer with some contempt. These are often the same managers who present esoteric works without making them accessible to a less than highly sophisticated audience, who write program notes in "artspeak," and who package their events to appeal to wealthy, status-seeking individuals.

A minor role is afforded to customer research. Without access to research results, the Buffalo Symphony would never have known that it had an image of stuffiness and formality that was keeping potential patrons away. Organizations cannot second-guess the attitudes or motivations of their publics, and they must devote both human and financial resources to challenge some managers' assumptions about their customers. Not only should research be used to analyze current audiences, but it should also be proactive in order to forecast and anticipate changes in customer needs.

Marketing is defined primarily as promotion. Promotion is only one aspect of the total marketing mix. To focus marketing efforts on promotional devices alone is to ignore the full range of benefits that marketing can provide. At best, it provides short-term solutions to long-term problems. A customer may be lured to attend a performance because of a two-for-one price offer, but, unless the product itself, the location, and the total experience are pleasing, it is unlikely that the customer will return for another production. Therefore, promotion must be considered as only one tool within the marketing field.

Marketing specialists are chosen for their product knowledge or their communication skills, rather than for their knowledge of marketing principles and methods and especially of consumer behavior analysis. While product and communication skills are useful, they do not represent a full range of marketing skills. The best marketing managers in the private sector are those who know their consumer markets and their competitors well. They make active use of consumer research and know how to develop and implement systematic marketing plans. In contrast, many nonprofit managers believe that marketing their organizations is so different that there can be little transferability of skills. Many business managers of arts organizations were once performers themselves or were formally trained in artistic disciplines. Their product orientation leads them to be more comfortable working with others who have a similar perspective. Indeed, arts marketing requires special sensitivity—sensitivity to the nonprofit point of view in which quality is maximized rather than the bottom line, and sensitivity to the artist's imperative for freedom of expression. An artist would never create a painting to appeal to the public in the way a brand manager designs the box for a consumer product (except Andy Warhol). However, a marketing professional knows how to "package" the artist's creation and the organization's vision in order to make them appealing to the audiences.

Some nonprofit arts organizations no longer insist on hiring performing arts product specialists for their marketing positions, but rather choose public relations specialists or advertising personnel. What these marketers have in common is that they are good communicators. Their communication skills and their emphasis on persuasion reflect a view of marketing only slightly different from that found in the product-oriented organization. This emphasis on persuasion, as may be obvious by now, is consistent with a "selling" approach to marketing.

One "best" marketing strategy is typically employed in approaching the market and is viewed as being all that is needed. An arts administrator may perceive the market as monolithic or at least as having only a few crudely defined market segments. This approach ignores subtle distinctions and bypasses major opportunities. A climate of managerial certainty precludes experimentation either with alternative strategies or with variations of a single strategy applied across a number of market subsegments.

Thus, marketing strategies are aimed at the most obvious market segments, usually determined by loyalty status to the organization in terms of past ticket purchases. It is crucial that managers consider a patron's total involvement with the organization— ticket purchases, donations, attendance at special events, and volunteerism—in order to obtain a true snapshot of his or her involvement with and commitment to the organization.

Many nonprofit managers believe that simple, consistent strategies are the best. They fear change and experimentation, and often do not realize that adaptive and creative marketing strategies are a necessity.

Generic competition is ignored or misunderstood. The managing director of a theater asked a marketing consultant which other theaters were its greatest competitors. The consultant answered that none of them were—that this theater, which attracted an elderly, middle-, and upper-middle-class audience, was mainly in competition with DVDs, dinners with friends, and winters in warmer climates. Organizations compete at many different levels. It takes a customer-focused perspective to understand that when a couple chooses to attend a play, the choice is not made solely in the context of what other plays are available, or what other plays are available within a certain distance from home, or what other live performances are available. Rather, the selection is likely to have been made among several forms of leisure-time pursuits.

Customer-Centered Marketing

A rationale that can help even the most reticent arts administrators accept the customer-centered approach to marketing is the fact that customers are hard to change—the organization is not. The organization is under management control; the customer is not. The process of changing the organization to accommodate customers ensures that consumer trends will be carefully monitored. Again, this does not mean that an organization must deny its artistic vision and present more programs that appeal to a broader audience. It means that the organization's approach to marketing the entire experience must give the customer a central focus. In a sophisticated marketing organization, all marketing analysis and planning begin and end with the customer. A customer-centered organization always asks:

- Who is our current audience? How can we define and categorize them?
- Who are our most likely potential markets for future development?
- What are their current perceptions, needs, and wants?
- How satisfied are our customers with our offerings? In what ways can we make them more satisfied? In what ways can we create and market satisfaction for other potential audiences as well?

The following features characterize a customer-centered organization.

It relies heavily on research and data. Even when an organization's product stays the same over time, as is the case with a Shakespeare festival or a company dedicated to presenting music of the Baroque, the customers themselves may change. Because the consumer is central, the management must have a profound understanding of consumer perceptions, needs, and wants, and must constantly track changes so that the organization can respond to even subtle shifts as they occur. An organization should be proactive, not just reactive, in its strategic planning, and should develop a forecasting capability that can anticipate changes from the audience's perspective. In order to determine how best to meet changing audience needs and take advantage of new opportunities for audience development, performing arts marketers should undertake consumer segmentation research and leverage the abundance of data available to them to help pinpoint and appeal to target markets.

It creatively and strategically segments the audience into target groups. Many marketing managers of arts organizations do think of segmentation when planning strategy, but all too often, they do so only from time to time and in the most general terms. Managers of symphony orchestras, for example, are well aware that their prospects are better in high- than in low-income households, among women than among men, among the well educated rather than the less educated, and among certain age groups. This understanding certainly helps to determine where budgets are most effectively concentrated, but all too often these budgets are spent on a single "best" program, usually aimed at upscale households, who, in marketing lingo, are known to be the "heavy users."

Yet, even within this market, many possibilities for more subtle segmentation exist. A study commissioned by the NEA revealed that, despite industry-wide intuition to the contrary, the best predictors of likely symphony attendance were not the traditional demographic characteristics, such as income and education, that are commonly researched on audience surveys, but lifestyle factors, attitudes toward actual attendance, past experience, and childhood training.[11] Considering only the lifestyle measure, the study clearly showed that there were two major lifestyle groups interested in symphony attendance. One group was the traditional *cultural lifestyle group,* who made cultural events the center of their leisure pursuits. They tended to patronize the theater, opera, and museums as well as the symphony. They were very much interested in the program content and the artists when selecting performances to attend, and tended to be relatively insensitive to atmospheric factors or prices. The cultural experience itself was their primary reason for attendance. This group is undoubtedly the one that many theater and symphony marketers have in mind when they design their "one best" strategy.

The research, however, identified a very different lifestyle group that also included excellent prospects for the symphony, which was named the *socially active group.* Respondents in this group went out a lot, not only to the symphony, but also to all sorts of nonclassical events, and they liked to give and attend parties. Their symphony attendance was largely a social experience, an opportunity to meet with friends, to plan a dinner before the concert and perhaps dessert or cocktails after. Going out was the point, and the content of the program was of far less interest than who among their friends was going or what restaurant they should attend.

Obviously, the appropriate strategies to reach these two groups are very different, and a strategy developed to appeal to one group may be unattractive to the other. It is the marketer's challenge to develop messages that appeal to each of these groups and specifically target them.

It develops strategies using all elements of the "marketing mix," not just communication. The marketing mix, one of the key concepts in modern marketing theory, is the set of marketing tools that the firm uses to pursue its marketing objectives in the target market. McCarthy popularized a four-factor classification of these tools called the four Ps: product, price, place, and promotion.[12] An important fifth "P" has been added to the list: people.

- *Product*: The choice of *core product*, the works performed, is primarily in the domain of the artistic director, not the marketer. The *augmented* product consists of marketer-designed offerings such as subscription packages, memberships, singles nights, special pricing offers, postperformance discussions, and loyalty-building programs.
- *Price*: Performing arts organizations set prices differentiated by many factors, such as seat location, timing of purchase relative to performance date, and audience member category (student and senior discounts). In recent years, many organizations have adopted dynamic pricing as an integral part of their pricing strategy, varying prices over the time period that tickets are sold. Developing appropriate pricing structures is an important part of planning the marketing mix.
- *Place*: Place refers to the channels or access points through which the product is made available to the public. Many variations are possible in performing venues and ticket distribution, and creative distribution concepts are being successfully enacted by many performing arts organizations.
- *Promotion*: Promotion consists of all efforts that communicate with the public, including advertising, public relations, direct mail, websites, email, social media, telemarketing, and personal selling. This final step in the marketing process encompasses the communication of the strategies and tactics developed through the other aspects of the marketing mix.
- *People*: This refers to the arts organization's managerial and artistic staff, box office personnel, board members, volunteers, and others who come in contact with customers. These people's attitudes and communication skills can help or hurt the organization's marketing effectiveness.

It nurtures customers. The Internet, other technological advances, and globalization have combined to create a new economy. The old economy was built on managing industries; the new economy is built on managing information. Previously, say Kotler, Jain, and Maesincee in their book *Marketing Moves*, "the company had been the hunter searching for customers; now the consumer has become the hunter."[13] Instead of acting as hunters, smart managers function as gardeners, nurturing their customers. Marketers can respond to this "reverse" marketing by paying attention to the customer's four Cs: enhanced customer value, lower costs (tangible and intangible; perceived and real), improved convenience, and better communications. Businesses need to shift their focus from products to customers. This requires shifting the marketing mind-set from make-and-sell to sense-and-respond, from mass markets to markets of one, and from seasonal marketing to real-time marketing. Using the Internet, the customer can readily tell the organization what it wants and the organization delivers. Thus, the customer changes roles from *consumer* to *prosumer.*

The emergence of the new economy does not mean the extinction of the old, but requires a new focus overlaying preexisting approaches. Therefore, it is crucial that marketers fully understand and adopt the customer-centered philosophy before realizing the collaborative, values-driven, humanistic approach that is becoming the all-pervasive marketing methodology as people become more comfortable, adept with, and reliant on the tools of the new technology.

THE CUSTOMER ORIENTATION: HOW FAR DOES IT GO?

Marketing is a tool—actually a process and a set of tools wrapped in a philosophy—for helping the organization do what it wants to do. Marketing is a way of thinking for the entire organization, whether it is a nonprofit or in the business sector. Says Steve McMillan, former CEO of the Sara Lee Corporation, "Business success is a function of how well the marketing mind-set is integrated in the entire company."[14] Management must decide which goals marketing can help achieve, and how. It is management's prerogative to say that certain decisions will be made with little or no attention to marketing concerns. Thus, a theater company may decide to choose the season's program on the basis of the interests of its artistic directors. Marketing may then be assigned the task of maximizing audience revenues for that given program. This of course does not mean that marketing should fall back upon a selling philosophy. It simply means that marketing planning must start with customers in deciding whom to target and how to describe, package, price, and distribute a given program.

At the other extreme, a theater director may decide to be very customer-driven. He or she may very carefully survey the potential audience, analyze past revenues and audience reactions, and consider which artists and works are available to maximize future attendance. This organization would then establish an offering that maximizes sales. These two approaches are equally customer-oriented. They differ only in the management goals they were designed to achieve.

Most managers operate between the two extremes. Artist-driven and customer-driven performances are often mixed over the season. (It has been estimated that presentations of the *Nutcracker* constitute more than 30 percent of dance performances in the United States each year. The revenue gained from these popular, well-attended productions help support each dance company's more artist-driven, experimental, and less-familiar productions that have narrower appeal.) Or an organization may present productions that are artist-driven with respect to content but audience-driven with respect to talent, place of performance, pricing, or special events. For example, people are as likely to go hear Itzhak Perlman perform whether he is playing Beethoven or a contemporary composer. A singles night at the theater is likely to draw a crowd, whatever the performance. A concert in the park will probably draw different and larger audiences than one in the symphony hall. Arts managers, therefore, have a great deal of flexibility in choosing how audience-driven the programming will be for a particular planning period, but the way they market the offering should always be customer-centered.

Writer Marya Mannes says, "The artist of today says to the public: 'If you don't understand this you are dumb.' I maintain that you are not. If you have to go the whole way to meet the artist, it's his fault."[15] In other words, the artist and the organization presenting the artist's work have a responsibility to the audience. Efforts to bring the arts to the public—and the public to the arts—do not represent pandering to lowly desires, but rather provide access to man's highest achievements and, more basically, to man's humanity. The marketer has the responsibility of packaging and communicating the artist's product to appeal to the broadest possible audience by meeting patrons' needs and preferences. The artistic experience exists in the communication between the performer(s) on stage and the patrons in the hall; it is the role of the manager, the marketers, and the artistic directors to help facilitate that communication, thereby fulfilling the mission of the organization. And that communication is made infinitely easier and much less costly than in the past, thanks to the pervasive use of the Internet, email, and various social media platforms.

The function of marketing is not to deny leadership or inspiration. Marketing exists in conjunction with other functions in order to find the most effective means for sharing inspiration with the public. The key is knowing how to make the product more accessible to the patron, not how to change the artist's vision or the organization's mission.

Innovations in sets, costumes, lighting, and supertitles in recent years have elevated the visual aspect of the opera-going experience to approach the musical aspect, creating a total experience that is highly appreciated by artists and audiences alike. The Metropolitan Opera, which was among the last holdouts in the American opera world against translated supertitles, "succumbed" because of the proven importance of this feature in increasing the audience's understanding and appreciation. An important by-product is that supertitles provide artistic decision-makers the freedom to venture beyond the traditional, well-known repertoire and produce unfamiliar works by well-known composers as well as works in such languages as Russian and Czech. Other visual aspects such as images on large screens enhance the concert-going experience and have been widely adopted for certain performances at many orchestras.

Say Pine and Gilmore in their book *The Experience Economy*, we live in a "new economic era in which every business is a stage and companies must design memorable events that engage customers in an inherently personal way."[16] Trendy cafes offer the ambience of cushy sofas, roaring fires, soft jazz, and wireless Internet so that their patrons are happy to pay a premium price for a cup of coffee; sports stores feature rock climbing walls for kids to keep parents browsing and returning to the store; but cultural institutions have the great advantage that by their very nature they provide authentic, unique, quality experiences with their core offering that is being performed on stage. Arts organizations must ascertain that every encounter the public has with them—from the organizations' marketing materials to the ticket purchase transaction and the experience in the lobby—anticipates and enhances the nature of the performance experience. It is the goal of artists and marketers to *engage* the audience.

Not every audience group will respond to postperformance discussions, social media tactics, or cocktail receptions. Most importantly, not all audience segments will be interested in all of an organization's artistic offerings. Offers need to be designed for and positioned to their target audiences. And there must be a real commitment to ongoing implementation of the strategies.

CHAPTER 3

UNDERSTANDING THE PERFORMING ARTS MARKET: HOW CONSUMERS THINK

DANIEL KAHNEMAN, THE NOBEL PRIZE–WINNING BEHAVIORAL ECONOMIST, shares the story of a man who told him that he listened to a gloriously beautiful symphony for 20 minutes, after which a terrible screeching sound ruined his experience. Kahneman replied to him that the experience was not ruined, as he enjoyed 20 wonderful minutes, but it was his *memory of the experience* that was ruined. Kahneman describes this as the riddle of experience versus memory. Says Kahneman, "We don't choose between experiences, we choose between memories of experiences. Even when we think about the future, we don't think of our future normally as experiences. We think of our future as anticipated memories. We don't tell stories, our memory tells our stories." Furthermore, "the remembering self is the one that makes decisions. The experiencing self has no voice in decisions. Look at this," says Kahneman, "as the tyranny of the remembering self."[1]

CREATING MEMORABLE EXPERIENCES

Arts managers need to focus on creating memorable experiences so that people will take away positive, even delighted, thoughts and feelings about their experiences at performances and develop lasting, fond memories. People choose to pursue or repeat an activity or not based on the feelings, impressions, and experiences that are packaged together in a memory.

Research conducted by WolfBrown consultants has found that anticipation is one of the lead indicators of impact and that impact increases the more familiar a person is with the work they are going to see. Preparing patrons for the experience they are about to have helps to set them up for a better, stronger memory of the show. Similarly, patrons who participate in postperformance activities such as discussions with artistic personnel or their own companions are more likely to have better and stronger future memories of the performance.

How often are bad memories playing into people's decisions not to return to a performing arts organization? Research conducted on behalf of the symphony orchestra industry found that of the patrons new to the symphony, dubbed the "unconverted trialists," most did not return the following season. As people have more experience attending the symphony, their "churn rate" decreases precipitously. If people have many good memories of experiences at the symphony, one bad experience won't play so important a role. It would be helpful for marketers to learn why new concert attenders select a certain performance. Do they have a special interest in or relationship with a performer, which accounts for sporadic attendance? Would they enjoy the experience more with different music? Or is it the social aspects of the evening out that disappointed? Research clearly demonstrates that most performing arts attenders are not primarily culture buffs but come for social reasons.

Diane Paulus, artistic director of A.R.T. in Cambridge, Massachusetts, offers another perspective on the experience. Says Paulus,

> We've got to open up the definition of what theater *is.* OBERON (A.R.T.'s Zero Arrow Street Theater) is accessing a new demographic: a younger, under-30 audience. This audience isn't one that *goes to the theater*—they go *out at night....* The theater needs to be something where you feel: "I have to experience it."... People are craving experience—they are *desperate* for experience.[2]

Paulus continues, "There's a syndrome in our profession—to blame the audience, especially young people.... We have to flip that analysis and say, 'Maybe it's *us*—maybe it's the arts producers... perhaps we have to do a better job of *inviting* this audience back to the theater."[3]

Overview of Consumer Behavior

At the core of effective marketing is an understanding of consumer behavior. Marketing planners must understand the motives, preferences, and behavior of their organizations' current and potential consumers. In recent years, sociologists, psychologists, economists, and arts researchers have been working in areas that assist organizations in understanding the performing arts audience. They seek to answer such questions as: What motivates a person to purchase a ticket? To buy or renew a subscription? To attend once and not return? What factors create satisfaction, even delight, and stimulate loyalty to a performing arts organization?

A study conducted by the Australian Opera indicated that the major factor in a person's decision to attend was the reputation of the opera itself.[4] *La Boheme* was more appealing than the less familiar *Lulu*. Yet, in another study of cultural patrons, researchers found that some audience segments have an appetite for new experiences, meaning both new and unfamiliar works.[5]

For many people, it is not the work being presented that informs the attendance decision, but other aspects of the total concert- or theatergoing experience. Those who complain that people do not like the performing arts and that nothing marketers say or do attracts people to come must look more carefully at how they are packaging and promoting their offers to the public. Often functional details prevent people from attending, not the artistic offerings.

When the Cherry Lane Theater in New York City's West Village presented *EVEolution*, a two-woman play chronicling motherhood's bumps and triumphs, managers couldn't understand why they weren't attracting young moms, the ideal audience for this show. It turned out that the performance schedule didn't work for moms who needed to be home when school lets out. So the theater changed its Wednesday matinee performance from 2:00 p.m. to 11:00 a.m., which allowed mothers to be out in plenty of time for the 3:00 p.m. school pickup. These performances attracted not only carloads and busloads of moms, but also older adults who enjoyed beating the late afternoon rush hour.[6]

Symphony administrators at the Saint Louis Symphony Orchestra (SLSO) thought people who didn't attend their concerts were not aware of their international tours, their awards, and their great reviews. When they hired a market research firm to find out what people thought about the orchestra, they got some surprises. People did know about the SLSO's fine reputation. But they were not attending for other reasons—largely because of factors that created anxiety for them. People wanted to know: What kind of people go to the symphony concerts? Are they people like me? Where do I park? In fact, the symphony's prestige actually contributed to people's anxiety.

So, in response to the research report, the orchestra ran a new sales campaign that soft-pedaled its strong reputation in favor of an emotional appeal. Rather than advertising, "Virtuoso musician coming to orchestra hall," they used such lines as "When Wynton Marsalis was here, killer wails were heard in the hall!" They even offered a money-back guarantee for the opening weekend, which stimulated a significant increase in ticket sales. The success of the campaign was proved by the fact that no one asked for a refund.[7]

Music critic Alex Ross believes the term "classical music"

> is a masterpiece of negative publicity, a tour de force of anti-hype. It cancels out the possibility that music in the spirit of Beethoven could still be created today.... When people hear "classical," they think "dead." The music is described in terms of its distance from the present, its resistance

> to the mass—what it is not. You see magazines with listings for Popular Music in one section and for Classical Music in another, so that the latter becomes by implication, Unpopular Music. No wonder that stories of its imminent demise are so commonplace.[8]

In a similar vein, the *Chicago Tribune*'s leisure section is entitled "Arts and Entertainment." Does this mean their editors think that arts are not entertaining? Does this title subconsciously induce people seeking an entertaining evening *not* to select a fine arts event?

The Marketing Mind-Set

A marketing mind-set requires that the organization systematically study customers' needs and desires, perceptions and attitudes, preferences and concerns. Because product value is defined by the consumer, managers must distinguish between the way they think about what they do and the way their target markets think about what they do. Arts managers are the experts on their products, so they see the world through biased eyes. It is important for arts managers to always keep in mind that the organization's success is a function of the thoughts and emotions the target patrons attach to the organization and its offerings.

What gives the marketing manager power is knowledge of the customer. Marketers need to approach people as whole human beings with minds, hearts, and spirits. People seek not only functional and emotional fulfillment, but also human spirit fulfillment in the products and services they choose.

Choice

In the last three or four decades of the twentieth century, people often subscribed eagerly without even knowing what was to be performed the next season, trusting the artistic director to choose for them. But more recently, many long-standing subscribers have dropped their subscriptions in favor of multiple single ticket purchases and it has become more and more difficult to attract new subscribers because people want to choose exactly which performances to attend.

Self-expression has become a critical element of our culture. People want to have more and more control over their lives and choice is an essential part of control. For decades, arts organizations have said "Subscribe now!" (In other words, have it our way.) Consider the completely opposite tag line "Have it your way!" from mass marketer Burger King, which understands what motivates contemporary consumers. This situation is problematic for arts marketers who must adapt to what has become a much more complex task: selling tickets on a show-by-show basis and attracting people to come back frequently.

In her book *The Art of Choosing*, Sheena Iyengar explains that our brains "respond more to rewards [that we] actively choose than to identical rewards that are passively received."[9] In other words, we tend to have more positive thoughts and feelings if we choose which performances to attend and which benefits we receive than if they are given to or imposed on us. This is confirmed by the numerous studies that have shown that even though subscribers may be the most frequent and loyal attenders, it is the single ticket buyers, the people who select shows one by one, who report having the highest satisfaction with the experience.

Psychologist Barry Schwartz, author of *The Paradox of Choice: Why More Is Less*, argues that contrary to the conventional wisdom that greater choice is for the greater good, the abundance of choice in today's Western world is actually making us miserable. Infinite choice is paralyzing and exhausting to the human psyche, says Schwartz. It leads us to set unreasonably high expectations, question our choices before we even make them, and blame our failures entirely on ourselves. With so many options, people find it hard to choose at all.[10] According to Schwartz, even if we manage to overcome the paralysis and make a choice, we end up less satisfied with our choice than we would be if we had fewer options to choose from. Schwartz tells us that constantly being asked to make choices, even about the simplest things, forces us to "invest time, energy, no small amount of self-doubt, and dread." There comes a point, he contends, at which choice becomes debilitating rather than liberating. Did I make the right choice? Can I ever make the right choice? So much choice produces paralysis rather than freedom. Fewer people participate when faced with too many options. The situation results in the escalation of satisfaction; multiple choices escalate people's view of what they expect. The more options there are, the easier it is to regret anything at all that is disappointing about the option that you choose. This is a peculiar problem of modern affluent society. Schwartz concludes, tongue in cheek, that everything was better back when everything was worse.[11]

Says Kent Greenfield in his book *The Myth of Choice*, "we like things, people, beliefs, and products that are familiar and like them even more when we feel threatened or stressed." Greenfield cites a study that asked people at a conference to establish their preferences for snacks over several days, picking one kind of snack per day. The participants anticipated that they would want variety, so they ordered a variety of snacks. Even though the attendees ordered in advance, each day they were offered the full range of snacks. What did they choose? They did not vary their choices day by day; rather they selected their same favorite every day. According to Greenfield, from the marketer's perspective, the perfect product is one that is purchased out of habit or compulsion, but which the purchaser feels he or she has exercised free will and rationality in choosing.[12] During the 20-year period from 1992 to 2012, nonprofit Broadway theaters offered 39 new works and 105 revivals. The mission of some of these theaters is to produce the classics, so it is expected that many revivals would be performed, but, nonetheless, the numbers clearly demonstrate an audience preference for the tried and true.[13]

Is it any wonder that people, especially those unfamiliar with symphonic repertoire and guest artists, are often confused by the brochure of a large symphony orchestra that may offer one hundred or more different programs over the course of a season? And when people have made a choice and then are unhappy with the program they selected, are they reluctant to select another performance to attend?

When choices are complicated, unfamiliar, numerous, or overwhelming, we need cognitive shortcuts and other aids in order to make decisions. This is where great marketing enters the scene.

Risk and Uncertainty

Arts marketers often ask how they can *attract* more people to attend their performances. But they should consider not only identifying what factors will entice people to attend, but identifying and breaking down the *barriers* that prevent people from attending.

Daniel Kahneman has done extensive research about people's propensity to dislike uncertainty and to be extremely risk-averse. He argues that people think in terms of gains and losses, but fear loss much more than they value gain, especially in the short term.[14] In other words, when people ask themselves "Will I be better off or worse off if I take this action?," the negative aspects loom larger than the potential positives. Furthermore, intuitive decisions are shaped by factors that determine the accessibility of different features of a situation. Highly accessible features influence decisions, while features of low accessibility will be largely ignored. Says Kahneman, "Unfortunately, there is no reason to believe that the most accessible features are also the most relevant to a good decision."[15]

Too often, advertisements feature the names of performers, composers, or choreographers that only the most die-hard fans would recognize. Says Welz Kauffman, executive director of the Ravinia Festival, the summer home of the Chicago Symphony Orchestra, "Rather than working to win over new audiences, the average newspaper advertisement for a classical event simply lists names that are probably unfamiliar to a majority of readers."[16]

Common risks or potential losses that people face when considering attending performing arts events are the fear that they will not understand a performance, that they might applaud at the wrong time, that they will be bored, or just not like the performance, incurring both the tangible and intangible costs of a bad experience. When considering a ticket purchase, infrequent or nonattenders often ask themselves: Will there be other people like me there? Will it look like I do not belong? Do I need a degree in music to appreciate the performance? Will I be able to get tickets? Good seats? Seats I can afford? Will I be able to find someone to accompany me?

People also fear that they will regret their choices. Says Kahneman,

> People expect to have stronger emotional reactions to an outcome that is produced by action than to the same outcome when it is produced by inaction. . . . The asymmetry in the risk of regret favors conventional and risk-averse choices. Consumers who are reminded that they may feel regret as a result of their choice show an increased preference for conventional options.[17]

Henry Fogel, former president and CEO of the League of American Orchestras, fixes blame firmly on the music establishment for the fact that classical music has been marginalized. "Complex program notes, musicians in white ties and tails, and dowagers who hiss if one claps at the wrong time all keep newcomers out of the concert hall," he noted. After seeing a conductor wag his finger at concertgoers who applauded too soon, Fogel wondered "how many more times those people will actually pay money for tickets so they can be humiliated."[18]

It is not only the new or infrequent attenders who face risks in attending performances. People who frequent full-length story ballets may not expect to appreciate modern dance repertory. Researchers conducting a study of dance attenders in Scotland found it difficult to recruit participants for their focus groups because even frequent attenders of ballet and contemporary dance became anxious when they realized that the topic would be dance. They felt uncomfortable talking about dance and were worried that they would seem stupid or ignorant. One respondent said: "I don't understand half of what they're trying to put over in dance but I just like freedom of movement." Many participants said that they were careful in their choice of programs to attend as they wanted to be sure they were going to enjoy an event before they committed to seeing it. The variability of the quality of the experience from one production to another, especially between different performing groups, adds another element of risk that creates a barrier to attendance.[19]

In a 2002 study of the attitudes of performing arts ticket buyers who have not made the leap to opera, about 30 percent said that they would be unlikely to attend because they do not understand the art form and would feel "intimidated" and "uncomfortable." Almost 40 percent said they would attend an opera if personally invited by a friend or family member whose opinion about opera they trusted.[20]

Not all of potential patrons' concerns can be addressed by arts marketers, but many of them can. Arts managers should put themselves inside the head of the reluctant attender and think in terms of dispelling myths and fears and breaking down other barriers to attendance.

A theater on the outskirts of Copenhagen, about a 45-minute drive from the city center, is facing great difficulty attracting people to its performances. The theater's director told me that his theater offers high-quality avant-garde productions and the drive should not be a barrier for so many people. I replied that maybe the physical distance was not the key factor keeping people away. If "avant garde" implies "on the edge," maybe people are risk-averse to attending a performance that may be too edgy for their taste; that may even go *over* the edge. I suggested to the theater director to consider whether the idea of "on the edge theater" serves to attract people or if it serves as a barrier. People want to *trust* the organization at which they purchase tickets. If trust is a key factor in the ticket purchase decision, what should the theater's primary message be to potential patrons?

In a similar vein, *Chicago Tribune* theater critic Chris Jones titled his "Fall Theater Guide 2012," "Take a Risk This Fall." Says Jones,

> Chicago is, above all, known as a tryout town, an experimental city, a boundary crosser.... Any theater lover in these parts should have a taste for adventure.... We've combed the fall performance schedules with an eye to offering some suggestions for the progressive theatergoer: the arts fan who likes to be surprised and challenged, who seeks out intellectual titillation and formative innovation and who does not necessarily want to bank on the tried and true or the kind of thing he or she has seen many times before.... On our Top 10 list for the start of the 2012–2013 theater season, you'll find solo tours de force, world premieres, Midwest premieres, Chicago premieres, experiments and all the risk-taking any audience member could want.[21]

How many people will this approach actually attract? How many will avoid the risk?

Decision-Making

Daniel Ariely's book title *Predictably Irrational* encapsulates his analysis of how our minds work when making decisions. According to Ariely, we are far less rational than conventional economics theory assumes. Moreover, these irrational behaviors of ours are neither random nor senseless. They are systematic and, since we repeat them again and again, predictable.[22]

Gerald Zaltman, author of *How Customers Think: Essential Insights into the Mind of the Market*, asserts that many managers still believe that consumers make decisions deliberately, by logically processing information to arrive at a judgment. Says Zaltman, "Consumer decision making sometimes does involve this so-called rational thinking,... however, it is the exception, rather than the rule. As it turns out, the selection process is relatively automatic, stems from habits and other unconscious forces, and is greatly influenced by the consumer's social and physical context." "More important still," claims Zaltman, "emotions contribute to, and are essential for, sound decision making.... In actuality, ninety-five percent of thinking takes place in our unconscious minds—that wonderful, if messy, stew of memories, emotions, thoughts, and other cognitive processes we're not aware of or that we can't articulate."[23]

Gary Klein, author of *Sources of Power: How People Make Decisions*, lends the validity of scientific research to everyday decision-making techniques such as intuition and use of metaphor. The use of story-telling and metaphor enables decision-makers to devise meaningful frameworks and compare their present situations to previous events. Intuition, says Klein, is not based on instantaneous insight, but rather on the rapid (perhaps even subconscious) interpretation of perceptual cues.[24]

Experience is a central factor in how we make decisions. A person who does not have experience to draw upon will rely heavily on traditional decision-making models—gathering information and weighing options. For such people, it is necessary to provide answers to the questions that will help them understand the nature of the experience. More experienced people can see a situation for what it is and quickly, even automatically, size it up and make a decision. For these people, building on their experience is an excellent way to draw them into an offer.

A common marketing principle is that past behavior is the best predictor of future behavior. If marketers know what people have done in the past and why they made the choices they did, this information can be used to help stimulate future purchases. People typically have a *confirmation bias*; we are prone to notice information that confirms what we already think and disregard facts that tend to disprove our preexisting notions. So, the marketer needs to learn about people's attitudes and then try to approach the people by starting from their perspectives and easing into modifications. Two factors make this task particularly challenging. First, people differ from one another and it is complicated to relate to them one by one. Second, people change over time. Lifecycle stage may affect a person's attendance behavior, to either more or less frequent attendance. Also, interests change. An opera company offering many of the most beloved operas the next season may find many new patrons attracted to the programs, may please the loyal patrons who prefer the great classics, and may deter those who have seen these operas many times and don't want to invest their money or time in yet another familiar production.

The Rational Decision-Making Process

It is crucial for marketers to consider how people make decisions when designing marketing offers and communications. Although behavioral economists and psychologists teach us that decision-making is largely emotional and intuitive, aspects of the traditional, rational approach to decision-making still play a role for people in many situations and should be understood by marketers.

In the model of rational decision-making, a person follows the steps of identifying a particular need, seeking a set of options that could meet that need, evaluating the pros and cons of each option, calculating the tangible and intangible costs of overall satisfaction per option, and then making a well-reasoned decision.

Low-Involvement to High-Involvement Decisions

For many performing arts attenders, the decision to purchase a ticket is stimulated by the mere receipt of an attractive announcement of an interesting performance.[25] A study conducted for the Theatre Royal Stratford East in England concluded that "respondents go to the theater because they want to see a particular performance or a particular actor."[26] For such people, arts attendance is often routine and familiar, and the buying decision is a relatively simple, low-involvement one.

The renewal of a theater subscription for the fifth consecutive year by a satisfied patron is a low-involvement decision. The decision has become so routine that there is little or no observable decision-making. This type of purchase decision is sometimes termed a *straight rebuy.* In a straight rebuy, the buyer does not need much information because he or she knows the organization, and the costs and benefits it confers, from previous experience. Requiring a higher level of involvement is the *modified rebuy*, such as the decision by a couple who are longtime subscribers to the symphony to add a subscription to the orchestra's new chamber music series. The couple is familiar with the organization and they are merely modifying a purchasing decision they have made in the past. This is a case of decision-making with experience, but one that requires more information than a straight rebuy.

In the case of *high-involvement* exchanges, the buying decision process is far more complex. High personal involvement typically occurs when the type of exchange is new to the decision-maker and the consumer does not know much about the product category. The consumer is also highly involved when the purchase decision is important, expensive, or risky; when it reflects upon the consumer's self-image; or when reference-group pressures to act in a particular way are strong.

The decision to attend an opera for the first time can be considered a high-involvement decision. This situation represents a new task for people who are faced with a new offer of an unfamiliar kind from an unfamiliar seller. Such potential ticket buyers typically undertake an extensive search for information about the opera, the performers, the venue, ticket prices, parking information, and more. In this context, it is easy to understand why organizations avidly promote subscription renewals: it is much easier for a marketer to sell a straight rebuy than a first-time purchase.

Theorists believe that even low-involvement decisions actually involve cognitive activity that is a simplified version of the process involved in more elaborate decision-making. Furthermore, it is likely that the same factors that make a decision relatively trivial at one time may make it more deliberate at another time—such as when a longtime subscriber becomes dissatisfied with the programming and considers not renewing his or her subscription, or when an economic setback requires that the patron consider the relative importance of this expense.

Stages of Decision-Making
The decision process proceeds through five stages:

1. *Need recognition*: We have a babysitter and no plans for the evening.
2. *Information search*: What entertainment options are available to us?
3. *Evaluation of alternatives*: Should we go to a play, a film, or to a concert?
4. *Purchase decision*: The play will be here for another few weeks and we will be able to stream the film on Netflix soon. Let's go to the concert; I hear the soloist is wonderful.
5. *Postpurchase behavior*: I really liked that soloist and the orchestra played beautifully. I want to get tickets for another performance soon.

Breaking down the buying decision process into five stages makes it clear that the decision process starts long before the actual purchase and has consequences long afterward. This insight encourages the marketer to focus on the factors and influences that can affect each stage of decision-making.

NEED RECOGNITION. The buying process starts when the consumer recognizes a problem or need, which may be triggered by either internal or external stimuli. An internal stimulus may be a desire to see a much-discussed show, a famous star, or the new work of a respected playwright. External stimuli include advertisements, reviews, email blasts, direct mail, websites and social media, and word-of-mouth recommendations from friends or colleagues. By gathering information from a number of consumers, the marketer can identify the most frequent stimuli that spark interest in a performance or subscription. The marketer can then develop marketing strategies that trigger consumer interest.

INFORMATION SEARCH. An interested consumer is inclined to search for more information. For example, someone driving home from work may be thinking about how to spend Saturday evening. That person will be likely to have heightened attention when the local theater reviewer comes on the radio. Upon arriving home, the person may begin an active information search, looking up theater schedules and prices, contacting friends, and checking for ticket availability. How much search the person undertakes depends on the amount of information he initially has, the value he places on additional information, and the satisfaction he gets from the search itself. Some people are content to respond to one seemingly viable offering; others need to investigate all their options before making a selection. In general, the amount of search activity increases as the consumer moves from limited problem solving (Shall we go to the jazz concert or to the movies tonight?) to extensive problem solving (Which one of these three worthwhile theaters shall we subscribe to next season?).

Most often, people do not compare one production to another when deciding whether or not to attend. If a particular dance company or play sparks their interest, they will seek information to help them decide if this is their "kind of thing."

The major information sources that the consumer will turn to and the relative influence each will have on the subsequent purchase decision are of key interest to the marketer. Consumer information sources may be personal (family, friends, and colleagues), commercial (advertising, email blasts, posters, etc.), public (mass media, reviews, award-granting institutions such as the "Tony"), or experiential (based on previous experience with similar products). Arts marketers often survey their audiences to determine which information sources play a part in their decisions and to assess the relative importance of each source. The patrons' responses help organizations to prepare effective communications for the target market. However, people often are exposed to a variety of message sources and it is hard to know which one, or which combination, was influential. Of course, it must be remembered that this approach only analyzes the responses of current patrons. If an organization would like to attract new target markets, the information-seeking habits and preferences of the target groups must be researched. This process is not necessarily as difficult as one might think. One arts organization arranged with a car repair shop in a target neighborhood to keep track of the radio station to which each incoming car was tuned. One could even limit such a search to certain brands of cars. Of course, in this age of sophisticated software and social media, it is easy to track people's online search behavior and learn a great deal about their interests and web-based activity.

Evaluation of Alternatives. If consumers were to evaluate alternatives based solely on the information they collect, the marketer's job would be relatively simple. But the process is complicated by all of the cultural, social, personal, and psychological factors that influence consumer behavior. The consumer sees each product as a bundle of attributes, which have varying capabilities of delivering the benefits sought and satisfying certain needs.

Consumers will differ as to which product attributes are relevant to them. One person may choose to attend a certain concert because of the symphony being performed, another because she likes a featured soloist, yet another because good friends invited her to join them. Other factors also play a part in the evaluation process, such as ease of parking or ticket prices, and their influence will vary with different consumers. There is a significant difference between the importance and the salience of product attributes to consumers. A symphony may make its awards and great reviews salient in its ads, but other attributes such as programming and ancillary social events may be more important to consumers.

Marketers should emphasize those attributes that are most important to the targeted audience.

There are several tactics a marketer may employ to influence the consumer's evaluation process. Consider a modern dance company performing at an outdoor festival in a new city. First, the marketer can attempt to change consumers' beliefs about modern dance (You think modern dance is for highbrows? Well, you'll be dancing in the streets after this performance!). Second, the marketer may attempt to increase the perceived weighted importance of certain factors (Listen to the worldwide acclaim the company has received). Third, the marketer may attempt to change beliefs about competitors (And you thought basketball is athletic and beautiful to watch!). Fourth, the marketer may call attention to neglected favorable consequences (Enjoy a picnic in the park under the stars before the show). Emotional appeals are seen as the most effective way to draw in potential ticket buyers. Says MOMIX on its website, "With nothing more than light, shadow, props, and the human body, MOMIX has astonished audiences on five continents for more than 30 years."[27]

PARTICIPANTS IN THE PURCHASE DECISION. The decision-making process in arts attendance is made more complex by the fact that the target consumer is often not an individual, but a group. Five different roles may be played by people involved in the decision process:

1. *Initiator*: The initiator is the person who first suggests the idea of becoming involved in a particular exchange.
2. *Influencer*: An influencer is a person who offers or is sought out for advice on the decision. The influencer may be a friend who has seen the show; people unknown to the potential buyer who offer their comments, usually online; or a professional critic.
3. *Decider*: The decider is the person who ultimately determines any or all parts of the decision to participate in the exchange: whether to take action; what action to take; how, when, or where to take action.
4. *Transactor*: The transactor is the person who completes the actual purchase.
5. *Consumer*: The attender may or may not participate in the actual purchase decision roles.

For example, while reading the 2012 season program book for the Ravinia Festival, I noticed that MOMIX was the featured dance company for the summer. I had heard of the company but had never seen them perform and didn't know much about them. I called my friend Madeleine who frequents Ravinia and is a dance aficionado. Madeleine said: "You've never seen them? Don't miss this show!" Intrigued, I looked up information about the company on its website and read a review in the *Philadelphia Inquirer* from a May 2012 performance. Said critic Marilyn Jackson, "MOMIX soars and enchants again." Other excerpts from the review say: "Genius stagecrafter and choreographer Moses Pendleton," "his inventiveness and artistry far surpass the popular Cirque du Soleil," "Pendleton brings beauty, mystery, emotion, and uproarious fun to the table, too."[28] One comment that particularly attracted me was the judgment that MOMIX surpasses Cirque du Soleil because I saw Cirque du Soleil's "O" and thought it was among the most inventive shows I have ever seen. Both Madeleine and the critic were highly effective influencers. Once I decided to attend, I invited others to accompany me and bought the tickets.

POSTPURCHASE BEHAVIOR. After experiencing a performance, a consumer will register some level of delight, satisfaction, or dissatisfaction. There are two major theories that explain whether consumers will be satisfied with a performance.

The *expectations-performance theory* holds that satisfaction is a function of the consumer's expectations and the perceived outcome.[29] If the outcome matches expectations, the consumer is satisfied; if it exceeds them, he or she is highly satisfied; if it falls short, he or she is dissatisfied. This theory suggests that the marketer should not overbuild expectations. If anything, the organization should underpromise and overdeliver so that consumers will experience higher-than-expected satisfaction. Consumers form their expectations on the basis of messages and claims by the organization, other communication sources, word-of-mouth recommendations, and past experience. If the consumer is satisfied, he or she will exhibit a higher probability of purchasing the product again. The satisfied customer will also tend to say good things about the experience to others, and marketers know that their best advertisement is a satisfied customer. The larger the gap between expectations and performance, the greater the consumer's dissatisfaction. Even if people liked a play, when it is touted as being "hilarious" and patrons find it mildly funny, there will be much dissatisfaction and people will be less likely to believe future claims by the organization.

The *cognitive dissonance theory* holds that making any choice always involves giving up other things, and therefore some postdecision regret, or cognitive dissonance, might be felt.[30] Dissonant consumers will resort to one of two courses of action. They may try to reduce the dissonance by vowing not to repeat the choice (I'll never go to modern dance/that theater again!). Or consumers may seek information to help them feel better about their choice (Now that I read the critic's review and explanations from the artistic director, I appreciate the play more).

Arts organizations strive to satisfy audience members over many productions and seasons. An important part of this effort is to take positive steps to reduce postpurchase dissonance and help patrons feel good about their choices. Warm letters of welcome should be sent to new subscribers. The organization can provide interesting information about the performances and human interest stories about the performers and staff. Managers can also solicit suggestions and complaints from the patrons by providing them with easy means to voice their interests and concerns. The biggest problem, of course, occurs when someone comes once and does not return. It requires a great deal of marketing skill to bring new people into the hall, and it is frustrating to lose patrons after just one experience. Maybe these people would like the next play of the season better? How can marketers entice people to come back and find out? All too often, desperate marketers use a strategy of deeply discounting prices to lure people into the hall, but if people don't want to come, it hardly matters what the price is.

UNDERSTANDING CONSUMER BEHAVIOR

Since the objectives of marketing are to create some level of satisfaction, commitment, and loyalty among target consumers, it is essential that all strategic planning start with understanding consumer behavior. This understanding allows the marketer to develop a marketing program that meets the needs of its publics.

BENEFITS SOUGHT

Marketers are typically most effective when they tap into outcome-based offerings—the benefits realized by the customer—rather than input-based offerings—descriptions of the program being presented. People want healthy teeth, not toothpaste; they want their cars to run, not a trip to the gas station. Similarly, people want an experience at a performing arts event that, depending on their preferences, is aesthetic, sociable, entertaining, familiar, new, educational, inspiring, and so on. For most people, a message that they will hear Mahler's magnificent Second Symphony or see a critically acclaimed play by even a well-known playwright will not generate a promise of what is in it for them. Descriptions of the offering should include, either directly or implicitly, promises of the benefits various consumer groups value.

Intrinsic and Extrinsic Benefits

The topic of what benefits people seek in attending a performance and what aspects of the experience they value was a focus of the ticket-buyer surveys in the John S. and James L. Knight Foundation Classical Music Consumer Segmentation Study. The insights derived from the survey responses focus on the fact that classical consumers derive "layers of value" around the concert experience that do not always relate to what is happening onstage. Benefits gained vary from the extrinsic to the intrinsic. Some people use concerts to entertain friends and family members ("occasion value") while others enjoy them as a means of nurturing and sustaining their personal relationships ("relationship enhancement value"). Respondents rate highly the "ritual/ambience" value of the concert experience. In focus groups, classical consumers discuss the more intrinsic "healing and therapeutic value" and the "spiritual or transformational value." Yet, all these layers of benefits and values surround the actual artistic and educational experience, which is what the orchestras sell.

While promoting anything "less" than the artistic and educational value of the performance experience is distasteful to some who work in the arts, understanding and interpreting the complex values that consumers draw from the experience and highlighting these benefits in marketing material is likely to stimulate demand, especially among those already predisposed to attending.

Self-Invention

To make their messages and offerings relevant to today's consumers, marketers must also respond to the concept of *self-invention,* which has blossomed in the era of the Internet, where extensive information is available a click away, empowering consumers to make better choices and to enrich their experiences. People have more confidence in their own abilities (72 percent) than they do in doctors (57 percent), police (39 percent), public schools (26 percent), the judicial system (18 percent), and advertising (7 percent), which ranks near the bottom of people's confidence level. Say Smith et al., "Expectations of self-invention are blurring the lines between consuming and marketing. In a world of self-invention, people want more, if not all, of the power and the control that marketers traditionally presumed was their alone."[31] As more and more people are configuring their iPods with the music they select for their own listening pleasure, few arts managers face the music that people are increasingly resistant to subscription packaging because they want to choose which performances to attend.

Social Factors

Social factors involved in the experience of attending an arts event play a significant role, especially for those who are not avid cultural attenders. Alan Brown, who directed a major survey of symphony audience attenders in 15 cities on behalf the John S. and James L. Knight Foundation, says, "A mounting body of research suggests that who invites you to a concert has as much to do with your decision to attend as other factors such as the program, guest artists, etc. . . . The vagaries of social context cast a long shadow on demand for arts programs."[32] Social factors such as reference groups, family, social roles, and status affect a consumer's mind-set and behavior.

Craving Connections

The Knight Foundation study also found that even though people connect with classical music more than ever, the traditional concert-hall experience is not the primary way these people relate to the art form. Rather they listen to music on their iPods, on satellite radio in their cars, or play CDs in their homes, although CD sales are dropping precipitously. When some people do attend concerts, they often come for social reasons: their friends are there, it is a good date or a spousal night out, they are entertaining visitors, or someone asks them to go.

Among respondents to the 2002 American Express opera study, one-third of attenders who did not buy their own tickets but were invited by a friend spoke warmly of the social aspects of the opera-going experience, such as a group dinner before the performance or the enjoyable "camaraderie" with friends. About 25 percent of nonattenders said that they did not attend opera as they did not want to attend alone, noting that their spouses or friends did not enjoy opera. On the other hand, regular opera goers reported that social aspects were not a major influence on their decision to attend. In fact, 25 percent commented that they often attended opera alone and enjoyed it a lot.[33]

Yankelovich Monitor consumer surveys show a nearly universal craving for connections and relationships. Of the respondents, 91 percent say that they are looking to find more time for the important people in their lives, whereas 73 percent say they are looking to do things that make them feel closer to others.[34]

Programs long in use by arts organizations, such as postconcert social gatherings, are highly limited as effective tools for appealing to people's social needs, as not much intermingling takes place at these events. Rather, says Brown,

> many individuals use arts programs as a means of investing in their personal relationships. While arts organizations sell artists and repertoire, consumers are buying spiritual journeys, emotional therapy and better relationships.... What's needed and quickly are new models for customer-centered marketing relationships with the goal of offering consumers a menu of involvement opportunities that fit into their lifestyles and reinforce their self-perceptions.[35]

Opinion Leaders

Consumers are also influenced by opinion leaders—those whose opinions they value. Opinion leaders are found in all strata of society. A specific person can be an opinion leader in certain areas and an opinion follower in others. The more highly a person esteems the opinion leader, the more influential the opinion leader will be in shaping that person's product and brand choices.

Initiators and Responders

How does a marketer define the customer? Is a customer someone who purchases a ticket or someone who attends a performance? For marketers, the difference is important. If the definition of "customer" is expanded to include people who enjoy concerts but do not attend without an invitation, then a fundamental realignment of marketing strategy is called for: a shift toward strategies that create and facilitate attendance in small social groups.

One survey of US adults clearly identified two types of consumers—the *initiators* and the *responders*. The initiators, who comprise 18 percent of culturally active adults (those who attended any arts events in the past year), strongly agree with the statement, "I'm the kind of person who likes to organize outings to cultural events for my friends." Responders, who comprise 56 percent of culturally active adults, report that "I'm much more likely to attend cultural outings if someone else invites me." Initiators are constantly scanning the media for activities (58 percent of initiators are "always looking for cultural activities to do" versus 16 percent of responders). They are more likely to be "extremely interested" in arts activities (24 percent versus 8 percent), and also more likely to say that arts activities play a "major role" in their lives (58 percent versus 23 percent). Initiators are more likely than responders to be single (36 percent versus 22 percent), younger (average age is 42 compared to 48), and female (60 percent versus 53 percent).

The 2002 Knight Foundation survey of 15 orchestra audiences found, importantly, that the incidence of initiators among orchestra subscribers is lower than that among culturally active adults. Subscribers make a major commitment far in advance to attend several programs and typically outlay significant money all at once. Getting other people to commit to the same is not an easy undertaking. Says Brown,

> Arts organizations must embrace the growing number of omnivorous and independent-minded single-ticket buyers who enjoy arts activities but won't subscribe. Arts groups spend a lot of money trying to convert Responders into Initiators. The research suggests, however, that as many as 85 out of 100 concert attendees do not have the psychological profile of an Initiator. . . . Would it not be more productive to find a new way to market to Responders indirectly through their Initiator friends?[36]

The benefits of arts participation accrue not only to the responders, but also to initiators. Focus group research suggests that some initiators derive meaning and satisfaction from the process of creating cultural experiences for their friends—the drama of putting it all together.

So, rather than making futile efforts to target responders directly, arts marketers should seek out the initiators and inspire and empower them to become "sales associates" for the organization. The highly effective program at Piccolo Teatro di Milano that leverages initiators' (whom they call "animators") goodwill and passion for the theater is described in chapter 15.

Reference Groups

People's reference groups usually have a direct influence on their attitudes or behavior. Since leisure-time activities have a strong social component, group affiliations are highly influential for involvement and attendance patterns. The more cohesive the group, the more effective its communication process, and therefore the greater its effect on the individual. Reference groups include informal primary groups, such as family, friends, neighbors, and coworkers, and more formal secondary groups, such as religious and professional groups. People are also influenced by groups to which they do not belong. Someone aspiring to be a member of the board of trustees of the local opera society may make large donations and attend special events to cultivate acceptance by the group's members.

Peer Group Influences

In a study of Cleveland's cultural patrons undertaken to compare the relative importance of peer group influences and childhood arts education for later arts attendance, it was found that mere exposure of children to culture appears to have little effect on later attendance habits. Performing arts attenders are about twice as likely as nonattenders to have friends who participate in the same cultural activities. Adult reference groups are so important that where they are absent, the effects of childhood exposure and education tend to dissipate.[37]

The implication for arts marketers is that attendance can be stimulated by promoting group sales among various membership groups and businesses, by offering gift tickets with subscriptions, and by encouraging the purchase of gift certificates. Marketing is not only an interaction between the organization and the customer, but also involves many exchanges of influence and information among the customer and the people in his or her network of friends, family, and associates.

CHAPTER 4

EXPLORING CHARACTERISTICS OF CURRENT AND POTENTIAL PERFORMING ARTS ATTENDERS

EFFECTIVE MARKETING COMMUNICATIONS APPEAL TO EACH TARGET CUSTOMER'S core values, lifestyle, and interests. It is crucial for arts marketers to stay up to date with the changing attitudes and expectations of their current patrons. Also, arts marketers must develop and implement strategies that attract new audiences while continuing to build loyalty and frequency among current audience segments.

In this chapter I will address age and life cycle status, gender, and ethnicity; personal, cultural, social, and psychological factors; and the ways these characteristics affect various audience segments' attitudes, behaviors, expectations, and preferences.

UNDERSTANDING DEMOGRAPHIC MARKET SEGMENTS

There are numerous segmentation variables that marketers can and should employ when analyzing and developing strategies to reach various audience groups. The demographic analysis that follows is just the beginning.

SEGMENTING BY AGE AND LIFE CYCLE

Arts organizations face the challenge of appealing to very different age and life cycle groups: the Traditionalists, also known as the "Silent Generation"; Baby Boomers, the largest generational segment; the 30- and 40-somethings, known as Generation X; and the younger set born before the year 2000, the members of Generation Y. Generation Z is composed of the children born since the turn of the current century. Each of these generations is characterized by a different level of exposure to and interest in the arts, by different factors that affect their propensity to attend arts events, and by differences in lifestyle that strongly affect which marketing messages and media are most effective in attracting them.

The Aging Population

There are now more Americans aged 65 and older than at any other time in US history. The nation's elderly population will more than double in size from 2005 through 2050, from 37 million to 81 million, as the Baby Boom generation enters the traditional retirement years. The number of working-age Americans and children will grow more slowly than the elderly population, and will shrink as a share of the total population.[1] In Great Britain, it is estimated that by 2035 the number of people between 16 and 50 years of age will decrease by 1.5 million and the number of people over 50 will increase by 6 million.

Traditionalists. Born between 1927 and 1945, the Traditionalists have among the highest rates of participation in live classical music, according to audience demographic research conducted by the League of American Orchestras (LAO). However, this generation has been aging out, with their overall population expected to decline by 45 percent, or 2.8 million people, by 2018.

To generate trial and repurchase, orchestras have created programs geared toward the senior segment, typically pure classical programming and pops series, and have scheduled concerts to better align with senior lifestyles, such as matinees and late morning "coffee and cookies" series. They also have offered discounts for senior subscription buyers and enhanced customer service experiences with high touch customer service for subscribers, including thank-you lunches. To target new audiences, orchestras have built awareness with free concerts in parks and senior living communities.

Orchestras have also partnered with social networks and businesses that have high concentrations of seniors, such as libraries, spas, insurance companies, estate planners, and community centers to promote their offerings. Some orchestras have also offered bus shuttles to their venue from senior homes.

Traditionalists prefer brick-and-mortar educational institutions and traditional lecture formats to online, web-based education and are much less technologically adept than the younger generations.

Baby Boomers. Born between 1946 and 1964, Baby Boomers are well established in their careers and often hold positions of power and authority. They welcome exciting, challenging projects and strive to make a difference. Baby Boomers have the greatest wealth of any generation, and their financial resources will grow as they age and as they inherit from their parents. The one in four Americans aged 50 and older control half of the nation's buying power and three-fourths of its assets. This mature market represents $150 billion in annual discretionary income. Current estimates suggest that in Britain people over 45 control 80 percent of the national wealth. US labor statistics indicate that nearly 80 million Baby Boomers will exit the workplace in the next decade, retiring at the rate of 8,000 per day.

According to the League of American Orchestras, nearly 2.5 million Boomer audience members stopped attending live classical concerts between 2002 and 2008. Further audience declines among Boomers are projected to be 1.7 million by 2018 if recent trends continue. Orchestras should focus on developing initiatives to increase Boomer attendance frequency.

MARKETING TO MATURE ADULTS. *American Demographics* magazine calls today's 50 and older population "the youngest, the wealthiest, the healthiest, best educated and most ambitious group of retirees ever." Their wealth will be much less important than the values they bring to the idea of growing older and the ways those values influence their spending. Says Ken Dychtwald, CEO of Age Wave,

> A totally new model of maturity is emerging—a vibrant landscape filled with new beginnings and personal reinvention, where people remain involved, productive, and connected. People don't like the idea of becoming out of touch, distant, or unnecessary—which are all too often descriptors for many of today's seniors. I don't think it's aging that frightens people. It's the fear of becoming uninspired and unwilling to try new things. In essence, opening yourself up to new experiences and making new friends is the ideal antiaging medicine.[2]

This is good news for arts and cultural organizations.

Mature adults are a substantial, vital, and loyal segment. They tend to have significant discretionary time and income and prefer *being* experiences of nonmaterialistic origin to purchasing and possessing products.

Mature adults tend to spend more tightly on goods and more loosely on lifestyle enhancement experiences. Some arts organizations offer discounts to seniors for all performances and in all sections of the hall, not considering that many people in this segment are fully able and willing to pay full price. Arts marketers should address the needs of low-income seniors by offering discounts for one or two performances per week or only for nonpremium seating areas.

Barriers to attendance for this group are likely to be functional factors, not the arts themselves. Mature adults are excellent candidates for group sales. Senior centers and senior residences often have their own program directors who are likely to be receptive to group sales offers from performing arts organizations. Typically, senior centers and homes provide their own transportation and their program directors would eagerly arrange with an arts marketing director to provide a pre- or postshow lecture and reception with refreshments.

Abraham Maslow describes the following personality attributes of mature, self-actualizing people:

- *Superior perception of reality*: This quality makes them more discerning when evaluating advertising claims.
- *Increased acceptance of self and others*: This characteristic results in a higher capacity for humor and for coping with stress.
- *Increased spontaneity*: Mature people are more likely to live day to day and respond well to the unexpected.
- *Increased detachment and desire for privacy*: Advertising images should acknowledge the unique individuality felt by mature adults.
- *Increased autonomy and resistance to enculturation*: Mature people know what they want, and they resist efforts by others to change their minds. Give them facts, not value-inflated, self-serving claims.
- *Greater freshness of appreciation and richness of emotional reaction*: Touch their hearts and they will allow you to enter their minds.
- *Greatly increased creativity*: First invoke the poet in the mature consumer, but be sure not to abandon the rational side of the process.[3]

Based on Maslow's theories, David Wolfe has developed a set of principles for positioning an offering to mature adults:

- *The Gateway-to-Experiences Principle.* Mature consumers tend to be motivated more by the capacity of a product or service to serve as a gateway to experiences than by the generic nature of the product or service itself.
- *The Age-Correlate Principle.* Age is a correlate, not a determinant, of consumer behavior in mature markets and hence should not be used to define and predict specific consumer behavior among older people. Stage of life, expressed in terms of personality maturity, has a far greater influence on consumer behavior than chronological age or social, economic, or career position.
- *The Price/Value Principle.* Consistent with the desire for quality, mature consumers tend to place a higher value on price in purchasing basics, but in their more discretionary purchases, they tend to place a higher value on the potential experience resulting from the purchase. That is, mature consumers tend to spend "tightly" on goods and services necessary to maintain lifestyles and more "loosely" on products and services that enhance lifestyle.
- *The Altruistic Factor.* Mature adults tend to respond more favorably to marketing messages that emphasize introspective or altruistic values and less favorably to marketing messages that emphasize selfish interests.[4]

Performing arts marketers can use these principles to develop relevant auxiliary programming and messages. The goal is to create multiple perceptions in consumers' minds of what the offering can do for them.

Arts managers and marketers who bemoan the graying of their audiences should consider not only the enormous value of mature adults as important audience members, but also these adults' ability to help develop audiences of the future.

MARKETING TO GRANDPARENTS. About one-third of American adults are grandparents. In 2010, it was estimated that there were 115 million grandparents in the United States, compared to 70 million in 2005. With more Baby Boomers approaching grandparent age, this number is expected to continue to swell. Today's grandparents have higher levels of education and income than their predecessors and are more likely to participate in arts and culture. They also play a larger role in their grandchildren's lives, including their education and cultural enrichment, especially now that many parents work long hours. As a group, grandparents now spend more than $30 billion a year on their grandchildren, a twofold increase over what was spent a decade ago. It is estimated that grandparents account for almost 17 percent of toy sales in the United States, yet today's grandparents are even more interested in spending money on structured activities they can do *with* their grandchildren.

In the year 2000, 6 million Americans reported vacationing with their grandchildren in a typical month. Grandtravel, grandparents traveling with grandchildren, has become so popular that tour operators, hotels, cruise lines, and even Elderhostel have developed programs and promotions tailored to this market. Arts marketers should capitalize on these trends by offering a special grandparent or child and grandparent memberships and promotions tailored to grandparents and their grandchildren.

Younger Generations

GENERATION X. The members of Generation X were born between 1965 and 1980. On the whole, they are more ethnically diverse and better educated than the Baby Boomers. They were the first generation to grow up with computers and portable music players.

Generation X is independent, resourceful, and self-sufficient and values both freedom and responsibility. Unlike previous generations, members of Generation X work to live rather than live to work. They appreciate fun in the workplace and espouse a work hard/play hard mentality.

Generation X participation rates in live classical music have averaged about 9 percent, compared to 14 percent for previous cohorts, reflecting a lower affinity for live classical music. Gen Xers have the highest penetration in listening to online radio and buying digital music, suggesting opportunities for orchestras to engage them through digital platforms. To increase trial, some orchestras offer discounted tickets to families and young professionals, present pre- and post-concert events, and tailor schedules around Gen Xer schedules, such as Sunday afternoon family series. The primary gap appears to occur in the repurchasing stage. Among Gen Xers, repurchasing of subscriptions with five or more concerts has declined nearly 25 percent more than other generations for live classical music.

Generation Y. Born in the 1980s and 1990s, Generation Y is a large cohort, rivaling the Baby Boomers in population, with about 80 million people. Growing up with technology and personalized attention, Gen Yers expect instant gratification and products tailored for them. Armed with smartphones, laptops, and other gadgets, Generation Y is constantly plugged in. This generation prefers to communicate through email, text messaging, and social media rather than face-to-face contact, and prefers webinars and other online technology to traditional lecture-based presentations. Generation Y has the highest level of engagement in online and mobile activities, including social networking, watching streaming videos online, downloading music, blogging, and instant messaging, compared to previous cohorts.

Generation Y is confident, ambitious, achievement-oriented, and family-centric. The fast-track has lost much of its appeal for Generation Y, whose members are willing to trade high pay for fewer billable hours, flexible schedules, and a better work/life balance.[5] Generation Y is highly cost-conscious; however a cheap ticket price will not make or break a Gen Yer's decision to buy a ticket. It is value, not price, that is most important to them.

Today's younger generation has a statistically higher propensity for self-reflection and search for truth than their parents. They are inspired by humanism and are not motivated by hype. Angst, stress, and disillusionment are often by-products of their mind-set.

Address this market using a style and sensitivity that supports the artistic and humanistic quality of the art form. Avoid overcommercialization. "Dumbing down" the message is not only unnecessary, it could in fact be a turnoff.

To generate awareness among Gen Yers, some orchestras offer initiatives such as free music lessons and ensemble concerts in local schools, youth symphonies to promote engagement, and a "Concert for Kids" series to encourage parents to bring children to concerts. Despite successful educational awareness programs, orchestras have struggled to convert Gen Yers to trial.

Teenagers. Another major demographic group with important buying power is youths aged 12–19. Teens are the most informed and media-aware consumers in history. They are especially tech savvy, rapidly adopt state-of-the-art products, and then adapt them to their ever-evolving lifestyles. Arts marketers have a great opportunity to target teenagers in their interactive environment; only those marketers who appeal to teens technologically and interactively are likely to succeed. Each arts organization should consider designing part of its website to be particularly appealing to teens.

Having been exposed to more than 1,200 advertising messages per day during their lifetimes, teens are more immune to traditional ad messages than any other generation before them. They don't want to be told by advertisers that something's cool; they would rather find it on their own or through their peer group. Word-of-mouth and viral marketing approaches, especially through social media, are the best ways to reach this generation. Marketers might encourage people in this age group to become participants in a teen club or teen membership program.

No matter how successful an arts organization may be, it must plan for the future by building audiences among today's youths. Arts marketers can follow the lead of Stéphane Lissner, former artistic director of La Scala, the world-famous opera house in Milan and long the exclusive playground of the rich, who announced in summer 2005 that he would welcome for free anyone under age 18 who is accompanied by an adult at a series of Sunday concerts.

A highly effective program for bringing teens to arts events is the High 5 program in New York City.

High 5 Tickets to the Arts in New York City

High 5 Tickets to the Arts was created in New York City in 1993 to develop new audiences for dance, theater, music, film, and visual art by making New York City's cultural life affordable and accessible to teenagers. The High 5 calendar presents incredible variety and selection, offering thousands of tickets each season to events at the most prestigious concert halls, off-Broadway theaters, avant-garde performance venues, and world-class museums in the New York City area. The arts organizations donate tickets to their performances and exhibits, and High 5 markets, promotes, and sells these tickets to teenagers for just $5. High 5 serves teenagers from every segment of the population, but makes a special effort to reach communities where teenagers are economically disadvantaged, culturally diverse, and at risk.

To market its programs, High 5 sends a weekly newsletter called *The Week in Arts* by email to its extensive mailing list of teens, schools, and community-based organizations. In addition, High 5 provides audiences with daily updates on social media platforms including Facebook and Twitter.

Several special programs enhance the High 5 arts ticket offer. Take 5 allows adults to take groups of teens to arts events at High 5 ticket prices. Through its partnerships with dozens of youth service organizations, High 5 is able to provide access to the arts for thousands of teens each year. The

Teen Reviewers and Critics program engages teens in critical thinking about the arts through their participation in a workshop series with professional arts critics.

Not all cities provide conditions ideal for a program such as this, but aspects of this program can be borrowed or adapted for use by arts organizations, arts service organizations, and community groups in cities and towns with various levels of cultural offerings.[6]

CHILDREN. Many arts organizations have been proactive and creative in educating youngsters about their art forms by presenting in-school programs, special student performances, and a wide range of educational material. Typically, the goal of special children's programs is to build the audiences of tomorrow. However, Peter Brosius, the artistic director of the Minneapolis Children's Theatre Company (CTC), has a more immediate goal, which he described on the occasion of the CTC's winning the first Tony award for excellence in regional theater ever given to a theater aimed entirely at kids: "Central to the mission of this theater is the belief in the power of young people—not as future subscribers or donors, but as an arts audience of today." Even so, CTC, which was founded in 1965, has developed several generations of adult audiences who started out by attending its performances.[7]

Following are examples of other highly successful programs for children.

Young People's Concerts at the New York Philharmonic

The Young People's Concerts, which began in 1922, achieved worldwide popularity when Leonard Bernstein hosted them on live television in the 1960s. The same spirit of learning and wonder animates the Young People's Concerts today, awakening children's musical curiosity and laying the foundation for a lifetime of enjoyment. Children are invited to come early to meet the musicians, try out some instruments, play musical games, explore the day's theme, and get ready to enjoy a great concert.

In recent years, the New York Philharmonic has also been offering Very Young People's Concerts, a junior version of the Young People's Concerts. Designed for children aged three to five, the one-hour concerts introduce youngsters to classical music through games, active listening, and hands-on music making with members of the Philharmonic and faculty from Columbia University's Teacher's College. Each concert has a host and a theme, such as "Fast and Slow," "Loud and Soft," or "High and Low."

Teddy Bear Concerts in Sweden

Sweden's Malmö Symphony Orchestra, like many Swedish arts organizations, has a strong commitment to children's concerts. This orchestra's version is called Nallekonsert (Teddy Bear Concert) and is aimed at children from the age of four upward. It is estimated that twenty-five thousand children attend the orchestra's school concerts every season. The orchestra's website has a special section that features music quizzes and games for kids.

Keeping Score in San Francisco

The San Francisco Symphony's *Keeping Score* program provides innovative, thought-provoking classical music content on PBS television, the radio, the web, and through an education program, which is a national model for classroom arts integration for K-12 teachers. The *Keeping Score* educational program trains teachers to integrate music effectively into core subjects such as science, math, English, history, and social studies. Teachers are trained by the symphony's educational staff, by musicians, and by arts educators and receive professional development training and assistance throughout the school year. The *Keeping Score* Education program has worked with partner sites in Fresno, Sonoma, and Santa Clara counties; Flagstaff, Arizona; and the Oklahoma A+ Network statewide program. An integral part of the program is the annual *Keeping Score* Summer Teacher Institute—a multifaceted professional development experience that builds teachers' understanding of both music and integrated curriculum design.[8]

FAMILIES. Some arts performances offer prime opportunities for marketing to families, such as productions of *A Christmas Carol* and *The Nutcracker*, which are typically advertised as family entertainment. Because nearly half of North American dance companies' annual earned revenue comes from holiday season productions of Tchaikovsky's *Nutcracker*, dance companies typically leverage every possible marketing opportunity to attract patrons to these performances and increase revenue with specially designed amenities. The New York City Ballet offers parents the opportunity to purchase a photograph of their child with one of the ballerinas; many *Nutcracker* performances nationwide are followed, for an additional fee, by after-concert receptions replete with tea and cookies and dancers milling about. Typically, it is the occasion of the Christmas-time celebration that induces families to attend and to spend more than they would ordinarily, not the cultural event in and of itself. Of course there are many opportunities for marketing to families throughout an organization's entire season of performances.

Family concerts are geared toward children, but parents are the ones who buy the tickets. So arts organizations market to parents, aiming to convince them that their children will have an enriching, entertaining experience. Furthermore, says Pam French Blaine, vice president for education and community programs at Orange County's Pacific Symphony, "We are educating the kids. Parents will put themselves in learning situations for the sake of their children in ways that they might never do for themselves."[9]

Unfortunately, the limited revenue potential of family programs makes them a low priority in marketing budgets. Yet arts managers and marketers need to carefully consider the long-term value of acquainting children and their parents with the organization's art form and building a familiarity that will translate into regular arts attendance when these individuals are older and have disposable income.

Family Offers at DCDC

In the 1990s, Dayton Contemporary Dance Company (DCDC) offered a Saturday afternoon family program with a special low price for two adults, plus any number of children at $5 each. The program was advertised in the local newspaper's Friday arts section. The program drew very few patrons.

I suggested that we try a new strategy to attract families. First of all, I shared that offering the program for two adults may be counterproductive since often just one adult will want to accompany a child or children, because others in the family may have conflicting activities. Also, there are many single-parent households with a divorced mom or dad looking for activities to do with his or her child. Furthermore, most people don't think to look in the arts section of the newspaper to search for ideas for activities, so we needed to be visible in the newspaper sections that our target markets frequented. Accordingly, we placed an ad in the Wednesday women's section with a little girl looking up at her mom saying, "Mom, let's do DCDC this weekend." Similarly, we placed an ad in the Monday sports section showing a boy looking up at his dad and saying, "Dad, let's do DCDC this weekend." The new pricing structure was for one adult or more and retained the $5 kids' price. The first weekend we sold out all the available family tickets.

Singles

Today, more than 50 percent of US residents are single, nearly a third of all households have just one resident, and five million adults younger than 35 live alone. By 2000, 62 percent of the widowed elderly were living by themselves. Percentages are similar in the United Kingdom, where, as of 2003, 37 percent of adult men and 41 percent of adult women were neither married nor cohabiting. In 2010, about a third of adults aged 46 through 64 were divorced, separated, or had never been married, compared with 13 percent in 1970. In Sweden, 55 percent of births are to unmarried women; almost a third of Japanese women in their early thirties are unmarried; and more than one-fifth of Taiwanese women in their late thirties are single.[10] Furthermore, 72 percent of US mothers are working, which does not bode well for attracting this segment of extremely busy people to performing arts events.[11]

Performing arts managers face challenges with this demographic. Single working parents with children at home rarely have the discretionary time to pursue interests like arts events. Also, although there are many places that people feel comfortable visiting alone, typically, people like to attend a performance with a companion. And arts organizations have not been good about appealing to singles with their offers. It is clear that many arts organizations are either unaware of the huge number of singles (and of odd-numbered groups) or simply do not take them into consideration when they offer "twofer" promotions (two tickets for the price of one). This problem can be resolved simply by offering half-price tickets instead. Singles nights for young adults have become an effective strategy to attract this market segment. Singles night subscriptions are often so popular that they sell out. One of the primary benefits of such events is that they appeal to a young crowd whose members enjoy participating when more people like them are in attendance. These events often feature special programs such as preshow casual dinners or postshow discussions about the performance, sometimes with a wine-and-cheese reception, that encourage socializing.

Too many marketers ignore the significant potential market of older singles, people who like to frequent the arts but often do not do so for lack of a companion with whom to attend or lack of access to transportation. Arts organizations can increase the chance that older singles will attend and will have a more satisfying, interactive experience by creating opportunities for conversation. They can successfully reach out to this market by, for example, reserving a block of "mature singles" seats for certain performances so that these patrons might have a like-minded person at hand with whom to chat before the show, during intermission, and at postshow events.

Segmenting by Gender

There is substantial evidence that gender affects how people process information, respond to messages, and make decisions. It is believed that these differences emerge in part due to socialization and in part due to biological reasons.

Marketing to Women

Women make more than 80 percent of purchasing decisions in the United States. They bring in at least half the income in a majority of US households and they own more than three-fourths of the nation's financial wealth. Women 50 and older control a net worth of $19 trillion.[12] This is estimated to grow to as much as $40 trillion over the next decade as Baby Boomers inherit from their parents, creating the largest wealth transfer in the United States in history. With successful careers, investments made during the "boom" years, and inheritances from parents or husbands, Baby Boomer women are more financially empowered than any previous generation of women.[13] Young single women purchase twice as many homes as young single men. Undoubtedly, women are the largest and most important market segment, and marketers should seriously consider what approaches are most effective in reaching out to them.

The more mature luxury consumers place the highest priority in making memories and experiences. They don't buy things just to have more things; they want the experience to go along with it. Luxury consumers expect superior quality and are extremely discerning.[14]

Since the late 1980s, women have earned two-thirds of all undergraduate degrees and 60 percent of master's degrees. Countless surveys over the decades have demonstrated that attainment of high education levels is strongly correlated with arts attendance.

Women's communal focus prompts them to consider both the self and others in making a decision. They want comprehensive information, want to compare more options, and often have many questions they want answered.

Women also like stories; they like to hear them and they like to tell them. Because women are predisposed to share information with their friends, especially information based in emotional content, marketers should create messages that not only influence the reader but also help her to transmit the message. Martha Barletta, author of *Marketing to Women*, suggests that it is much easier for women than for men to recall and recount an ad that contains sound bites, strong visuals, and stories.[15] Women enjoy ads that include lifestyle images and that portray a "slice of life." Such ads could stimulate them to imagine their experience at the performing arts venue.[16] Women will bond over brands, and form communities around brands. What better catalyst is there for connecting people than the performing arts, which bring people together and stimulate ideas, curiosity, and conversation?

Tell your genuine story is trend expert Faith Popcorn's advice to anyone trying to attract people to their performances. Says Popcorn, "The story of your organization's genesis is crucial in communicating genuineness to people. Identify your company's narrative and what values it represents and then communicate how those formative ethics continue to drive the mission."

Marketing to Women at the Joffrey Ballet

"We have to market toward women," said Jon Teeuwisen, former executive director of the Joffrey Ballet. "Our audience is 60 percent female, and most of the men who come do so because women bring them." The Joffrey's ads were changed to show less athleticism, which appeals to men, and more grace and inspiration, which attracts female balletomanes. "We wanted to reach the inner ballerina in women who have taken classes," Mr. Teeuwisen explained. "We also started a very successful women's board, and organized affinity groups which featured professional women as keynote speakers before performances."[17]

Marketing to Men

Men tend to be highly goal directed and prefer messages that help them make a quick and simple decision.[18] Unlike women, they respond better to messages that focus directly on a single benefit, and are likely to enjoy ads that are funny or have a unique creative style. Men are not browsers and they tend to hone in, more quickly than women, on what they're looking for.[19] They make decisions on products by the process of elimination—deciding which aspects of a product matter most and eliminating products lacking those attributes. For marketing purposes, men eliminate the competition before everything is considered, while women only reach a decision when looking at the overall picture.

Since women make about 80 percent of the art-going decisions and ticket purchases, I am giving men just a short overview in this context.

Marketing to Gays and Lesbians

It is estimated that 4.8 percent of the population in the United States, or about fourteen million people, are gay. Nineteen percent of both gay men and women have postgraduate degrees, compared with 14 percent of heterosexual men and 12 percent of heterosexual women.[20] "Gays are twice as likely to have graduated from college, to have an individual income over $60,000, and to have a household income of $250,000 or more. Moreover, gays and lesbians comprise a $600 billion market that is fiercely loyal to brands that advertise directly to them."[21] In fact, research shows that 94 percent of gays and lesbians "would go out of their way to purchase products and services marketed directly to them."[22]

Gays and lesbians represent a unique demographic in that they include people of all races, incomes, and ages—all with different buying habits. Because the market is so diverse, the organization should decide which cross-section of the gay community is best to reach.

In focus groups nationally, gay men and lesbians express a definite preference for advertising that specifically reflects their mind-set and sensibilities. Extending an organization's reach among this audience requires understanding the community infrastructure available to access them. The result is often a sophisticated mix of advertising, direct marketing, community presence, and Internet promotions.[23]

Consider advertising in local gay media, including gay websites, a directory of which can be found at Yahoo! The gay population is remarkable for its presence as a social community. Marketers can achieve success with gay consumers, who are moving further into the mainstream every day, by focusing more on their daily lives—their financial and family concerns, for example—and less on their political identities. Because fewer gays and lesbians have children than heterosexuals, they tend to have more discretionary time and disposable income. This makes them ideal targets for arts events. Yet, gay parents represent a unique opportunity for marketers of family events, as approximately 33 percent of lesbian couples and 22 percent of gay male couples have children in their households.

Segmenting by Ethnicity

According to the Pew Research Center, if current trends continue, the population of the United States will rise from 296 million in 2005 to 438 million in 2050, and 82 percent of the increase will be due to the arrival of immigrants and their US-born descendants. The non-Hispanic white population, which comprised 67 percent of the population in 2005, will become a minority (47 percent) by 2050. Hispanics will grow from 42 million (14 percent of the population) in 2005 to 128 million (29 percent) in 2050. African Americans made up 13 percent of the population in 2005 and will remain roughly the same proportion in 2050. Asians, who made up 5 percent of the population in 2005, will comprise 9 percent in 2050. The combined spending power of Hispanics, African Americans, and Asians is in excess of $600 billion a year, and it will only rise as their populations grow at a faster than average rate.[24]

The search for new patrons for now and into the future means looking to these nontraditional markets. Arts organizations must learn how to include members of diverse minorities among their audiences, boards of directors, donors, and volunteers. Says David Brooks in the *New York Times*, "Communications technology hasn't brought people closer together; it has led to greater cultural segmentation, across the world and even within the United States.... As people are empowered by greater wealth and education, cultural differences become more pronounced, not less, as different groups chase different visions of the good life."[25] This places new demands on arts marketers to understand how to break down physical, psychological, and cultural barriers to participation and how to help people believe that the marketers' offerings are for them.

Too often, cultural organizations bring in ethnic programming without making connections with the affiliated ethnic groups in the community. The only way to develop compelling communication strategies is to invest a great deal of staff time in locating business, religious, social service, media, and educational leaders in target ethnic communities; setting up an advisory group composed of people of the target ethnicity; and soliciting as much information as possible to learn cultural biases and preferences and how best to meet the needs and interests of each group. Grassroots community efforts require an intensive time investment, which pays off in understanding, learning, and long-term audience development.

Hispanics

Hispanics are the largest minority group in the United States, and this population is expected to continue to grow at a more rapid rate than other population segments. Hispanic household income has also been increasing at a rapid rate and college enrollment among members of this segment has shown a steady increase. Moreover, Hispanic parents are encouraging educational opportunities for their children.[26]

Traditional ideas about family retain a strong hold on Latinos in the United States, whether they are foreign born or US born. According to a Pew Hispanic Center/Kaiser Family Foundation survey, 89 percent of Latinos believe that relatives are more important than friends, compared with 67 percent of non-Hispanic whites and 68 percent of African Americans. The results of this research indicate that arts organizations could most effectively reach out to potential Hispanic audiences with family-oriented programming, pricing, and scheduling.

This niche market of today is becoming a mass market in its own right, segmented not only by nationality (e.g., Mexican or Guatemalan) but also by spending behavior and other psychographic characteristics. Hispanic identity and a connection to Hispanic culture have remained remarkably strong across generations of Latinos now living in the United States, so marketers must learn how to create campaigns that resonate with Latino identity and culture.[27]

A number of firms have found that bilingual direct-mail pieces draw better responses than English-only or Spanish-only pieces. However, direct mail from arts organizations to Hispanics must be extremely well targeted in order to garner a response rate that is worth the cost to the marketer.

African Americans

The Census Bureau reports that in 2010, US non-Hispanic blacks numbered an estimated 39 million people, or 12.6 percent of the US population. Since the mid-1980s, the buying power of African American households has more than doubled and has grown 50 percent faster than that of the US population as a whole. Other key social and economic indicators—such as home ownership and college enrollment—are also improving at above-average rates. In 2009, 2.9 million blacks were enrolled in college, compared to 1.4 million in 1990. Another factor in the rising affluence of African Americans is a noticeable increase in the number of high-income, married-couple African American families. As a result of these long-term trends, more and more African American households are achieving middle- and upper-income status.[28]

Despite economic changes in the African American community, values and attitudes remain the same. Home and family are extremely important to this segment, as are church, religion, education, and community involvement. In a study done by Yankelovich/Burrell, 39 percent of African Americans say the home provides all, or almost all, of their satisfaction, whereas only 25 percent of nonminorities found their home so satisfying. Religion ranks very high among African Americans, with 73 percent citing it as being very important, compared with 47 percent of nonminority populations.[29] In order to effectively market to and ultimately win the African American customer, it is important that organizations demonstrate an understanding of and respect for African American culture in their communications, marketing programs, and sponsorships. African American consumers want to see themselves in advertisements; they want to know they are being invited to select a product or service. Organizations should place ads in media that this community trusts and, when possible, feature members of the community, whether they be artists or audience members, in the advertisements. Building relationships within the African American community through channels such as events and community-based programs is important, not only to generate business but also to maintain loyalty.

The significant number of African American theaters and dance companies across the United States attests to the strong interest among this market segment in the performing arts. Many predominately white organizations are working to attract African Americans to their performances by incorporating works by African Americans on the stages, by casting people of color in their productions, by diversifying their boards and staff, and by reaching out to members of this segment in their local churches and community centers.

Asian Americans

According to the 2010 census, there were more than 17 million Asian Americans the United States, nearly double the number in 2000. The Asian American population is the most affluent demographic segment in the United States and presents a more favorable economic profile than the country's non-Hispanic white population, earning more on average than any other racial group. Half of Asian Americans have a bachelor's degree or higher, compared to 30 percent of native-born Caucasians.

The Asian American market consists of numerous subgroups with distinct languages and cultural frames of reference. Although the great majority of

Asians in the United States have English language facility, they respond best to messages delivered in their native language and via their ethnic media. Furthermore, advertisements need to reflect an Asian cultural context, and some of the most effective marketing campaigns have been tied in with Asian festivals and holidays.[30]

MARKETING TO INDIVIDUAL SEGMENTS

People link with others who are similar to them. The implications, says Emanuel Rosen in his book *The Anatomy of Buzz*, are that first, the more similar your employees are to your customers, the easier the communication between them will be. Second, people who are similar to each other tend to form clusters.[31] Identifying these clusters and reaching out to them effectively is a prime challenge for marketers. Because of generational differences and the diversity of ethnic subgroups, locating the media outlets and messages that will target these various markets is quite difficult. As a general rule, marketers can best reach diverse groups where and how they live. "A lot of it really is guerilla marketing," said Erica D. Zielinski, general manager of the Lincoln Center Festival.

For example, for the US premiere of *I La Galigo*, avant-garde director Robert Wilson's adaptation of an Indonesian epic, Lincoln Center marketing interns approached Indonesian mosques in Long Island City and Indonesian restaurants in Park Slope. They dropped flyers at an Indonesian-owned bank on Wall Street, a South Asian martial arts studio in the Flower District, and even a couple of yoga studios in SoHo.[32]

MAJOR FACTORS INFLUENCING CONSUMER BEHAVIOR

Most factors that affect a consumer's arts attendance decisions, interests, needs, and satisfaction levels go beyond issues specific to the artistic offering. Factors as broad as macroenvironmental trends and as specific as an individual's own psychology play a major role in influencing people's attitudes about what they purchase, how much they spend, which leisure activities they pursue, and which needs they try to fulfill.

MACROENVIRONMENTAL TRENDS

Large social, political, economic, and technological forces influence our attitudes, values, important decisions (education, career and job choices, and investment decisions), and day-to-day choices, including how to spend our leisure time. In a study commissioned by Dance/USA, researcher David Meer observes the impact of people's changing values on their leisure-time activities.[33] He notes a hunger for the classic family and a focus among young adults on child-centered activities, especially as people work longer hours. There is also a greater focus on productive leisure: staying healthy and in shape, other self-improvement activities, and building relationships. Conversely, one of the latest "crazes" is to "veg out" and have large periods of unplanned time.

Trend analyst Faith Popcorn recommends thinking of trends as a kind of database about consumers' moods, a rich source that can be tapped to help solve marketing problems.[34]

Popcorn says that *cocooning*, the tendency to look for a haven at home, has manifested itself in mushrooming home video centers, Internet and social media usage, take-out food, online-order merchandise, and in the fact that millions of Americans are working out of home offices. Says Popcorn, "Don't expect consumers to come to you anymore. You'll have to reach them in the cocoon itself." As a result, people may crave more experiences outside the home, and performing arts marketers can capitalize upon that desire.

Popcorn identifies the trend *fantasy adventure*, which refers to the desire to have exciting exploits, heartfelt experiences, exotic adventures, and broad sensory and emotional experiences—but to undertake them in the safest possible ways. This trend represents a wonderful opportunity for performing arts marketers, who offer adventure by association, and a range of emotional and aesthetic experiences from the comfort and security of a theater seat.

Small indulgences is a trend that developed in response to the economic crises of the first decade of the twenty-first century. Today, people buy accessories to update last year's outfit or go on a long weekend trip instead of a two-week jaunt to Europe. This trend is not just about cutting back, but also about buying the best there is in one or two lifestyle categories. For arts marketers, it may translate into the opportunity to encourage current patrons to upgrade to the best seats in the house, or to encourage infrequent attenders to subscribe for a full season rather than taking a weekend away. ("It's something to look forward to all year long.")

Egonomics refers to the consumer's need for personalization. As market niches grow smaller and software makes customization not only possible, but also expected for many products, each consumer's preferences for product features, packaging, distribution, and pricing grow in importance. The performing arts marketer who personalizes and customizes communications to various audience members will gain a competitive edge.

Ninety-nine lives is the trend to do it all and be it all (work, parent, volunteer, travel, stay fit, achieve self-fulfillment, socialize, read, and on and on). What people really want is to buy back time. What this means to marketers is that whatever can be done to speed service, increase the ease of ticket purchase, and fit performing arts events into people's lifestyles and time constraints will help increase performing arts attendance.

CULTURAL FACTORS

Of the factors that affect consumer behavior, cultural factors—from national identity to membership in small social groups—exert the broadest and deepest influence. A growing child acquires a set of values, perceptions, preferences, and behaviors through the process of socialization into his or her culture(s).

Nationality

It is frequently asserted that Europeans are more predisposed to arts attendance than are Americans. Europe has a longer cultural history, and Europeans are more habituated to arts attendance. In Europe, the arts receive significant government subsidy, and performances are often very affordable, attracting working-class people as well as the upper classes.

Western music is seemingly more popular in Asia than it is in the West. In Tokyo, for example, there are seven full-time symphony orchestras.

In the United States, by contrast, the performing arts are considered as more elitist. All too frequently, people feel they must be highly educated and capable of making sophisticated responses in order to appreciate opera, ballet, symphony, and chamber music. In one survey of adult Americans, more than 90 percent of the 1,059 people questioned said they felt the arts and humanities were important to freedom of expression, were life enriching, and were a means of self-fulfillment. However, 57 percent said that the arts and humanities played only a minor role in their lives as a whole and little or no role in their daily lives.[35] To J. Carter Brown, former chairman of the National Cultural Alliance, the survey indicated that Americans are "an enormously receptive population" for the arts, and that arts managers must find ways to provide opportunities for more involvement. If Americans could be as comfortable attending the theater or dance as they are attending a movie, both the public and the arts organizations would benefit.

Subcultures

Each culture consists of subcultures—religious, racial, and regional groups—that provide more specific identification and socialization for their members. These subcultures influence a person's tastes, preferences, and lifestyles, as well as the nature and extent of his or her interest in performing arts. Even when programming is directed to a specific subculture, various factors come into play. When a Central American play, performed in Spanish, was presented in a city's central theater district, no native Spanish-speaking people attended. However, when the production was moved to the Mexican Cultural Center, the Latino community filled the house.

Social Class

A person's social class also affects his or her behavior and attitudes. Social classes are hierarchically ordered, relatively homogeneous, and enduring divisions in a society. Their members share similar values, interests, and behavior.[36] These classes show distinct product and brand preferences in their leisure activities as well as in their choice of such consumer goods as clothing and automobiles. The advertiser has the challenge of composing copy that rings true to the targeted social class and presenting it through the media that are most likely to reach that target group.

The characteristics of each class's members give marketers clues as to how to reach them. Working-class consumers may be encouraged to celebrate a birthday or anniversary with an inexpensive dinner and theater package, while members of the upper middle class may be encouraged to entertain their clients with box seats and cocktails in the "members only" room.

Innovativeness

People differ greatly in their readiness to try new products. An innovative person is quicker to adopt new ideas than other members of his or her social system. Everett Rogers has identified five categories that are defined by how long people take to adopt innovations. *Innovators*, who represent 2.5 percent of the population, are venturesome and willing to try new ideas. *Early adopters*, who make up 13.5 percent of the population, are opinion leaders in their community and adopt new ideas early but carefully. The *early majority*, comprising 34 percent of the population, are deliberate; they adopt new ideas before the average person, though they rarely are leaders. The *late majority*, comprising another 34 percent of the population, is skeptical and tends to adopt an innovation only after a majority of people have tried it. Finally, *laggards*, who make up the last 16 percent of the population, are tradition bound and adopt an innovation only when it takes on a measure of tradition itself.[37]

The adopter classification has many implications for arts marketers. It suggests that an avant-garde theater should research the demographic and psychographic characteristics and media preferences of cultural innovators and early adopters and direct communications specifically to them. The majority of the population will wait for the cue of opinion leaders and jump on the bandwagon only after the theater has received repeated acclaim. A company that performs the classics with original-style sets and costumes is likely to appeal to the late majority and the laggards, among others.

Psychological Factors

A variety of personality traits, self-concept issues, and emotions also affect people's attitudes and behavior.

Personality

Personality may be described in terms of such traits as self-confidence, dominance, autonomy, deference, sociability, and adaptability. Cultural innovators and opinion leaders are likely to be more self-confident, dominant, and autonomous than less innovative people, who may be more deferential and more interested in the social benefits of arts attendance.

Self-concept is composed of how one views oneself (actual self-concept), how one would like to view oneself (ideal self-concept), and how one thinks he or she is viewed by others (others' self-concept). A message such as "You belong there!" for the opening night at the symphony may reach all of these groups, including those who believe the message and those who want to believe it.

A person may undergo psychological passages or transformations during which his or her needs or tastes change. The factors that motivate someone to purchase a ticket to a play or that actually restrain him or her from doing so may operate at a subconscious or unconscious level.

It is often assumed that behavior follows what Michael Ray has called the "think-feel-do" model of consumer behavior.[38] That is, consumers presumably take in information, form some emotional response, and then act when the appropriate resources are available. However, Ray suggests that in many situations, people are influenced to take action by their feelings, following the "feel-do-think" model. This alternative model for influencing emotions has received growing attention in the private sector. Through messages that communicate few if any actual facts, advertisers attempt to create feelings or moods and provoke positive associations with an organization's offering. Coca-Cola would "like to teach the world to sing in perfect harmony," creating warm feelings and sudden thirst in its listeners. The National Basketball Association, in its ads during the playoffs and finals, shows exciting "dunks" followed by the single phrase: "I love this game." One can readily see how this approach is likely to be more emotionally appealing than the logo "Where acclaimed theater begins."

Beliefs and Attitudes

Through socialization and the learning process, people acquire beliefs and attitudes that influence their buying behavior and that lead them to behave fairly consistently. They allow people to economize on energy and thought, rather than having to interpret and react to every object or situation in a fresh way. Some arts organizations are trying to change people's attitudes about what kind of experience they will have attending a performance. For example, Lincoln Center has a "Serious Fun" series, and the Milwaukee Chamber Theatre asks, on the cover of its season brochure, "Wanna Play?"

By teaching the target individual that a particular action will lead to a desired reward, the probability of the action is increased. Thus, if a dance company offers a special singles night including a wine-and-cheese reception after the performance, people who have never been interested in attending dance performances may attend (change their behavior) in order to gain a desired reward unassociated with the performance itself (the social hour) without changing their attitudes about dance. Some of the attendees may find that the performance was far more accessible and entertaining than they had anticipated and that, as a result, their change in behavior has stimulated a change in attitude.

Motivation

A person has many needs at any given time. Some needs are biogenic, arising from physiological states of tension such as hunger, thirst, or discomfort. Other needs, which are less immediate, are psychogenic, arising from psychological states of tension such as the need for recognition, esteem, or belonging. A motive is a need that is sufficiently pressing to drive a person to act. Two of the best-known theories of human motivation—those of Abraham Maslow and Frederick Herzberg—offer implications for consumer analysis and marketing strategy.

According to Abraham Maslow, motivation is driven by particular needs at particular times, from the most pressing to the least pressing.[39] Maslow proposed that the more basic needs require gratification before a person is able to achieve substantial gratification of higher-level needs. Gratification at each level contributes to the person's maturation. Maslow characterized the first four needs as "deficit-driven." He diagrammed motivations in the shape of a pyramid, with the first level of needs at the broadest place, the bottom of the pyramid. People must be able to satisfy the basic physiological needs for food, sleep, and sexual expression; otherwise they will not be able to move to higher need levels. Next they must satisfy safety and security needs, which include both physical security and a sense of psychological well-being. Social needs—for belonging and love—must then be met. People will also seek gratification of esteem needs. An individual's substantial success in gratifying these four deficit-driven needs lessens the potency of the drives behind those needs, thus freeing up psychic energies to meet growth needs.

Performing arts organizations can meet these needs in a variety of creative ways. For example, social needs are met when a theater offers a special night once a month for gays and lesbians or for young singles. Esteem needs are fulfilled when arts organizations recognize donors on plaques and in program listings or invite them to special events with the artists.

Maslow envisioned self-actualization as a level of maturity at which a person is beyond striving, beyond basic psychological fear, beyond a need to demonstrate who he or she is, beyond shaping his or her own life largely around the expectations and views of others. It is a stage of life characterized by a high level of well-being, healthy self-esteem, and a positive outlook on life, in which a person is "being all he or she can be."[40] Self-actualizing people represent an important target market for performing arts attendance.

It is interesting to note that before Maslow died, he regretted what he had said earlier as he felt that fulfillment of self-actualization is a prime need of all human beings.[41]

Psychologist Frederick Herzberg developed a "two-factor theory" of motivation that distinguishes *dissatisfiers* (factors that cause dissatisfaction) and *satisfiers* (factors that cause satisfaction).[42] For example, lack of parking adjacent to a performing arts center may be a dissatisfier, keeping away senior citizens who are concerned about safety, inclement weather, or who are challenged by long walks. If an orchestra does not offer ticket exchange privileges, this policy may be a dissatisfier that keeps frequent business travelers from making advance purchases. Arts managers must anticipate and counteract dissatisfiers. At the same time, appealing programming and other satisfiers are necessary to actually stimulate sales.

PERSONAL FACTORS

A variety of other personal characteristics affect a consumer's preferences and behavior. These factors include the person's occupation, economic circumstances, lifestyle, and life cycle stage.

Occupation

It is useful for marketers to identify the occupational groups that have above-average interest in the arts. The Los Angeles Philharmonic (LA Phil) found that among subscribers to its classical series, 54 percent were professionals; 29 percent held executive, administrative, or managerial positions; and 9 percent held technical, sales, or clerical positions. The LA Phil also found that various occupational groups have different programming preferences. College-level teachers represented 7 percent of professionals at classical concerts but only 2 percent of professionals at pops concerts, while 11 percent of classical audiences and 5 percent of pops audiences were physicians and dentists.

With this kind of information, marketers can reach potential patrons through their professional publications and they can place stories and advertisements in the media read by high-interest occupation groups. Messages can also be designed to reach people directly in their workplaces. One arts organization arranged with a sponsoring corporation to promote the performances as a heading on all of its interoffice memos and citywide messages.

Economic Circumstances

Marketers of income-sensitive goods and services pay constant attention to trends in personal income, interest rates, and attitudes toward the economic climate. If economic indicators point to a recession, arts marketers may choose to take steps to redesign, reposition, and reprice their products in an effort to retain and build their audiences. Miniseries, discount coupons, and blocks of lower-priced seating are among the offers that have been developed by arts organizations in order to attract patrons with less discretionary income. Yet, through the recent recession, as well as in better economic times, the highest price seats remain the best sellers.

Lifestyle

People coming from the same subculture, social class, and occupation may nonetheless lead very different lifestyles. One person may choose a "belonging" lifestyle, spending a lot of time with family and voluntary organizations. Another may choose an "achiever" lifestyle, working long hours on challenging projects and playing hard at travel and sports.

Lifestyle refers to a person's pattern of living in the world as expressed in his or her activities, interests, and opinions. It is a dynamic factor—each person may pursue different lifestyles at different times in life.

According to Harper Boyd and Sidney Levy, people are artists of their own lifestyles, and marketers have the opportunity to provide them with the pieces of the mosaic from which they can pick and choose to develop their own composition. "The marketer who thinks about his products in this way will seek to understand their potential relationships to other parts of consumer lifestyles, and thereby to increase the number of ways they fit meaningfully into the pattern."[43]

Experiential Life Stages

In contrast to family life cycle stages, David Wolfe has defined three experiential stages in adult life that largely define consumer behavior, especially in terms of discretionary purchase behavior. First is the *possession* experience stage, which usually develops in young adulthood, reflecting the stage of life when the acquisition of possessions, both animate and inanimate (house, car, jewelry, spouse, child), drives consumer behavior. Possession aspirations are strongly linked to the establishment and maintenance of the image one wishes to project to others. By age 40 and the onset of the midlife years, possession aspirations normally move beyond their peak potency, although they continue to be important, and *catered* experience aspirations begin their ascent. These experiences involve the purchase of services, rather than things: season tickets to the theater, a weekend getaway to a spa, a fine meal in an elegant restaurant. The more possession and catered experiences a consumer has had, the more those things that once mattered so much decline in importance. This is the time when *being* experiences, essentially of nonmaterialistic origin, take center stage. A being experience may consist of watching a beautiful sunset, learning something exciting, or listening to music. Being experiences enhance a sense of connectedness, sharpen one's sense of reality, increase one's appreciation for life, and contribute to inner personal growth.[44] Because art often creates an environment and serves as a catalyst for self-actualization, people who seek being experiences are a naturally fruitful target group for performing arts marketers.

In my own experience of teaching arts management to hundreds of college and master's degree students and interacting with young children, I have found that many young people focus deeply and strongly on being experiences, and that sensitivities to these interests often appear at a very early age.

There is a complex interplay among the cultural, social, psychological, and other personal factors that shape consumer behavior. The marketer must develop a thorough understanding of the factors that affect behavior and take them into account when selecting the target audience as well as positioning, pricing, and communicating the offering.

CHAPTER 5

PLANNING STRATEGY AND APPLYING THE STRATEGIC MARKETING PROCESS

CENTRAL TO EFFECTIVE MANAGEMENT AND MARKETING IS CAREFUL AND THOROUGH planning. Developing new tactics that are relevant to today's consumers requires more than some good ideas; it requires thinking strategically. Strategic market planning necessitates a broad understanding of the organization's current needs and opportunities, and it entails determining if ideas old and new are a good fit with the organization's core strengths and resources.

STRATEGIC PLANNING OVERVIEW

Strategic planning addresses all the issues of an organization, of which marketing is a central focus. If the core purpose of the organization is to bring art to the public and public to the arts, then basically all the efforts of the organization relate to the marketing function.

THE COMPLEXITIES OF PLANNING

The managerial environment and the marketplace are increasingly complex and changing, so it is difficult to make predictions about funding levels, audience development, and even programming opportunities. Also, in some organizations, there is no clear consensus as to which strategies will best carry out the mission. Whereas profit-making firms can pursue a well-defined goal of profit maximization, nonprofit organizations pursue a more complex set of goals and must apply different measures of success.

Stakeholder Roles

Diverse stakeholders in the organization have different perspectives on what they most value. Due to the very nature of their role as "owners" of the organization, board members prioritize fiscal responsibility, while artistic directors, the keepers of the organization's vision, prioritize artistic excellence and exploration. Some key stakeholders may prefer an emphasis on highly sophisticated programming, while others may place higher value on maximizing audience size or providing a broad spectrum of educational outreach programs in schools. Goals can often clash when, for example, the artistic director would like to present more new works while the board would prefer to "play it safe" and present familiar works, or when the board wants to earmark fundraising efforts to grow the endowment for the organization's long-term security, while the choreographer is eager to raise additional funds to provide for live orchestral music at performances. Meanwhile, managers often prioritize growing their administrative budget in order to increase staff size, improve staff quality, and to have additional funds for marketing and fundraising activities so they can best fulfill their objectives of maximizing both earned and contributed revenue.

Benefits of Planning

Paradoxically, such conditions warrant even stronger strategic planning, which promises several benefits. Strategic planning (1) directs the arts organization to identify long-term trends and their implications; (2) helps define the key strategic issues facing the organization; (3) opens better channels of communication among the key players in the organization; and (4) improves management control by setting objectives and providing measures of performance. Strategic planning helps an organization to develop a shared vision of its policies, goals, objectives, and activities. It defines the organization's planned trajectory.

The Planning Mind-Set

Both the organization and the environment with which it interacts are perpetually changing, and it is up to the marketing manager, other upper management personnel, and the board of directors to be aware of the current pulse and sensitive to opportunities and threats for the future. The key to an effective marketing strategy is to maximize *relevance* to the consumer. Relevance is not a constant; as people adapt to changes in their environment, they develop new expectations.

Adrian Slywotzky, who coined the expression *value migration*, says, "As customers' priorities change and new designs present customers with new options, they make new choices. They reallocate value."[1] Therefore, given a primary allegiance to the organization's mission, the strategic plan should be as flexible, adaptable, and changeable as the environment in which it exists.

THE STRATEGIC PLANNING PROCESS

Strategic planning is the process through which arts managers, artistic directors, and board members define their objectives (where they want to go), determine their strategy (how they will get there), identify the necessary resources (what it will take to get there), and evaluate their results (how they know if they got there).

It is important to consider, says Jay Conrad Levinson, that "marketing is not an event, but a process. It has a beginning, a middle, but never an end. You improve it, perfect, change it, even pause it. But you never stop it completely."[2]

To be effective, the strategic planning process requires the input of all the organization's stakeholders and collaboration between the board of directors, the business managers, and the artistic personnel. Typically, the strategic planning committee will consist of key board members, including the committee's chairperson and the chairs of the marketing, development, and finance committees. The committee will also include the artistic director or his or her representative and top-level managers, usually including the executive director and directors of marketing, development, and finance. Other staff and board members who specialize in various areas are brought into the process to speak to their particular interest. Once the committee members have formed a consensus, the plan is brought before the full board for approval, after which the executive director disseminates the strategic vision and plan throughout the organization. Then the marketing director develops and implements an annual marketing plan and special campaign plans to achieve the target objectives and goals.

Often it is helpful for an outside consultant to facilitate strategic planning sessions, as managers and board members may be unaware of some elements of the analysis, may not recognize the extent to which these factors do or could affect operations, or may be unsure of how to correct their problems.

STEPS IN STRATEGIC PLANNING

The strategic market planning process consists of the following four steps:

1. *Strategic analysis*: (a) Assess the organization's strengths, weaknesses, opportunities, and threats, and (b) analyze the organization-wide mission, objectives, and goals.
2. *Market planning*: (a) Determine the objectives and specific goals for the relevant planning period, (b) formulate the core marketing strategy to achieve the specified goals, and (c) establish detailed programs and tactics to carry out the core strategy.
3. *Marketing plan implementation*: Put the plan into action.
4. *Control*: Measure performance and adjust the core strategy, tactical details, or both as needed.

These components constitute an integrated set of steps. The control process is used to reformulate current strategies and to help plan future strategies. In this sense, planning is an ongoing process without a beginning or an end.

Integral to strategic planning is organizational self-assessment. Peter Drucker recommends that managers consider these five questions:

1. *What is our business (mission)?* What results does the organization seek to achieve? What are its priorities? Its strengths and weaknesses? To what extent does the mission statement currently reflect the organization's goals and competencies?
2. *Who is our customer?* Who are our current and prospective customers? Who are the primary and supporting customers? What are their levels of awareness of and satisfaction with the organization's service? Which segment(s) are good prospects for future development as patrons?
3. *What does our customer value?* For each primary customer group, what specific needs does the organization fulfill? What satisfaction and benefits does it provide? How well is it providing value? Is the same value available from other sources?
4. *What have been our results?* What criteria do we use to measure success? To what extent has the organization achieved the desired results?
5. *What is our plan?* In what areas should we focus our efforts? What new results should be achieved? What activities should be abandoned, expanded, or outsourced?[3]

Internal and External Environment Analysis

The first step is to analyze the organization's internal environment (i.e., strengths and weaknesses) and its external environment (i.e., opportunities and threats). This is known as a SWOT analysis.

It is extraordinarily illuminating for board and staff members to hear one another's views about the organization by listing everything that comes to people's minds in each of these four categories and then discussing all the factors. Not only are important issues revealed, but in most cases, the process helps participants arrive at a consensus as to future directions.

Internal Analysis: Strengths and Weaknesses

Every organization has strengths that can be exploited in the pursuit of its goals. Prahalad and Hamel refer to an organization's real strengths as its core competencies.[4]

A great opportunity is no opportunity if the organization lacks the required resources and capabilities. Performing arts organizations need the following competencies:

- *Program quality*: The higher the program quality, especially relative to direct competitors, the greater the organization's strength.
- *Efficiency level*: The more efficient the organization is at producing a program, renewing its subscribers, attracting new patrons, attracting adequate contributions, and managing administratively, the greater its organizational strength.
- *Market knowledge*: The more the organization knows about its current and potential customers, the greater its market strength.
- *Marketing effectiveness*: The more proficient the organization is at marketing, the greater its effectiveness in audience building and fund-raising.

In evaluating the organization's strengths and weaknesses, managers should rate each factor as to whether it is a major strength, minor strength, neutral factor, minor weakness, or major weakness. It is also necessary to rate the importance of each factor—high, medium, or low. Not all factors are equally important for succeeding as an organization.

This analysis tells us that even when an organization has a major strength in a certain factor (i.e., a distinctive competence), that strength does not necessarily create a meaningful advantage, as it may not be important to the organization's customers. Also, an organization does not have to remedy all of its weaknesses, since some are unimportant to its operations or customers. The strategic question is whether an organization should limit itself to those opportunities where it now possesses the required strengths, or consider better opportunities where it might have to acquire or develop new strengths.

One small but highly regarded Chicago theater performed on the second floor of a building where the only access was up a long, steep flight of stairs and the only parking facility was two blocks away. The theater attracted a sizable audience considering these limitations, but board and staff members wanted to grow their audience and be able to accommodate people who were unable to climb their stairs. So the theater decided to move to a new venue that was being built in an adjacent community. The new space seated about four times as many people and was extremely expensive for them to rent. The company was unable to grow its audience significantly and raise the money necessary to pay the high rent, so it had to shut down completely. Board members and managers had been unrealistic about this theater's potential.

External Analysis: Opportunities and Threats

An organization operates in a constantly changing and often turbulent external environment. The macro environment consists of large-scale fundamental forces—demographic, social, cultural, economic, and political—that shape opportunities and pose threats to the organization. These forces are largely uncontrollable but must be monitored for purposes of both short- and long-term planning. Opportunities can be seen as a combination of the organization's core competencies or strengths with an unfilled niche in the external environment. Common threats are anticipated or realized funding decreases and reduced prospects for future audience vitality due to an aging patron base.

The most successful organizations maximize opportunities by identifying attractive markets and developing the organizational strengths required to appeal to those markets. Four types of opportunities can be distinguished:

1. *Build opportunities* are those current activities that warrant more investment because they increase audience interest and participation, attract more funds, increase community involvement, and so on.
2. *Hold opportunities* are those that warrant maintaining the present level of investment. Although full-subscription purchases and renewals have slackened in recent years, organizations are maintaining their strong subscription efforts while offering new, flexible packages to encourage frequent attendance at and loyalty to the organization.
3. *Harvest and divest opportunities* are activities that should be reduced or eliminated in order to free up resources that can be better used elsewhere. Thus, some dance companies have reduced the number of *Nutcracker* performances they present to better fill the hall for the remaining performances. Due to waning audience interest, some organizations have discontinued offering flex plans.
4. *New product opportunities* are programs or services that might be added to the organization's current offerings, such as Rush Hour and Saturday afternoon casual concerts, singles nights, and other events planned to accommodate the lifestyle preferences of special interest groups.

These opportunities should be classified according to their attractiveness and their probable success. Market attractiveness is made up of factors such as the ones listed here:

- *Market size*: Large markets are more attractive than small markets.
- *Market growth rate*: High-growth markets are more attractive than low-growth markets.
- *Surplus building*: An organization may wish to present widely attractive programs to help subsidize its artistically driven programs that have narrower appeal.
- *Competitive intensity*: Markets with fewer or weaker competitors are more attractive than markets that include many or strong competitors. This does

not mean that weak competition justifies entry; the market must also be attractive on other grounds.

- *Cyclicality*: Cyclically stable art forms and art organizations enjoy relatively consistent levels of demand over long periods of time. For example, a theater that presents the classics is likely to enjoy far more stable audience interest than a theater that presents new plays by unknown playwrights.
- *Seasonality*: Seasonality provides opportunities for special programming (*A Christmas Carol*, summer music festivals). Most often, these programs garner high revenue and help support the costs of other programs throughout the year.
- *Scale economies*: Programs for which unit costs fall with large-volume production and marketing or that build on previous efforts are more attractive than constant-cost programs.
- *Learning economies*: Programs that get better or more efficient with each performance or production are more attractive than those that do not benefit from past experience.

Product/Market Opportunity Analysis

Opportunities can also be identified through the use of a product/market opportunity analysis. An organization may look at its present offerings in terms of (1) building on what already exists, (2) making modifications to current offerings, or (3) developing entirely new products. Markets may be analyzed in the same way: (1) building current markets that have potential for growth, (2) modifying current markets, or (3) developing new markets.

Market penetration consists of broadening the organization's infiltration of current markets with current products. This is often the easiest strategy, as it requires the fewest changes. This strategy is most workable when there is significant room for growth in currently sizable markets with existing offerings. Offering better marketing programs and increasing advertising and promotion to the target market will help the organization deepen its market penetration. For example, a chamber music group gave four free tickets to board members, as a means of encouraging them to introduce their friends and business associates to the organization. The promotion included a postperformance reception with the artists to build involvement and enthusiasm.

Geographical expansion consists of taking the organization's current offerings into additional geographic markets by performing in alternative venues in outlying areas, going on tours, and distributing recordings and videos through various media. Many symphonies, dance companies, and a few theaters, such as Piccolo Teatro di Milano, regularly tour internationally. Many organizations offer outreach programs in local schools, neighborhood parks, and other community locations.

New markets involves finding new groups to attract with current offerings, such as new age or ethnic groups and special interest groups. It also includes attracting more distant audiences by offering special transportation, dinner-theater packages, group plans, and price incentives. The marketing director of Hubbard Street Dance Chicago once reserved a car on the commuter train from downtown Chicago, where many young professionals live, to the suburban Ravinia Festival where the company was performing. More than one hundred singles who lived in the downtown area responded to the targeted mail offer that included refreshments on the train, the dance program, and a postconcert party with some of the performers and staff.

Offer modification may involve changes that make programming more accessible to various target markets. The time or length of the program may be geared to specific audience segments, or a concert may be advertised as casual in order to encourage spontaneous attendance. The New York Philharmonic has added performances at the lunch hour and the rush hour, appealing to people who work in Manhattan, but who live in outlying areas. The Los Angeles Philharmonic offers casual concerts at which the musicians wear jeans or khakis to appeal to the younger crowd. Casual, hour-long, Saturday afternoon concerts offered by many orchestras provide an appropriate setting at a convenient time for parents to share fine music with their children.

Offer innovation involves the development of new product offerings, such as the creation of a membership or a group sales plan where none previously existed. Other examples include the special children's series at the New York Philharmonic and the jazz series at Lincoln Center.

Geographical innovation involves offering a new program in a new geographical area. A theater may develop a program of plays by African American playwrights and perform them in African American neighborhoods. Theaters may air select productions on radio stations.

Total innovation refers to creating new offerings for new markets. Dancing in the Streets offers adventurous, free public performances in unexpected places and creates new choreography for each location in which it performs.

Thus, the product/market opportunity analysis stimulates thinking about new opportunities in a systematic way. New opportunities are evaluated, and the better ones are pursued.

Threats and Opportunities for Collaboration

Along with opportunities, organizations face various threats. An environmental threat is a challenge posed by an unfavorable trend or development in the environment that may harm the organization. Although such changes are largely uncontrollable, an organization should prepare for contingencies. In the case of a recession, the organization should think about how to develop other revenue sources and/or cut costs. If the audience is aging, the organization should focus on attracting younger patrons. When fewer patrons are interested in full-season subscriptions, the organization should offer alternatives that appeal to its customers' needs for lower commitment levels and more spontaneity.

Threats can also come from groups and organizations that compete for the same audience. Yet, many opportunities exist for arts organizations to develop mutually beneficial collaborations with one another and with leisure-oriented businesses such as restaurants, hotels, and other tourism providers.

Greater Boston Theatre Expo: Adapting to a Changing Market

In September 2013, 55 Boston-area theater groups participated in a free expo, intended to give theatergoers a chance to meet, greet, and get to know them. Participating groups ranged from big companies such as the American Repertory Theater and the Huntington Theatre Company to newer small groups such as Fresh Ink Theatre and Theatre on Fire. Companies were represented by marketing staff, performers, and/or other creative staff. The program did not include performances—just conversation, special discount offers, and some raffles.

The expo was a collaborative effort by the companies, who lured their own theatergoers to the event. This event gives small companies access to traditional theatergoers, and gives the traditional companies exposure to the avid theatergoers who comprise the strong audience base for the smaller companies. It is also a good thing for people involved in Boston's cultural landscape to see so many theaters all in one place, thereby asserting their collective value.[5]

ANALYZING THE ORGANIZATION'S MISSION, OBJECTIVES, AND GOALS

Once the organization has completed its SWOT analysis, the next step in strategic market planning is to analyze and define the organization's mission, objectives, and goals.

MISSION

Effective nonprofit organizations have four characteristics: (1) a clearly articulated sense of mission, (2) strong leaders and a culture that motivates the organization to fulfill its mission, (3) an involved and committed volunteer board that provides a bridge to the larger community, and (4) an ongoing capacity to attract sufficient financial and human resources.[6]

An organization's mission comprises its purpose and the results it wants to achieve. Says management strategist Peter Drucker,

> The mission focuses the organization on action. It defines the specific strategies needed to attain the crucial goals. It creates a disciplined organization. It alone can prevent the most common degenerative disease of organizations, especially large ones: splintering their always limited resources on things that are "interesting" or look "profitable" rather than concentrating them on a very small number of productive efforts.[7]

The Guthrie Theater: A Mission Recaptured

The highly respected Guthrie Theater in Minneapolis, Minnesota, thrived for many years with a mission of presenting fine-quality performances of the classics. When a new director arrived, he tried to be more things to more people and changed the programming by adding some new and experimental plays and popular entertainment shows. The number of subscribers dropped from 23,000 to 13,000. The Guthrie had lost its grip on its purpose and its audience. For the next several years, the board of directors struggled with the Guthrie's identity—what kind of theater to be. It finally returned to its role as a theater offering vital productions of the classics. By the end of the second year, subscriptions were up 70 percent to 22,000, the highest since the inaugural season.[8]

Defining the Mission

Although the question "What is our mission?" sounds simple, it is really the most challenging question an organization can ask. Different members will have different views of what the organization is about and should be about. The mission statement should describe what the organization does, whom it serves, and what it intends to accomplish. It should be broad enough not to need frequent revision and yet specific enough to provide clear objectives and to guide programming. It should be understandable to the general public and should be stated as succinctly as possible, with a forward-looking approach. [9] Organizations may need to hold numerous meetings over several months to clarify their ideas and to develop consensus.

In defining the mission, it is helpful to establish the organization's scope on three dimensions. The first is *consumer groups*, namely, *who* is to be served and satisfied. An organization's consumers include all the parties that it intends to serve, including the artistic director, the performers, and the various audience segments. The second dimension is *consumer needs*, namely, *what* is to be satisfied. The goal may be, for example, to inspire or educate the audience—but how to accomplish each goal depends on the tastes and levels of sophistication of the audience members, combined with the intent of the organization's leadership. The third is *approaches*, namely, *how* consumer needs are to be met. A summer music festival may be created to share the joy of music in the context of a casual ambience. Performances held in neighborhood settings may provide convenience and a level of comfort that leads people of particular groups to enjoyment of the art form. A theater performing the classics is exposing its audiences to great plays through the ages.

An organization should strive for a mission that is *feasible*, *motivating*, and *distinctive*. Feasibility means that the mission should be realistic in terms of the organization's financial and human resources. The organization should avoid a "mission impossible." An institution should aim high, but not so high as to create an unachievable goal or to produce incredulity among its publics. The mission should also be motivating: it should embrace goals that excite the organization's staff, board, and artists, who should feel they are worthwhile members of a worthwhile organization. A mission works best when it is *distinctive*. If all theaters in a city resembled each other, there would be little loyalty to or pride in any one company. Also, certain consumer needs would be overserved while others wouldn't be met at all. By cultivating a distinctive mission and personality, an organization stands out more and attracts more loyal patrons.

The Mission Statement

Following are examples of mission statement development at the Illinois Symphony Orchestra and the San Francisco Ballet.

Mission Development at the Illinois Symphony Orchestra

In 2008, I was invited to consult with the board and staff of the Illinois Symphony Orchestra (ISO) with a primary focus on strategic planning. We began the strategic planning process by reviewing the ISO's mission statement. Here is the mission statement as it existed before I began my project:

> The mission of the Illinois Symphony Orchestra is to provide live professional orchestral music to the communities of central Illinois; to perform at the highest levels of musical excellence and with the highest standards of professionalism; to serve people of all ages, races and economic means; to introduce classical music to children in central Illinois through education and outreach initiatives; to serve as a vital community member as we strive to be one of the premiere professional orchestras in the State of Illinois.

After several meetings with key stakeholders in the organization and intensive review of many documents, I developed a new mission statement that, I thought, better addressed the organization's strengths, its potential, and its limitations. Here is my revised mission statement:

> The Illinois Symphony Orchestra is dedicated to providing inspiring professional orchestral concerts at high levels of musical excellence to the widest possible audiences in the communities of central Illinois. The Illinois Symphony enhances its role as a vital cultural resource in the region by offering enriching and entertaining outreach and educational programs to families and school children.

The board approved this mission statement. My rationale for the changes I made is as follows:

- We do not provide music to communities, we provide it to *people*. The word "audiences" gets at this idea better.
- Many of the musicians are not regulars with the symphony and different musicians need to be sought and hired on a per-concert basis. Therefore, it is not possible for the ISO to perform at the "highest levels of musical excellence," as do major orchestras in large cities, whose top-quality musicians are employed on a full-time basis and play together constantly. So it made sense to me to change the mission in this regard to "high levels," instead of "highest levels."
- Although the ISO wants to clearly show in its mission statement that it is all-inclusive in the people it wants to reach, some of the claims in the original mission statement are not feasible. The ISO does not really want to provide the concerts to "all ages." Just imagine two-year-olds in the audience, and, as much as the ISO would love to serve underprivileged people, it is not in the financial position to give away tickets to those who cannot afford to buy them. Unless the ISO plans to do these things, they are not realizing their mission as stated. Therefore, I recommended that "widest possible audiences" would cover what is realistic.
- Why is it a goal to be one of the premiere orchestras in the state? The ISO cannot begin to compete with the world-renowned Chicago Symphony Orchestra. Other Illinois orchestras also present very stiff competition to the ISO in the quality dimension. Claiming a mission of being a premiere orchestra in the state is actually a demotivator to artists, staff, and audiences alike who would rightfully question this aspect of the mission.

Revising the Mission at the San Francisco Ballet

The first step in our strategic planning process was to review the San Francisco Ballet's mission statement to determine if it still met the organization's needs and aspirations. The existing mission was:

> San Francisco Ballet Association is dedicated to producing superior performances of classical and contemporary ballet, to providing the highest caliber of training for dancers aspiring to professional careers, and to becoming America's model of excellence and innovation in ballet artistic direction and administration.

The planning committee recognized immediately that this mission statement was outdated. For several years, the company had garnered worldwide recognition for the quality of its dancing and innovative programming; the dance school was considered one of the country's finest and was in high demand by fine, young dancers; the artistic and administrative functions were extremely well run by high-caliber personnel. The mission had been accomplished; a new mission that would be motivating, distinctive, and feasible needed to be crafted.

Over the course of many meetings among artistic staff, management staff, and the board, key elements for a new mission statement were determined. The ballet wanted to recognize its status as a superior dance company and school, yet motivate its staff, dancers, and students to continually improve. They wanted to expand the audiences they serve, both locally and on tour, and expose as many people as possible—young and old, dance aficionados and newcomers alike—to the joys of dance. To express the characteristics that make San Francisco Ballet distinctive and that are central to its very essence, *vitality* and *diversity* were added to the ballet's long-standing focus on innovation, although it was crucial to state that no matter how contemporary and creative the choreography, it was based in the tradition of classical ballet.

The unanimously agreed-upon new mission statement read as follows:

> The mission of San Francisco Ballet is to share our joy of dance with the widest possible audience in our community and around the globe and to provide the highest caliber of dance training in our school. We seek to enhance our position as one of the world's finest dance companies through our vitality, innovativeness, and diversity and through our uncompromising commitment to artistic excellence based in the classical ballet tradition.

Setting Objectives and Goals

While the mission statement comprises an organization's purpose and broadly states the results it wants to achieve, each organization must also develop specific objectives and goals for the coming period that are consistent with its mission statement. *Objectives* are usually intended to have a three- to five-year life span. The objectives of a midsized theater company planning to move into a new, larger venue in three years include raising $10 million in a capital campaign for the new building, increasing annual giving to accommodate the larger operating budget required in the new venue, increasing the subscriber base in anticipation of having 50 percent more seats to fill in the new space, building awareness of and interest in the theater to attract new patrons, and upgrading box office software so that customer-centered data can be easily accessed and to facilitate real-time online ticket ordering.

In order for objectives to be achieved, they need to be restated in operational and measurable terms, called *goals*. The objective of increasing the number of subscribers may be turned into a goal of a 10 percent increase the first year and a 5 percent increase each of the next two years. Goal setting leads managers to consider all the necessary planning, programming, and control aspects: Is a 10 percent increase feasible? What strategies will be used? What human and financial resources will it require? What activities will have to be carried out? Who will be responsible and accountable? How will achievement be tracked? All of these questions must be answered before goals are adopted.

As it celebrated a major anniversary, a dance company well known in the Midwest, stated its long-term goal as being one of the top five dance companies in the United States. In aspiring to this goal, management needed to address several issues. What does it mean to be a top-five company? Does top-five status refer to the performers, choreography, variety of repertoire, size of the company, benefits to customers, rave reviews—or all of the above? As to the goal's feasibility, where does the company now rank in the hearts and minds of its patrons? In how many areas and by what percentage in each area must the company improve in order to attain this goal? What is the competition that must be challenged in order for the goal to be achieved? How is progress or success to be measured—by increases in audience size, audience growth relative to competitors, quality of reviews? How will the relevant information be obtained? Finally, what benefits will the company gain by achieving this goal? Will it help to sustain current audiences and build new ones? Will it attract better dancers and choreographers? Will it attract more funding? Who is the goal for?

Completing the Strategic Planning Process

After conducting a strategic analysis, an organization must undertake three major steps to complete the strategic market planning process: market planning, implementation, and control.

Market Planning

Market planning is the process of determining specific objectives and goals for the relevant planning period; formulating the marketing strategy to achieve the goals; and establishing detailed programs, tactics, and budgets. Goals indicate where an organization wants to go; strategy answers how to get there. Each strategy must be refined into specific programs. In determining specific strategies, it is important that the organization correctly identify the issues that affect each target market.

Strategic Decision-Making

Good choices are dependent on the quality of the decision-making process. There is general agreement that better decisions are based on in-depth information. Yet many managers grab at the first available straw, only to discover that it's a short one. For example, an organization copies another art group's promotional idea without researching its effectiveness and without adapting the idea to its own specific situation; a marketing director judges a new ad campaign by her "gut" reaction rather than by conducting market testing to verify her intuitions. Similarly, when faced with a new and desirable community involvement program, a manager may recite the all too familiar "no money" response, rather than seek ways to implement the program using community-based collaborations. Or, a marketing director requests a budget that is a percentage higher or lower than last year's budget, rather than modifying it to reflect new opportunities and to reduce or eliminate funds designated for previously unsuccessful efforts. Sometimes decision-makers select a solution that is acceptable or reasonable, though not necessarily the best choice.[10]

The following model outlines the steps in strategic decision-making.[11]

1. *Define the problem.* It takes good judgment to accurately define the problem. Too often the problem is defined in terms of symptoms rather than underlying causes. It has been observed that a problem well defined is a problem half-answered.
2. *Identify the criteria.* Managers must identify the appropriate criteria for judging alternative courses of action. For example, in advertising a new production, managers must evaluate the alternative media (radio, newspapers, direct mail, the Internet) not only in terms of their cost but also in terms of their effectiveness.
3. *Weigh the criteria.* Importance weights must be attached to the respective criteria. For example, how much weight should be given to audience satisfaction versus performers' satisfaction versus financial results?
4. *Generate alternatives.* The next step is to generate alternative solutions. Here some creative brainstorming may greatly enlarge the number of realized alternatives.
5. *Make a choice.* Select the alternative that best meets the defined criteria.

This model of decision-making requires significant levels of information gathering and analysis. What happens when fast decisions must be made? Kathleen Eisenhardt contends that fast decision-makers use simple, powerful tactics to make effective and efficient choices.[12] They rely on regular—daily or weekly—performance reports, such as ticket sales information. They may require frequent staff meetings—two or three per week. Such frequent communication helps managers stay aware of what is happening daily and enables them to spot problems and opportunities, which, in turn, helps them react quickly and accurately to changing events.

Another factor in efficient decision-making is seeking advice from a colleague or consultants with broad experience in and knowledge of the field. Seeking advice from members of the executive board fosters a climate of trust and demonstrates discretion.

THE MARKETING PLAN

Once strategies are formulated, the next task is to plan marketing programs. This process consists of making basic decisions on marketing expenditures, marketing mix, and marketing allocation. It requires effectively combining the organization's resources to produce annual plans and budgets that work toward achieving the organization's overall goals. For the marketing manager, it also requires a deep understanding of how best to meet the target market's needs and preferences.

A structured marketing plan serves several important purposes. First, when a plan is written down, inconsistencies, unknowns, gaps, and implausibilities can be readily identified and dealt with. Second, it provides an anchor—stability in the midst of change. It is a basis on which to measure progress and to incorporate change as the market and the organization's own goals change. Third, it helps focus management on the organization's annual goals, changing market conditions, and key marketing issues. This is a crucial factor since a common complaint among marketing professionals is that their focus on day-to-day affairs causes them to lose sight of the big picture. Fourth, it leads to an implementation timetable that ascertains that tasks and goals are completed by certain dates.[13]

Formulating Marketing Plans

Marketing plans may take on different formats depending on managers' needs, but each plan should focus on identifying key issues, mobilizing resources, and measuring results. The marketing plan will approach each aspect of the marketing mix (product, price, promotion, and place) in a way that will position the offering effectively for the target market. The fifth "P," people—who will be targeted for the plan and who will be responsible for the work involved—is also crucial to address.

Robert Lauterborn suggests that marketers think in terms of the four Cs instead: *customer value* (not product), *customer costs* (not price alone), *convenience* (not place), and *communication* (not simply promotion).[14] In the same vein, Mohanibir Sawhney suggests that marketing should be organized around processes—processes for understanding, defining, realizing, delivering, communicating, and sustaining value.[15] Philip Kotler says that once the marketer thinks through the four Cs for the target customer, it becomes much easier to set the four Ps.[16]

To create a plan, the manager will (1) identify potential alternatives for achieving the organization's goals, (2) determine the financial implications of these alternatives, (3) decide which alternatives to pursue, and (4) decide on the marketing mix and marketing expenditures for each objective and/or program.

Marketing plans will have several sections, including an executive summary; current marketing situation; opportunity and issue analysis; objectives; marketing strategy; action programs; projected budgets, including statements of revenues and expenditures; and controls.

Executive summary. The planning document should open with a short summary of the main goals and recommendations to be found in the body of the plan. This permits upper management and board members to grasp quickly the major thrust of the plan. Each stated objective should refer to an element of the mission statement that it is designed to help realize. A table of contents should follow the executive summary.

Current marketing situation. This section of the plan presents relevant background data on the market, products, competition, distribution, and macroenvironmental factors. Data are presented on the size and growth (or decline) of the total market and on market segments over the past several years. Data are also presented on customer needs, perceptions, and purchase behavior trends. Each product is described in terms of target market(s), sales, prices, and distribution. When appropriate, major competitors are identified and described in terms of their size, goals, product quality, marketing strategies, and other characteristics that are helpful in understanding their intentions and behavior. Broad trends affecting each product and key target market are identified.

Opportunity and issue analysis. On the basis of the data describing the current marketing situation, the marketing manager identifies the major opportunities/threats, strengths/weaknesses, and issues the organization faces for each of its offerings during the time frame of the plan.

Objectives. At this point, management knows the opportunities and issues and is faced with making basic decisions about the plan's objectives. Two types of objectives should be set: *marketing* (such as reattracting lapsed subscribers or upgrading the organization's image through a branding campaign) and *financial.*

When I facilitated the long-range strategic plan for the San Francisco Ballet, we identified six objectives, which we called *strategic imperatives,* to make them compelling to potential funders, board members, and staff. These imperatives were divided into two categories: program imperatives and resource imperatives. The program imperatives—new works, touring, and outreach and education—are central to realizing the mission. The resource imperatives—major gifts, brand identity development, and board development—focus on the key resources necessary to implement these programs.

Marketing strategy. The manager now outlines the broad marketing strategy or "game plan." The basic strategy can be presented in paragraph form or list form, covering the major marketing tools. Following is a hypothetical strategy statement based on a theater's marketing objectives.

Target market: Long-term subscribers whose subscriptions have lapsed in the past two years.

Product: Engaging programming by well-known playwrights, selected by a committee of the theater's renowned founding members, those who made the theater great.

Price: No increase from previous season, discounts for early subscriptions, special perks.

Place: Cosmetic changes in the lobby to make it more inviting and comfortable.

Research: Allot expenditures for patron interviews to increase knowledge of customers' attitudes and behavior and to monitor changes in their attitudes and perceptions of the theater's image, product offerings, and reasons for attrition.

Sales force: Increase telemarketing efforts by 20 percent for personal contact with lapsed subscribers.

Service: Quick and courteous. Listen well. Create a response mechanism for all customer complaints.

Promotion: Create a new direct-mail campaign with a personalized letter for each target market and an email campaign with individually personalized messages. Increase advertising and direct mail budget by 20 percent.

Positioning: A bold and powerful theater company, one of the country's best, most highly respected, and most reliable. Our productions will engage and inspire you. We create flexible packages to meet your needs. (Positioning to be modified based on research results.)

In developing the strategy, the manager needs to discuss the plans with others whose cooperation and support are necessary for implementation. The financial officer can indicate whether enough funds are available; if not, trade-offs must be made to prioritize this effort over other expenditures; the executive director must be willing to dedicate the necessary resources and to support the concept of the program.

Action programs

The strategy statement represents the broad marketing thrusts that the manager will use to achieve specific objectives. Each element of the marketing strategy must now be expanded in detail to answer: What will be done? When will it be done? Who will do it? How much will it cost? What kind of benefit and how much benefit will it generate for the organization?

Budgets

A budget is an organization's plan of action expressed in monetary terms. Budgeting is a complex process that will be discussed in depth later in this chapter.

Controls

The last section of the plan outlines the controls that will be applied to monitor the plan. Typically the goals and budgets are spelled out for each month or quarter so that results can be reviewed each period and progress can be tracked. A contingency plan may be included to outline the steps that management would implement in response to specific developments. This planning will encourage managers to give prior thought to some challenges that might lie ahead.

Implementation

A brilliant strategic marketing plan counts for little if it is not implemented properly. While strategy addresses the *what* and *why* of marketing activities, implementation addresses the *who*, *where*, *when*, and *how*. Marketing implementation is the process that turns marketing plans into action assignments and ensures that such assignments are executed in a manner that accomplishes the plan's stated objectives.

Managers must have diagnostic skills to help determine whether or not a strategy is implementable. Allocation skills are used in budgeting resources (time, money, and personnel) to various functions and programs according to standards of efficiency and effectiveness. Monitoring skills are used in managing a system of controls to evaluate the results of marketing actions. Organizing skills are used in developing an effective working organization. Understanding the informal as well as the formal organization is important for effective implementation. Interacting skills refer to the ability of managers to get things done by influencing others. Interacting skills include applying effective behavior patterns during the implementation process, and effective use of the organization's power structure and of its leadership.

Many "best-laid plans" fail to see the light of day. Plans to innovate may fizzle out when management or board members are reluctant to commit funds to recommended projects. Plans to improve quality may get no further than some airy rhetoric. Without successful implementation, a strategy represents merely lost time and frustration to those who develop and support it and want to see the organization progress. Thus, a strategy is not truly well conceived if it is not implementable. The factors that allow for successful implementation must be considered during the formulation process itself. A strategist must be able to look ahead and ask, "Is this strategy workable? Can I make it happen?" If an assessment leads to the answer "no" or "only at an unacceptable risk," then the formulation process must continue.

In determining if a strategy is implementable, the strategist must consider the complexity, uncertainty, instability, and hostility of the environment. Internal organizational factors to be taken into account include leadership continuity, quality of decision-making, quality of management information systems, and the nature of the organization's reward systems.

A plan is a commitment to action—a commitment by people with their own ideas, attitudes, preferences, concerns, and needs. People must see the strategy's relevance, feel they are capable of implementing it, understand the required behaviors, know whether they have achieved the objective, and be rewarded for doing so. Important activities for implementing the plan should be prioritized or the organization may find that extraneous activities are draining valuable human and financial resources from the challenges at hand.

Strong leadership is crucial when new strategies are being implemented. Even though old methods may have proven to be ineffective, employees do not feel responsible if those efforts do not go well because, they say, "this is the way we've always done it." When staff and/or board members take this backward view, it is the role of the leader to take charge of making sure the new strategy is implemented and to take responsibility if it is not successful. It is important for the leader to be integral to the process for small efforts, as well as for large ones. In fact, staff is more likely to follow the leader on big changes after seeing the leader roll up his or her sleeves on small initiatives.

A Different Venue for the Lake Forest Symphony

The Lake Forest Symphony (LFS) performed in a hall in its eponymous town near Chicago for more than 30 years. I started consulting with the LFS one season after its long-standing performance hall was sold and demolished. Since there were no other venues in Lake Forest that were appropriate for professional orchestra performances, the symphony's decision-makers moved their performances to a beautiful facility with excellent acoustics, but in a different town about a 20-minute drive from Lake Forest. Most Lake Forest residents, who had been attending and supporting the symphony for up to 30 years, could not relate to "their" orchestra performing outside of Lake Forest, and stopped attending. Symphony managers offered bus service from a central location in Lake Forest to the new venue, but no one took them up on this offer.

After analyzing the situation, I first told management that the physical distance to the hall was not the problem; in fact many Lake Forest residents attend the Chicago Symphony Orchestra and Lyric Opera of Chicago, which are about an hour's drive away. Offering bus service made the new hall seem to be much farther away than it actually was. The distance these people felt was a psychological one and we needed to try to break through this barrier. I suggested that we offer complimentary tickets for one performance to people who had subscribed to the symphony in the former hall but had not attended the new facility. Staff members were reluctant to pursue my idea, as they believed that a moderate discount on the ticket price made sense but free tickets would cheapen the product and indicate our desperation. I countered that a discount would not be enough of a motivator to attract these people and I would take full responsibility of dealing with any possible backlash from my plan. The staff agreed, so I wrote a letter that said, in effect: "We miss you! We know that you love the Lake Forest Symphony and you'll love us even more in our state-of-the-art venue with fabulous acoustics, comfortable seats, and ample free parking. In fact, we miss you so much that we would like to offer you two complimentary tickets to an upcoming performance this season, according to availability." Nearly half of the people to whom we sent the letter took us up on the offer. Afterward, many of them wrote to us, thanked us for the tickets to the great concert, and said that the drive to the new venue wasn't bad at all. Most important, many of them started subscribing again.

CONTROL

Performance is the ultimate test of any organization. Says Peter Drucker, "The discipline of thinking through what results will be demanded . . . will protect the organization from squandering its resources."[17] This last step of the strategic market planning process, reviewing and adjusting the strategy and tactics and measuring performance, should actually be taken into consideration throughout the process. As we embark on various strategies, our information is necessarily somewhat limited. While experience and new knowledge are being gained, strategies and tactics should be reevaluated in light of fresh information. Frequent evaluations can serve to either validate a current approach, stimulate the modification of approach or of the goals themselves, or encourage the abandonment of a strategy altogether.

An organization should establish three levels of performance control: short-range or periodic, to assess progress and the effectiveness of operational marketing plans on a daily, weekly, monthly, or quarterly basis; midrange, usually annual; and long-range or strategic control, usually every three to five years. Many of the control processes and tools used for annual plan control are also used for shorter-term control.

Periodic and Annual Control

The short- and midrange control processes are driven by an approach called *management by objectives.* Four steps are involved. First, management sets monthly or quarterly goals. Second, management monitors the organization's performance in the marketplace. Third, management determines the causes of serious performance deviations. Fourth, management takes corrective action to close the gaps between the organization's goals and performance. This could require changing the action programs or the goals.

Sometimes a strategy should be abandoned altogether. Too often, management keeps investing additional resources, both human and financial, in order to salvage what has proved to be an ineffective project, thereby draining resources from more productive efforts. It is far better to accept a loss than to continue on an unproductive course of action. The time and expense already invested are "sunk costs," that is, they are historical, not recoverable, and should not be considered when evaluating the best course of action.

Management starts the control process by developing aggregate goals for the planning period, such as increasing the number of single ticket buyers who return for another performance. It is then up to the marketing manager to determine the meaningful criteria and the prerequisites for success. For example, the marketing department may have specific goals and tactics for reattracting patrons with special offers and engaging messages. During the relevant period, managers receive reports that allow them to determine whether their goals are being reached, and if not, to take appropriate actions.

Various control tools are available for managers to use in this process. Three common quantitative control tools are sales and revenue analysis, expense analysis, and ratio analysis. Two useful qualitative measures are customer satisfaction tracking and a marketing effectiveness/efficiency review.

Sales and revenue analysis. Sales and revenue analysis is the effort to measure and evaluate the actual sales and revenue being achieved in relation to the goals set by season, specific performance, audience type, and so on. If goals are not being met, the marketing manager may choose to modify various aspects of the offer.

Expense analysis. The budget can help to determine whether the organization is setting reasonable expense targets and to track actual spending and receipts against planned targets. Variances between actual and budgeted performance serve as a signal to management and the board that attention is required, whether the variance is positive or negative. Positive variance (where actual spending is lower than planned amounts) may either be the result of laggard performance in program implementation or may indicate cost savings in relation to anticipated expenditures. A negative variance may indicate either cost overruns or lower-than-expected revenues. This latter condition, if not dealt with in a timely manner, can create serious financial problems for the organization.

The following example dates back to the first edition of this book, but it continues to be relevant in demonstrating how to analyze expenses.

Analyzing Expenses at the Steppenwolf Theatre

At the Steppenwolf Theatre, subscription drive expenses are broken down by various categories. In the 1994–1995 season, 13,929 subscribers were recruited at a total cost of $248,681 or $17.85 per subscriber. The breakdown is exhibited in table 5.1.

New subscribers are important to Steppenwolf, even though they are costly to recruit. The organization increased its new subscriber budget for the 1995–1996 season to $85,050 for a goal of 1,800 subscribers. In other words, it planned to spend $30,000 more to capture an additional 317 subscribers, at a cost of nearly $100 per additional new subscriber. (As you will see in chapter 15, the lifetime value of subscribers makes it attractive to spend a considerable amount to attract them for their first season.) As interim results of the subscription campaign were tracked, revisions to the budget were made. By the end of July 1995, the campaign had recruited 2,100 new subscribers, exceeding the goal by 300 subscribers. This reduced the cost per new subscriber considerably, to $49.

In the 1994–1995 season, the subscription campaign yielded revenues of $1,423,730 for a cost of $248,680, an expense-to-revenue ratio of 17 percent. For the following season, projections were for revenues of $1,621,200 and expenses of $295,000, an expense-to-revenue ratio of 18 percent. (Printing and postage costs had increased considerably, accounting for the 1 percent increase in costs.) By the end of July 1995, expenses remained as budgeted, but revenues of $1,810,000 exceeded expectations, which brought the expense-to-revenue ratio down to 16 percent.

Table 5.1 Subscription drive expenses at the Steppenwolf Theatre

Category	Cost ($)	Number of subscribers	Cost/subscriber ($)
Early renewal	10,952	3,400	3.22
Official renewal	10,593	4,200	2.52
Telemarketing	61,391	3,157	19.45
New subscriptions	54,771	1,483	36.93
Flex plans	7,000	606	11.55

Ratio analysis. Analysis of ratios concerning various functions and programs can be used to compare progress by production, by season, or on a year-to-year basis, and determine their relative effectiveness and efficiency. An organization may analyze whether, for example, total ticket sales revenue for the year increased or decreased as a percentage of total combined earned revenue and contributed support. This is known as the ratio of net total earned revenue to total revenue and support.

A key ratio to watch is marketing expense to sales. For example, a marketing manager may track ratios of telemarketing to sales, advertising to sales, marketing research to sales, and promotions to sales. The marketing manager will be interested in the expense-to-revenue ratios of various marketing programs, as shown in the Steppenwolf Theatre example. An arts organization may also wish to compare the expenses and revenues associated with various programs to the organization's total expenses and revenues. This helps determine whether the programs are receiving a growing or dwindling share of the budget. The numerator is the category of expense, such as outreach programs, a ballet school, special events, or operating expenses. The denominator is the organization's total expense budget. The marketing manager should analyze such expenses as print advertising, promotions, direct mail, telemarketing, and so on as a portion of each program budget or of the entire marketing budget. It is useful to also analyze the much lower costs of email marketing, website management, and social media compared to the results they deliver. It is commonly accepted that social media is far more effective at building relationships than at providing any quantifiable measures, but the organization should still measure, the best it can, its hard costs and personnel costs for these efforts.

Customer satisfaction tracking. The preceding control measures are largely financial and quantitative in character. They are important but not sufficient. Also needed are qualitative measures that monitor changing levels of customer preference and satisfaction so that management can take action before such changes affect sales. Customer service is so central and crucial to the marketing function that I will address it in a separate chapter.

Marketing Efficiency and Effectiveness Review

Marketing effectiveness is not necessarily revealed by current sales and revenue performance. Good results could be attributed to featuring a well-known performer in a highly respected show, rather than to effective marketing management. Marketing improvements might boost results from good to excellent. Conversely, a marketing department might have poor results in spite of excellent marketing planning.

Peter Drucker recommends that managers ask the following questions to evaluate the effectiveness of each major program or activity:

- If we weren't already doing this—if we were not already committed to this—would we start doing it now?

- Are we working in the right areas? Do we need to change our focus?
- What have we learned and what do we recommend?
- What, if anything, should we do differently?[18]

To evaluate the efficiency of specific projects, managers should raise such questions as:

- Are higher revenue projections achievable?
- What additional sources of revenue are possible—raising fees, lowering fees to attract more patrons, devoting extra efforts to existing activities, initiating new programs, providing new services?
- Can expenditures be reduced by cutting certain costs?
- Can alternative approaches be used to improve efficiency and thereby lower costs without affecting quality?

In evaluating one program activity against another, the board and staff should be careful about using the criterion of cost-effectiveness. This criterion is useful for analyzing such factors as direct mail versus print advertising, for example. But from a larger perspective, the organization's mission may dictate that managers should carry on activities and programs that are not cost-effective, such as educational programs or performances of new music in a half-empty hall. The primary criterion should be a program's relationship to the mission of the organization, its purposes, and its goals.[19]

THE ORGANIZATION'S PUBLICS

Publics are individuals, groups, and organizations that take an interest in the organization. Various publics can help or harm the organization at different

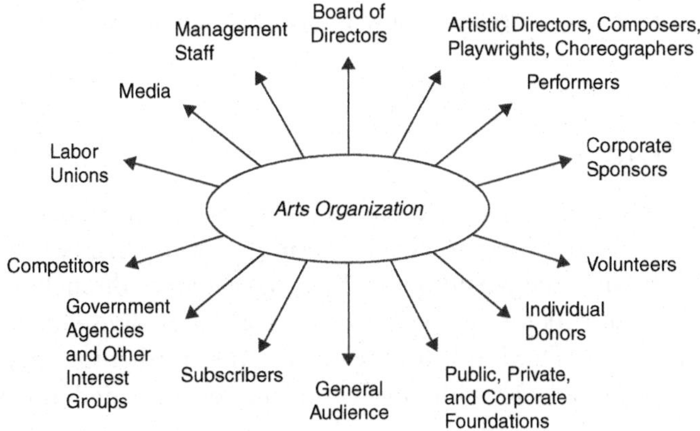

Exhibit 5.1 An arts organization's publics.

points in time, and their needs and interests must be served or accommodated. Exhibit 5.1 shows the many publics of a performing arts organization.

An organization is really a coalition of several groups, each giving different things to and seeking different things from the organization. Publics can be classified by their functional relation to the organization. An organization can be viewed as a resource-conversion machine in which certain input publics (playwrights, choreographers, composers, donors, suppliers) supply resources that are converted by internal publics (performers, staff, board of directors, volunteers) into useful goods and services (performances, subscription packages, benefits, educational programming) that are carried by intermediary publics (performing halls, advertising agencies, reporters, critics) to consuming publics (audience members, local residents, the media).

Not all publics are equally active or important to an organization. Some are central to its functioning, such as the artists, staff, donors, and volunteers. Certain publics may at times represent the organization's strengths and opportunities, and at other times may be the source of its weaknesses or threats. The common purpose may break down when one group pursues its own interest to the detriment of the organization. For example, orchestral musicians are a symphony's lifeblood, but they may put the organization at risk when they go on strike. The media may help bolster an organization with extensive promotional material or it may harm an organization's image by exposing internal strife. More peripheral groups, such as governmental agencies, may become particularly active when issues of general interest arise.

An organization must pay special attention to its internal publics. Internal marketing involves motivating and educating the organization's personnel, who define, refine, and carry out the organization's strategy. Effective marketing requires that all the key internal publics—the management, artistic staff, board of directors, support staff, and volunteers—understand and internalize the marketing mind-set. Often, it is the lowest-paid employees—the box office personnel, ushers, and parking lot attendants—who have the most direct contact with the patrons. Each must be trained to deal competently and courteously with the patrons and must learn how to create goodwill even during difficult encounters.

BUDGETING

A budget has three basic functions. First, a budget records in monetary terms the organization's realistic objectives for the coming year or years. Second, it serves as a tool to monitor the organization's financial activities throughout the year, providing a benchmark for managers and the board to analyze whether financial goals are being met. Third, it helps the organization predict the effects of marketing strategies and tactics with more confidence over a period of years.

The budgeting process is integral to the program planning process as the organization's resources are allocated and programs are dropped or added, weakened or strengthened. The budget operates on two levels. On the policy level, a budget defines in detail how the organization's financial resources will be used to accomplish the strategic plan. It requires that managers ask: How much will the program cost? How much money must be raised? How much can be earned? What will be the schedule of disbursements and income over the life of the program?[20]

COMMON BUDGETING PROBLEMS

Budgeting in arts organizations can be characterized as a balancing act of meeting the organization's objectives to the fullest possible extent within the limits of its financial capacity. But while attempting to maximize the use of their limited resources, managers commonly make certain mistakes in carrying out the budgeting process.

Excessive Emphasis on New Programs

A common budgeting problem in arts organizations involves attitudes toward existing versus new programs. Existing programs are assumed to be a fundamental part of the organization's operations and therefore are accepted virtually unexamined. In contrast, proposed new programs, often modest in comparison to existing programs, are subject to close scrutiny and negotiated endlessly by staff and the board. New programs are trying to find budgetary room within the confines of the organization's total resources, which most likely are already stretched to capacity. With all attention directed to proposed new programs, the unbalanced emphasis serves to institutionalize existing programs.

To ensure that knowledgeable, up-to-date decisions are made regarding resource allocation, both existing and new program activities should be thoroughly assessed in the budgeting process.

Incremental/Decremental Budgeting Policy

Normally the budgeting process begins with an estimate of revenues. For example, an organization's financial officer estimates that because of decreased grant support and because the current year's ticket sales did not meet projections, the total funds available next year will be 10 percent below the current level. As a result, all attention is focused on how to cut 10 percent from expenses. A common response is that the 10 percent reduction (or 10 percent increase in good times) should be shared equally by all budgetary units in the organization. This approach to budgeting is flawed because it assumes that all items in the budget should have equal weight. Here the focus is on the decrement or the increment rather than on the needs and importance of each budgeted function. Equal cuts are likely to foster overall mediocrity.

Late Preparation
Budgets are often prepared at about the time the new fiscal year begins, long after key decisions, such as committing to next season's shows or hiring new staff, are made. To use the budget merely to record actions already taken is to obtain only a portion of the benefits that can be derived from the budgeting process. Budgets should be prepared and approved by the board as new programming decisions are being made.

Lack of Participation
Frequently, one member of the professional staff, usually the business officer or the director, prepares budgets. Limited discussion occurs regarding programmatic needs and changing priorities and, as a result, decisions are made without full information. All key staff and board members should participate in the budgeting process to ensure that the budget accurately reflects the organization's goals and plans. Full participation also encourages commitment to accomplishing those plans.

Failure to Take Key Variables into Account
When preparing budgets, arts organizations often fail to plan for the impact of variables both within and outside of its control. For example, a symphony may hire a relatively unknown soloist to save money. However, the budgeters must consider that fewer tickets are likely to be sold in this case than if a famous performer were hired, so anticipated revenue from the event should be revised downward. Also, environmental factors such as a rainy season for an outdoor summer festival or the NBA playoffs taking place at the same time as a jazz festival are contingencies that should be allowed for in the budget.

Lack of Consistency and Direction
For some arts organizations, budgetary practices and procedures are not clearly defined. Budget forms and instructions should be standardized and understood by all who have responsibility for budget items. Once budgets are submitted, the director, the business manager, and appropriate board members should review them. Meetings should be held between budget reviewers and submitters to discuss and explain any necessary adjustments. Sometimes program managers set revenue budgets lower than expected to avoid the appearance of failure. They may also set expense budgets high to attract more resources. It is up to the director and board executives to create an environment in which all program managers will be motivated to project budgets as accurately as possible.

Lack of Periodic Review
In some arts organizations, the budget is "put to rest" once it is completed and approved by the board. However, the budget should be used regularly as a tool to review how well the organization is doing according to its budgeted plan. By

comparing actual expenses and revenues on a monthly or quarterly basis with those projected in the budget, the organization can determine possible problem areas before they get out of hand.

View of Budget as Inflexible
Budgets should not be considered as carved in stone or unchangeable once written. Rather, they need to be flexible and adjusted in consideration of changing circumstances. The plan itself must be taken seriously—or else it is of no value—but change must be a built-in factor.

TYPES OF BUDGETS

An organization may use one or more types of budgets, the most common being the traditional line-item budget, the program budget, and a PPBES (planning, programming, budgeting, and evaluation system).

The Traditional (Line-Item) Budget
The line-item budget lists all expenses by their object or source, such as salaries, benefits, travel, supplies, phone, rent, postage, printing, and so on. Revenues are treated in a similar manner and are listed in such categories as ticket sales, concessions, contributions, and grants. This type of budget is flawed in that it does not indicate how expenses and revenues are allocated across the organization's various functions.

The Program Budget
A program or project budget distributes revenue and expenses across the organization's functional areas and programs, such as regular season, touring, special events, education, and fund-raising. This budget gives managers much more information than the line-item budget. For example, it may show that touring is operating at a deficit while special events are operating at a surplus. Or it may show that fund-raising expenses are too high for the amount of revenue they generate. This can lead to a reprioritization of various activities and programs and an analysis of how to increase the cost-effectiveness of various projects.

For the program budget, a portion of staff salaries, office rent, utilities, and other fixed expenses may be allocated to various regular season productions, educational activities, tours, and other projects. Staff members should estimate the percentage of time they spend on each activity to allocate salary costs. Care must be taken not to overdo the monitoring of small expenses such as postage stamps, which causes frustration and wastes time. Such monitoring may be useful for short periods of time to get a rough approximation of what percentage should be applied to each program. The expenses that cannot be reasonably applied to specific programs should be placed in a general "administration" category. The program budget can be extremely helpful with fund-raising because many donors prefer supporting specific programs to providing general operating support. With a program budget, general operating expenses can be attributed directly to specific funded programs.

Combining Line-Item and Program Budgets

Table 5.2 shows how a line-item budget and a program budget may be combined, displaying the distribution of costs among various items for each program. For example, a line-item budget will show how much the organization spends each year on printing and mailing; a program budget will show how much is spent each year for the season's opening event; and the combined budget will show how much is spent for printing and mailing for the opening. This enables management to understand and evaluate each program's costs in detail.

PPBES

The planning, programming, budgeting, and evaluation system is an adaptation of the program budget, allowing a budget to be planned and evaluated in relation to multiple objectives. Often, expense budgets are provided for programs that generate no direct revenue. For example, a symphony may introduce a new concert series designed to attract younger audiences, to develop broader interest among current patrons in new music, and to provide educational opportunities to enhance the patrons' experiences. The criteria for the program's success may be the amount of audience crossover from more traditional programs and/or the number of patrons new to the symphony, not necessarily the amount of revenue earned. A PPBES system allows managers to evaluate programs on the basis of such varied objectives.

THE BUDGET AS A CONTROL TOOL

In order for a budget to be a useful control mechanism, it must be broken down into periodic increments corresponding to the financial statements. Comparisons should be made throughout the year between the budget and the financial statements. Most organizations do not have resources to fall back on

Table 5.2 Example of a program budget and line-item budget combined

	Salaries/ benefits	**Rent/ utilities**	**Supplies/ materials**	**Travel**	**Printing/ postage**	**Total**
Subscriptions						
Workshops						
Showcase						
Touring						
Fund-raising						
Total						

when it is determined that budgeted revenues have not been met or that expenses have been excessive. So the board and staff must be prepared to take action when the financial statements indicate a significant deviation from the budget. If the board and staff fail to take aggressive, corrective action, the board is shirking its responsibility and the budget has become a meaningless formality.

BUDGETING PRINCIPLES

Several principles serve to guide managers in formulating budgets.

Balanced Budget

An organization needs at least a balanced budget in order to remain solvent. A balanced budget means that the total budgeted revenue equals the amount budgeted for expenses. Managers may knowingly choose to operate certain programs at a financial loss. But when they do so, they must plan for other programs to compensate for expected shortfalls so that the overall budget is in balance.

Budgeting a Deficit

Some nonprofit organizations incorporate a deficit—where total expenses exceed total revenues—into the budget. In such cases, funds for operations are provided by indebtedness in the form of loans, unpaid bills, using advance ticket sales receipts, or "borrowing" from restricted funds. These are all obligations that the organization will eventually have to meet if it is to avoid bankruptcy. Since the budget reflects the organization's goals, operating under a deficit means that the goals are not being fully realized.

Some nonprofit organizations create a deficit as a strategy for raising funds. They intentionally overspend and create a crisis situation that is intended to attract the needed financial support. In the current era of accountability, good planning and fiscal prudence are highly valued, and even long-standing, loyal donors are not sympathetic to this approach.

Contingency/Reserve Fund

A contingency/reserve fund should be included in the budgeted expenses. It is recommended that between 5 and 10 percent of the organization's budget be set aside for this account. Larger organizations can allocate a smaller percentage; the less predictable the income, the larger the contingency/reserve fund should be.

This fund can be used in two ways. First, it can be used as a backup for unexpected circumstances such as a huge snowstorm the week of a major event, resulting in single ticket sales far below projected levels; unforeseen expenses; or cash flow problems that occur when the organization must pay bills before it has the income to do so. Second, at the end of the fiscal year, any money remaining in the fund becomes a reserve to be invested for future growth and to help carry the organization through tough times.

Determining Costs

There are two common methods for determining program and activity costs. The first approach, called *incremental budgeting*, is used for activities the organization has carried out in the past. It relies heavily on previous years' actual expenses and income. The manager investigates changes in the line-item costs of each program. If administrative costs are built in to program budgets, salary and rent increases should be included. A percentage should be added for other inflationary costs. Similarly, costs may be reduced incrementally.

The second approach, called *zero-based budgeting*, requires that each line item of the budget be calculated from scratch. This is not only necessary for new projects, but is also useful for reevaluating ongoing expenses. With a zero-based budget, staff members start with a budget of zero and must justify each amount they request.

The most effective method to determine costs is a combination of these two approaches. Previous expenses are extremely useful in planning the upcoming budget. Yet each budget item should be carefully examined to see if it can be reduced or if anticipated benefits can be achieved more efficiently some other way. Budgeters find it is most realistic to estimate costs on the high side. It is also important to account for the fact that new programs and activities will add to administrative costs such as staff, space, and office equipment.[21]

Budgeters must also take into account both their fixed and variable expenses. Since variable costs are more adaptable than fixed costs, managers often balance their budgets by varying their costs per production, service, or program. However, arts managers must always recognize the interdependence between the income and expense sides of the variable budget. The most accurate and revealing method of budgeting variable costs is to compare them to projected income on a precise event-by-event basis.[22]

In the next several chapters, I will explore marketing principles, strategies, and tools, and describe communication theories and methods. This will provide the foundation on which management can build and implement practical and effective plans.

CHAPTER 6

IDENTIFYING MARKET SEGMENTS, SELECTING TARGET MARKETS, AND POSITIONING THE OFFER

MARKETING PLANNING MUST START WITH STRATEGIC MARKETING—NAMELY, *segmenting, targeting,* and *positioning.* The marketer first identifies a variety of dimensions along which to segment the market and develops profiles of the resulting market segments. Then the marketer selects those segments that represent the best targets for its efforts. Finally, the marketer designs marketing strategies and positions the organization and its products in ways that are expected to have the greatest appeal to the target markets.

Customers vary greatly in their needs, attitudes, interests, and buying requirements. Since no organization can satisfy all consumers, each organization should identify the most attractive market segments that it can serve effectively. Strategic marketing enables organizations to spot market opportunities and to develop or adjust their offerings to meet the needs of the potential markets.

SEGMENTATION

Segmenting is the crucial first step. All customers are not the same and a single marketing strategy will do a poor job of serving many different customers. It is not unusual for arts managers to say, for example, that their typical customer is a 52-year-old college-educated woman. But averages say very little. How many of the patrons are 32 or 72? Are 51 percent of the patrons women or 75 percent? Even the actual ages, gender, and education levels of all current attenders convey little or nothing about their interests, habits, lifestyles, personality, preferences, and purchase behavior—factors that are far more significant for marketers in designing and promoting their offerings than demographic data.

Other important bases for segmenting the market are buyer behavior factors: their usage rates, the benefits they seek from attendance, their occasions for use, their loyalty status to an organization, and their buyer readiness stage. Benefits sought may vary by numerous psychological, social, demographic, lifestyle, and other factors.

The marketer's first responsibility is to segment its consumer base by aggregating consumers into groupings of similar characteristics. Various market groupings, or segments, exhibit different responses to offerings; what appeals to one group may be unattractive to another. The aim of segmentation is to identify groups within a heterogeneous market that share distinctive needs, preferences, and/or behaviors. Although, ultimately, each member of a market is unique, segmentation aims to identify broad groups for whom specific offers can be developed.

In their two-year study measuring the intrinsic impact of live theater on attenders, the principals of WolfBrown identified that the top three motivations of attendance are "to relax and escape," "to be emotionally moved," and "to discover something new." Young respondents are socially motivated and also are the most likely to attend "for educational purposes." High frequency patrons, 89 percent of whom are subscribers, are much more likely to cite emotional and intellectual reasons for attending, whereas low frequency attenders, 87 percent of whom are single ticket buyers, are motivated by production-specific factors, such as "to see the work of a specific artist." Among those attending the host theater for the first time in a year or more, 35 percent came because someone else invited them, illustrating the power of social context to drive attendance among infrequent attenders. The researchers found that, overall, motivations can vary dramatically from production to production, suggesting a need to carefully align marketing messages with motivations on a production by production basis.[1]

Alternative Bases of Segmentation

There are usually several ways to segment a given market, and forming meaningful segments is as much an art of insight as it is a science. The segmentation variables most commonly used by arts organizations are demographic: age, gender, income, education, occupation, religion, race, family size, position in family life cycle, and geographical factors—variables that were addressed in chapter 4. Even when the target market is described in terms of nondemographic factors, such as personality type, the link back to demographic characteristics is often necessary in order to determine the size of the target market and how to reach it efficiently. Following are other key segmentation variables.

Lifestyle Segmentation

Lifestyle segmentation is based on the notion that we do what we do because it fits into the kind of life we are living or want to live. Lifestyle has been identified as a better explanatory variable for arts attendance than any traditional socioeconomic characteristic, such as income or education. Lifestyle segmentation is also more dynamic than segmentation by personality. Whereas personality is seen to be an enduring, perhaps lifelong characteristic, lifestyle is more transient and is likely to change many times in one's lifetime.

Segmentation Categories for Arts and Culture: Values and Motivations

The British firm of Morris Hargreaves McIntyre, named Lateral Thinkers,[2] conducted extensive research with a goal of understanding audiences to be able to target them more accurately, engage them more deeply, and build lasting relationships. As a result, they have identified eight segments within the market for arts, culture, and heritage, which are based on people's cultural values and motivations, and which define the person and frame his or her attitudes, lifestyle choices, and behavior. The segments are distinguished from one another by deeply held beliefs about the role that arts and culture play in their lives, enabling marketers to get to the heart of what motivates people and develop strategies to engage them more deeply. The segments and their descriptive characteristics are:

1. *Affirmation*: self-identity, aspirational, quality time, improvement, "it's part of me"
2. *Enrichment*: mature, traditional, heritage, nostalgia, "way of life"
3. *Entertainment*: consumers, popularist, leisure, mainstream, socializing
4. *Essence*: discerning, spontaneous, independent, sophisticated, like to be inspired
5. *Expression*: receptive, confident, community-minded, expressive, fun-loving
6. *Perspective*: settled, self-sufficient, focused, contented, reflective, committed
7. *Release*: busy, ambitious, prioritizing
8. *Stimulation*: active, experiential, discovery, contemporary, likes new and exciting

Through interviews, the researchers learned, for example, that members of the *entertainment* segment are not likely candidates for the arts and that *essence* segment members will attend if the program seems especially worthwhile. Members of the *expression* segment like to "sit and soak up" what is exciting, thrilling, relaxing, and creates a feeling of awe, which provides good direction to arts marketers in how to reach out to these people.

Another lifestyle segmentation scheme was developed in a study of the cultural market in Philadelphia. Five segments were identified, each representing about one-fifth of cultural participants.[3] The most active cultural participants tended to be people who see themselves as sociable, who like to seek self-improvement and challenge, and who like to relax in their leisure-time activities. They place more importance on quality than on cost, they like to plan ahead, and they are confident by nature.

From these descriptions, and from those discussed in chapters 3 and 4, we can see how consumer lifestyles capture many influences—cultural, demographic, social, and family—that present opportunities for arts managers to create complementary situational influences.

Usage Segmentation

It has been said that the best predictor of future behavior is past behavior. Consider the so-called 80/20 rule: that 80 percent of the purchases in a category are made by 20 percent of the consumers. This group is usually referred to as *frequent* or *heavy users*. The remaining 80 percent are the *light users* and *nonusers*. Marketers know that it is much easier to stimulate increased attendance by current patrons than to attract those who have never attended. Most marketing resources have been devoted to efforts to increase the frequency and variety of attendance among the culturally active segments of the population. In recent years, however, the participation of the culturally active group has stagnated or slightly declined, and marketers are now enlarging their targets in efforts to increase audience size. At symphony orchestras, arts marketers have been successful in recent years in attracting many new people to their performances. However, approximately 80 percent of these people do not return, creating a situation known as *churn,* as described in chapter 2. Although a 20 percent response rate is quite strong for many offers, organizations hope to discover ways to bring back more of the other 80 percent.

Frequent users, light users, and nonusers share common characteristics with other members of their category. To develop a profile of these three segments, the Cleveland Foundation and the Pew Charitable Trust in Philadelphia commissioned studies of their area populations.

Frequent users were found to place a high value on leisure-time activities that spark the imagination or are new and different. They accept the arts as an important part of their lives, are predisposed toward active participation, and generally attend a variety of cultural events. For this group, satisfaction improves with more diversified use; the highest satisfaction levels are reported by those who attend six or more cultural organizations. They attach little importance to activities associated with performing arts events that allow them to see and meet other people. The only barriers indicated by these frequent patrons are cost and convenience.

Light users are defined as those who attend one performing arts organization only or who sporadically attend a few organizations. They select certain programs that interest them greatly, and cost, comfort, and convenience are central to their decision-making process. Meeting and seeing a variety of other people is also very important, as is self-improvement.

Nonusers report that it is essential that their leisure-time activities be fun and entertaining, convenient and inexpensive, feel relaxed and informal, and involve family and friends. They believe that cultural participation could involve family and friends, but few expect that a visit would be relaxed and informal, fun and entertaining, or inexpensive. (Interestingly, frequent users do tend to find performing arts attendance comfortable, relaxing, and enjoyable to share with friends. Performing arts managers can work to emphasize this perspective when reaching out to nonusers.[4])

Nonparticipants surprisingly report fewer demands on their time than participants and do not feel that time constraints limit their participation in cultural activities. It was actually found that in almost every case, cultural participants engage in competing activities involving sports, television, or DVDs as much as, if not more than, nonparticipants.

Most nonparticipants have consciously or unconsciously eliminated the arts as being of any possible interest or value in their lives. They have drawn a "cultural curtain" and ignore anything that is written or said about arts activities. Thus, converting nonparticipants is difficult and will succeed gradually at best, because it often involves changing basic attitudes.[5]

Bradley Morison and Julie Dalgleish believe that arts marketers have overfocused on regular attenders and should explore innovative ways to communicate with more elements of society. They suggest that marketers turn their attention to attracting a group they call the "Maybes," who represent the greatest potential for future audience growth and development. The Maybes tend to be uncertain about whether or not the arts are important to them. Some Maybes are intimidated by what they perceive as the formality of performance halls and the crowds who gather there. They are also insecure as to whether they know enough to be able to appreciate the arts. They will try an arts event that appears accessible and unintimidating, and will continue to attend on occasion as long as their experiences are positive. According to Sidney Levy, marketers must address nonusers' inhibitions and stereotypes. Says Levy:

> Some nonparticipants harbor many inhibiting images of the arts as relatively austere and effete, effeminate, esoteric, inaccessible, too demanding of study and concentration, arrogant, et cetera. Coping with these attitudes is not easy, but progress is made when experience shows the contrary or reorients the negative value. To bring about the experience, marketing usually recommends incentives, free samples, easy trial, and starting with

examples that most contradict the opposed imagery. Personalities help in this endeavor—men ballet dancers who are masculine and opera singers who are not foreign divas. English translations help greatly to make opera more accessible, as in the supertitles that are offered by many opera companies.[6]

The arts marketer should begin the process of attracting Maybes at a point of entry, a place that is familiar and accessible aesthetically and/or geographically to prospective patrons. Development of commitment to the organization is arranged in progressive stages, providing people with the opportunity to learn about the art form and increase their commitment to the organization at a gradual pace. Rather than encouraging subscriptions from the start, audience development of Maybes should be a long-term process of encouraging and assisting an audience member to become increasingly more involved in the life of an arts institution.[7]

Segmentation by Aesthetics

Aesthetic interest is a key factor in arts attendance. Some people perceive the arts to be hopelessly beyond them; they feel unequipped to participate, even to the point of hostility toward the arts. Others who feel alienated from the arts usually express flat disinterest, a lack of experience, or ignorance, or they may emphasize the esoteric nature of the arts.

Sidney Levy has observed, however, that many of these people have significant aesthetic interest that has found expression through other means. He claims that attitudes and behavior toward the arts are derived from fundamental factors such as excitement, realism, sexual identity, performance type preferences, social status, and seriousness. These factors offer clues to marketers as to what underlies the differences among individuals and groups regarding their aesthetic dispositions.[8]

Benefit Segmentation

People often base their performing arts attendance decisions on anticipated benefits. *Quality buyers* seek out the best-reputed offerings: critically acclaimed plays; performances by superstars; or works by renowned composers, playwrights, and choreographers. *Service buyers* are sensitive to the services provided by the organization, such as ticket exchange privileges, a convenient location, adequate parking, or educational and social events to complement the performances. *Economy buyers* favor the least expensive offers such as the low-priced community orchestra, free concerts and plays in the park, and half-price tickets. Benefit segmentation works best when people's benefit preferences are correlated with their demographic and media characteristics, making it easier to reach them efficiently.

Occasions

Buyers can be distinguished according to the occasions when they develop a need for or purchase a product. For example, the "need" for attending a production of the *Nutcracker* may be triggered by the desire for holiday entertainment. The purchase decision may be triggered when out-of-town guests arrive or the children are on vacation from school. People may attend the arts to celebrate birthdays and anniversaries or to entertain clients and friends. Marketers should consider the many possible occasions when people might attend performances and create messages and incentives to promote attendance on those occasions.

Loyalty Status

A market can be segmented by consumer loyalty status. Sometimes people are loyal to a specific type of offering and will attend musicals, Shakespeare plays, Beethoven symphonies, certain ethnic productions, or appearances by their favorite performers. Organizations that present specific repertoires, such as Shakespeare, classical ballet, or Baroque music, are likely to develop customer loyalty based on the offerings, as long as a satisfying level of quality and service is maintained. Some people may be loyal to an orchestra or theater to which they have subscribed for years, and will attend no matter what the production. Organizations typically segment the loyalty status of their patrons by subscription level, frequency of attendance, donation level, and volunteerism. Development and maintenance of customer loyalty will be discussed in depth in chapter 15.

Buyer-Readiness Stage

At any given time, people are in different stages of readiness to buy a product. Some people are unaware of the product; some are aware; some are informed; some are interested; some are desirous of buying; and some intend to buy. Their relative numbers make a big difference in how the marketing program should be designed. In general, the marketing program must be adjusted to the changing number of people in each stage of buyer readiness and to attract people in various readiness stages.

Criteria for Segmentation

Each organization must decide which approaches to segmentation would be the most useful for its own needs. Yet, any optimal segmentation scheme should possess the following characteristics:

- *Exhaustiveness*: Every potential target member should be placeable in some segment. If segmentation is done according to household status, there should be categories to cover relationships such as unmarried couples, single parents, empty nesters, communal-style livers, and so on.

- *Substantiality*: Each segment should be substantial in that it has a large enough potential membership to be worth pursuing. For example, although single parents are a growing segment of the population, they are unlikely to be frequent performing arts attenders while they have children to care for. On the other hand, a divorced parent is an ideal target segment member for family-oriented programs presented on weekend afternoons.
- *Actionability*: Actionability refers to the degree to which the segments can be effectively reached and served. Sometimes there may be no efficient advertising medium for reaching a segment specifically. And even if an advertising medium were available, its cost may exceed the available budget.

By analyzing the attractiveness of each segment, managers can prepare to make the following strategic decisions:

- *Quantity decisions*: How much of the organization's financial and human resources (if any) are to be devoted to each segment?
- *Quality decisions*: How should each segment be approached in terms of specific product offerings, communications, place of offering, and prices?
- *Timing decisions*: When should specific marketing efforts be directed at particular segments?

Assume that a dance company wants to attract both men and women, using different ads and appeals. The marketer knows that women primarily enjoy the grace, beauty, and creativity of the dancers, and that men respond most strongly to their athleticism. A single ad that tried to combine the two approaches could result in a confusing message, so the more "graceful" ad could be placed in the lifestyles section of the newspaper, and the more "athletic" ad could be placed in the sports section. The dance marketer might find that the added cost of placing two such ads is more than covered by the additional audience members the performances attract.

TARGETING

The decision of which and how many segments to serve is the task of *target market selection.* A target market consists of a set of buyers having common needs or characteristics that the organization decides to serve. Before selecting target segments, an organization should learn as much as possible about each segment under consideration to determine whether and how it can meet that segment's needs, interests, and desires.

PATTERNS OF TARGET MARKET SELECTION

Various patterns of target market selection may be considered by an organization: *single-segment concentration, product specialization,* and *selective specialization.*

Single-Segment Concentration

An organization such as a children's theater or a theater that produces plays for, by, and about gays, African Americans, or any other specific group, is basically a single-segment provider. Through concentrated marketing, the organization is likely to achieve a strong market profile owing to its greater knowledge of the segment's needs and the special reputation it builds. However, concentrated marketing carries high risks. The particular segment may withdraw its support for a variety of reasons, or a competitor may decide to pursue the same market segment. When an organization is a single-segment specialist, it should focus its efforts on penetrating that segment as deeply as possible.

An Unintentional Single-Segment Organization

The board members of a small chamber music society in a wealthy, quiet, suburban neighborhood became increasingly concerned about the fact that their audience consisted almost exclusively of gray- and white-haired people and that, over the years, the audience had been aging with few new or younger people attending. They engaged my services as a marketing consultant to determine ways to attract younger patrons. After conducting audience surveys and making observations at concerts, I developed some strategies that I thought would make the organization more appealing to younger people. I recommended to the board that the society (1) change the timing of its programs from 2:00 p.m. on Sundays, which is normally a family time, to later afternoon or evening when babysitters are more readily available; (2) change the name of the organization to eliminate the elitist term "society"; (3) add some adventuresome music to the currently traditional programming; (4) include some performing groups other than the traditional trio or quartet; and (5) present some performances in communities where more younger people work and reside.

The board members were unwilling to adopt any of these suggestions, since they felt such changes would drastically contrast with their own vision of the organization. They understood that their audience appeal would remain limited. So, understanding and accepting their clearly defined direction, I worked with them to develop strategies for enlarging their current market. We offered "bring a friend" promotions, gift certificates ("Give your friends, colleagues, and family members the gift of music!"), and other tactics to capitalize on the "birds of a feather stick together" principle that people already attending are the most likely ones to know others like them who would enjoy the concert. Within three years, the audience—which remained quite homogeneous—doubled in size. However, several years later, because the audience was becoming ever more elderly with no younger people joining them, the audience shrank precipitously and the organization finally closed its doors.

Product Specialization

A Shakespeare festival, an early music group, and a Latino dance company each concentrates on a specific product that appeals to certain market segments. This strategy creates a strong identity and a potentially loyal following among those who have a particular interest in the offering. The downside is that if demand for the product slackens, the organization may be poorly equipped to alter its offerings. Some organizations enhance their specialized product with complementary offerings that add interest and variety to the experience. Chicago Shakespeare enhances its core repertory, which consists of only 36 plays, with contemporary classics, such as works by Stephen Sondheim. Music of the Baroque performs in several Chicago-area churches, allowing patrons to experience Baroque music in varied theme-appropriate and convenient settings throughout the season.

Selective Specialization

Most commonly, arts organizations select a number of segments, each of which is attractive and matches the organization's objectives and resources. The strategy of multisegment coverage has the advantage of attracting a broader base of the art-going public and of diversifying the organization's risk. For example, if a large proportion of a theater's younger audience members begin to have children and focus on more home-centered activities, the theater can continue to attract an older (or younger) audience that has more discretionary time. Arts organizations should study trends and environmental factors that signal opportunities for targeting new segments or for concentrating their efforts on building certain current segments.

Choosing among Market Selection Strategies

The actual choice of target markets depends on specific factors facing the organization. The more limited the organization's resources, the more likely it is to concentrate on a few segments. The more homogeneous the market, the less the organization needs to differentiate its offerings. Each organization has to evaluate the best segment(s) to serve in terms of its relative attractiveness, the requirements for success within the segment, and the organization's strengths and weaknesses in competing effectively. The organization should focus on market segments that it has a differential advantage in serving.

POSITIONING

Once an organization has segmented the population and selected viable target market segments, its next responsibility is to promote those aspects of its offerings that appeal most strongly to its target market. The process of developing a focused positioning strategy is the act of designing the organization's image and offer so that it occupies a distinct and valued place in the target customers' minds. Positioning involves (1) creating a real differentiation and (2) making it known to others. The marketer's expertise lies in augmenting, promoting, and delivering the core product creatively and sensitively to meet the needs and preferences of target markets.

In the following example, two similar organizations attracted completely different audiences for the same generic product because they adopted different positioning strategies.

One Core Product—Different Positions

In 1993, both the Chicago Symphony Orchestra (CSO) and the National Symphony Orchestra in Washington, DC, opened their respective seasons with a performance of the Verdi *Requiem.* The core product was the same, but it was in the way that each positioned and promoted its concert that the two organizations defined their products and their target audiences.

In Chicago, the evening was an elegant, exclusive fund-raising event. Tickets were priced at a premium for both the concert and the gala benefit dinner that followed the performance. In contrast, the National Symphony presented Verdi's *Requiem* as a eulogy for the children killed by random violence in the city, contextualizing the music in such a way as to make it personally meaningful to new audiences. Seats were popularly priced to attract a broad range of Washington's diverse population.

In fall 2013, the CSO again presented Verdi's *Requiem* for its opening concert, timed to be in celebration of the two hundredth anniversary of Verdi's birth. This time, the CSO shared this sold-out concert with music lovers everywhere for free by offering a live video webcast on a giant screen at Millennium Park, at the Benito Juarez Community Academy Performing Arts Center in the primarily Latino Pilsen neighborhood, and online through several sites.

A product or organization evokes many associations, which combine to form a total impression. The positioning decision involves selecting which associations to build on and emphasize and which to remove or deemphasize. The positioning decision is central to influencing customers' perceptions and choices. A clear positioning strategy also ensures that the elements of the marketing program are consistent and mutually supportive.[9]

An arts organization can base its positioning on various attributes. Some attributes apply to the organization itself, such as the director's charisma, the programming focus, the organization's reputation, or the fine performance hall; others apply to an individual production, such as a star performer or a famous composer. The arts organization can also base its positioning on a set of attributes. Generally speaking, however, a marketer will select one or two attributes that appear to be the most attractive to the target audience, whether for the season brochure or to advertise an individual production.

The specific positioning the organization undertakes will depend largely on its analysis of its targeted market segments, its own strengths and weaknesses, and its competition. Many organizations will develop a "niche" or specialty. An organization can specialize according to customer, product, or marketing-mix (product, price, place, promotion) variables. A children's theater is a customer specialist. A Shakespeare company is a product specialist. A free community orchestra is a price specialist, while its high-priced competitor, the city's symphony orchestra, is a quality specialist. The Stratford Festival in Ontario, Canada, is a place or geographic specialist in that an entire community of hotels, restaurants, and shops has been developed to serve tourists who vacation in Stratford to attend the performances. In a large city like Chicago where there is a multitude of theaters, it is important for each to set itself apart from the others. For example, both Steppenwolf Theatre and Writers' Theatre are well known for their extremely high-quality productions. Steppenwolf focuses its positioning on its cutting-edge plays, often by well-known contemporary playwrights; Writers' Theatre focuses on the intimacy of the experience in its two venues, one with 50 seats and the other with 108 seats, and on its inventive interpretations of classic works.

The advantage of defining a position in a target market is that it almost dictates the appropriate mix of marketing strategies. If an arts organization adopts a "high-quality" positioning, then its programs should be of uniformly high quality, including its performers, sets, costumes, and venue; its prices may justifiably be above average, and all communications materials should be of consistently high quality, including the quality of the paper that is used for brochures. If any element is lacking, the desired marketing outcome may be sabotaged. A high-quality position can be undermined by seemingly minor factors such as low-quality stationery, hand-lettered signage, or inarticulate contact personnel. Conversely, an organization seeking contributions to correct a budget deficit should indicate frugality by using inexpensive mailers and conveying a no-frills attitude.

Some organizations prefer to leave their position less well defined in the hope of attracting more disparate market segments. Although such an approach might succeed on occasion, especially when there is little competition, the organization that establishes a distinct identity increases its chances of survival and prosperity in the long run. Positioning errors that managers should avoid are (1) entering a crowded marketplace—taking a position that is already held by other organizations, thereby gaining no distinctiveness; (2) focusing on an unimportant attribute, a factor that is of relatively little importance to the target audience; and (3) myopic positioning, focusing more on the seller's offer than on what the consumer wants. Positioning, then, is a matter of both substance and perception. The organization must truly have a set of offerings that differentiates it from the competition and that gives it a clear identity in the mind of the consumer.

According to advertising executives Al Ries and Jack Trout, positioning *starts* with a product. But positioning is not what you do to a product; it is what you do to the mind of the prospect.[10] Thus a children's theater may choose to position itself as an educational experience or an entertaining and even magical experience in trying to attract an audience. Or consider what the Atlanta Ballet did to reposition its status in the minds of its potential audiences.

Positioning the Atlanta Ballet

Some years ago, the Atlanta Ballet, which had always attracted traditional ballet-goers, was seeking to attract a broader audience. The ad agency Ogilvy and Mather donated time to the company to help shed light on how people viewed the ballet and to help the company reposition its message. After conducting research, Ogilvy and Mather recommended the following: "The Atlanta Ballet should focus on its entertainment appeal. We want to make Atlanta Ballet fans of people who may never have seen a dance performance—people who have the money and the interest to discover new kinds of entertainment. Our positioning should be: The Atlanta Ballet is entertainment everyone can enjoy."

By the following season, the company's theme had become "What makes the Atlanta Ballet the most exciting show in town? It's athletic! funny! sensuous! chilling! The Atlanta Ballet is different." In response to this campaign, the subscription audience quickly doubled. The new language worked because it changed Atlanta's *perception* of what was already on stage.[11]

Today, the ballet's tag line is: "Bold! Innovative! Exquisite!," which updates the previous emotional, exciting appeal.

Positioning Strategies

Every product and organization needs a focused positioning strategy so that its intended place in the total market—and in the consumer's mind—is clearly reflected in its communications. The more the performing arts organization knows about its target audience, the more effective it can be in positioning itself. The strategy requires coordinating all the attributes of the marketing mix to support the chosen position.[12]

Some alternative bases for constructing a positioning strategy are listed and illustrated here.

Positioning by Programming

Organizations like the Oregon Shakespeare Festival, the Baroque Band, the Old Town School of Folk Music, and Mostly Mozart are so clearly program-focused that their programming is positioned right in the name of the organization. By contrast, those organizations that produce a wide variety of works cannot develop a strong, clear position based on their diverse programming, so some develop a theme for each season. In its 2012–2013 season, the Steppenwolf Theatre promoted its season of five plays as "The Reckoning," saying on its website: "There comes a moment when we are called to account. Will our deeds be repaid? Will our secrets be revealed? Will we get what we deserve? Five stories about what lies in store when the past comes knocking." For its 2013–2014 season, the Atlanta Ballet describes a "season of power and passion." Such themes are typically developed by creative artistic and marketing directors working together.

Positioning on Specific Product Features

A theater may position itself as presenting the classics or a chamber music group may position on the Baroque period. The fact that a play or concerto has endured for decades or even centuries indirectly implies a benefit, namely, a highly satisfying, high-quality theater- or concert-going experience. Conversely, adventuresome arts goers may seek new work that is based on contemporary themes and styles. Of course, a multitude of product features are fair game for positioning on product features.

Positioning by Performers

Star guest performers are such an important audience draw that arts organizations will sometimes knowingly take a financial loss in order to present them. Star performers for even a single event can serve to elevate the image of the organization and stimulate subscription sales. "If you want to guarantee your seats to see Renee Fleming in *Streetcar Named Desire*, subscribe for the season," said a Lyric Opera ad.

Positioning by Reputation and Image

What other field besides the performing arts presents awards and recognition to its members so publicly? These awards stimulate demand; people like to see a winner. Arts organizations usually take advantage of the recognition they receive by quoting critical praise and listing awards in their promotional material. Image positioning can also be used to establish people's expectations about the spirit of the experience they will have at a given performance. Some classical music organizations are updating their musicians' attire to make the experience appear less stuffy. Promotional material may show raptly engaged audience members.

Positioning by Multiple Attributes

When presenting an entire series that includes widely varied programming, an arts organization may position itself as having something for everyone. Teatro Piccolo positions its artistic choices on multigenre and international content for each theater season, including classical and contemporary drama, music, and dance. Some organizations have websites that make it easy to present information by several attributes, such as genre, performer, composer, date, subscription cycle, and festival. This makes it easy for patrons to find exactly what interests them, when they want it.

Positioning on Benefits, Problem Solution, or Needs

The Atlanta Ballet's position focuses on the benefits the consumer will receive by attending—namely, having an "exciting, fun, sensuous, chilling" or "powerful, passionate" experience. Problem solution may address people's interest in celebrating an occasion with a special event or sharing the arts-going experience with a special person. A need may be as straightforward as exposure to an art form or artist or the desire to see one's favorite dance performed again.

Positioning for Specific Usage Occasions

Organizations may position certain performances on the user's own occasions. For example, "Give your mom a special and memorable musical event on Mother's Day." Or, "If watching football players during the Superbowl isn't your thing, come watch our beautiful dancers for the afternoon."

Some of the most popular and beloved shows are positioned for the holiday season. People have occasions to celebrate all year long and organizations would benefit by helping people decide where and how to celebrate.

Positioning for User Category

Probably the most common example of positioning by product user is children's or family theater, such as the Children's Theatre Company in Minneapolis. The Arena Stage in Washington, DC, reaches out to gay and lesbian patrons with such programs as its 2012 "Engage@arenastage: Aging with AIDS." At this program, ticket holders were invited to take part in a conversation with guest panelists following select performances of *The Normal Heart.* The panels explored how HIV/AIDS is represented in the media and in the arts as well as how the virus continues to affect gay men, women, senior citizens, youth, and people of color both nationally and in the Washington metro area.

Positioning against or Associating with Another Product

By claiming it is "different, athletic, and funny," the Atlanta Ballet is differentiating itself from other ballet companies, or as different from what people expect ballet companies to be like. It is also associating itself with other leisure pursuits that people tend to consider entertaining. Such association can strengthen the product's position in the minds of consumers. Say Ries and Trout, "The basic approach of positioning is not to create something new and different, but to manipulate what's already up there in the mind, to retie the connections that already exist."[13]

In the business sector, it is not unusual for companies to position against a major competitor (Avis: "we're number two, we try harder." Avis's Twitter handle is @AvisWeTryHarder), but this approach is rarely taken in the arts, except for a lighthearted comment like "The home team hasn't sounded this good in years," which was a critic's quote referring to the New York Philharmonic in an ad that ran during the football championships.

Since arts attenders are generally more satisfied the more variety of events and organizations they attend, arts organizations benefit from a mentality of collaboration, not competition.

Positioning as Number One

People tend to remember number one and to value it much more highly than any other offering or person that may be a close runner-up. For example, when asked, "Who was the first person to step on the moon?" we answer "Neil Armstrong." When asked, "Who was the second person on the moon?" many of us draw a blank. This is why managers fight for their organizations, productions, and performers to have the number-one position, either as the "largest" or "oldest" or "most famous." Such a position can be held by only one organization, production, or person. What counts is to achieve it with respect to some valued attribute.

The Exclusive Club Strategy

Sometimes a number-one position in terms of a meaningful attribute cannot be achieved, or it may otherwise benefit an organization to associate with other organizations of its kind. For example, some organizations emphasize their quality level by stating in their communications materials that their organization is "one of the world's finest."

Positioning by Director's Charisma

Organizations that have the benefit of a well-known, highly respected, and charismatic director often position largely on that one person. The Los Angeles Philharmonic strongly promotes itself based on its young, vibrant conductor

Gustavo Dudamel, and Michael Tilson Thomas is a major draw for patrons of the San Francisco Symphony. Sometimes a strong emphasis on one person can backfire for the organization. This is the case if, for example, a beloved conductor falls ill for an extended period of time and patrons need to be assured that they are not being cheated by having a substitute perform. The American Repertory Theater in Cambridge, Massachusetts enjoyed, for many years, dedicated audiences that were loyal to longtime director Robert Brustein. After Brustein retired, audiences shrank for a few years until Diane Paulus, an exciting, new, vibrant director, was brought on board and attracted new audiences to the theater.

Positioning by Location or Facilities

Some performing halls have such a reputation for quality and prestige that just being there is an event in itself. The most familiar example of such a place is Carnegie Hall. Capitalizing on this benefit, one brochure stated in part: "Climbers dream of Mount Everest; Divers dream of the Great Coral Reef; Music lovers dream of Carnegie Hall." Canada's Stratford Festival calls itself "one of life's great stages." The Sydney Opera House in Sydney, Australia, because of its iconic architecture, has attracted millions of visitors to concerts irrespective of what program might be performed. The Walt Disney Concert Hall, designed by world-famous architect Frank Gehry, is, according to the LA Phil's website, one of the most acoustically sophisticated concert halls in the world, providing both visual and aural intimacy for an unparalleled musical experience.

Repositioning

When an organization wishes to change the way it is perceived in the marketplace, it can reposition its offerings in a variety of ways—from altering its basic, core product; to changing the way the product is packaged, priced, or promoted; to simply changing the way the product is presented to the public so that customer perceptions are changed.

The starting point for any repositioning strategy is developing an understanding of how the organization is perceived by its current or potential audience. This process involves measuring the organization's image, which is the sum of beliefs, ideas, and impressions people have of its offerings.

An image differs from a stereotype. A stereotype suggests a widely held image that is highly distorted and simplistic and that is associated with a favorable or unfavorable attitude toward the object. An image, on the other hand, is a more personal perception of an object that can vary from person to person. Approaches for investigating people's views of an organization's image can be found in the next chapter on market research.

CREATING A CUSTOMER SCENARIO

Patricia Seybold refers to the broad context in which a customer does business as the *customer scenario*. By building a detailed understanding of common customer scenarios, an organization can often find creative ways to expand its reach into the lives of buyers.[14] Seybold observes that many managers "think they are taking on the customer perspective, but they're really only focused on the point at which the customer comes into contact with their company. That touch point is certainly important, but it's rarely the center of the customer's experience." To create customer scenarios, Seybold suggests the following steps:

Select a target customer set. Be as explicit as possible. For example, select someone with little or no previous exposure to classical music who is interested in learning about the art form.

Select a goal(s) that the customer needs to fulfill. For example, help the customer develop familiarity with and enjoyment of classical music.

Envision a particular situation for the customer. Offer a series of concerts with readily accessible music, with preconcert lectures, and with the conductor speaking from the podium about the music before each piece is performed. Enhance accessibility and knowledge with information about the concert-going experience, sources for learning about the music, composers, and performers; offer the opportunity to purchase recordings or download music; make special offers for upcoming programs that are likely to be of interest; offer receptions where patrons can mingle with the musicians.

Determine a start and an end point for the scenario. The key is to focus on the total customer experience, from the time an interest is identified through post-concert contacts to build an ongoing relationship with the customer. Consider likely opportunities and media for reaching out to these potential customers to make the initial contact and approaches for maintaining communication with them. Throughout this process, the marketer should map out as many variations of each scenario as she can think of and mentally walk through each step as the customer would. To accomplish this, think of the individual activities the customer performs and the information needed at every step. What can an organization do to support those activities and supply that information? In what ways can the marketer increase convenience to the customer by streamlining the decision-making and purchase processes? The exercise should be undertaken for a number of different target customers, customer goals, and customer situations.

THINKING LATERALLY

Philip Kotler and Fernando Trias de Bes describe traditional segmentation, targeting, and positioning as *vertical marketing*, which consists of creating modulations within a given market. On the other hand, *lateral marketing* restructures markets by creating a new category through new uses, situations, or targets.[15]

Lateral marketing comes up with answers to such questions as: What other needs can I satisfy with my product if I change it? What nonpotential consumers could I reach by changing my product? What other things can I offer to my current patrons? In what other situations can my product be used? Once the lateral marketing process comes up with a valid new offer, then vertical marketing comes into play developing segments, targets, and positioning.

One beneficial lateral marketing strategy may be to reverse the process by which the patron and the box office make contact. Typically, the organization distributes marketing information through a variety of media and *hopes* ticket buyers will contact them. With today's sophisticated databases, the organization knows what types of events people prefer (such as classical versus jazz, chamber versus symphonic music, full-length versus modern dance, and what day of the week and seating location people tend to prefer). With this information, the box office personnel can reach out directly to individual patrons with phone calls or with email messages saying, in effect, "Because you came to X and Y, we thought you'd be interested in seeing our upcoming production. We have center-section seats for you for the matinee performance, as you have preferred in the past. May I hold these tickets for you while you check your calendar?"

Lateral thinking is helping orchestras to innovate. The Memphis Symphony Orchestra collaborated with hip-hop artist Al Kapone for a program of orchestral-backed rap as part of its Opus One series performed in alternative spaces. The Detroit Symphony Orchestra put on a concert called Mix@TheMax, featuring the young Brooklyn chamber orchestra, the Knights, performing Copland's *Appalachian Spring* in a cabaret-style setting, followed by cocktails with the musicians. In San Diego, hour-long programs in the orchestra's Symphony Exposé series start with a host explaining the story behind a particular piece in a multimedia presentation, and then the orchestra performs it.

The New World Symphony's entry into the young audience–building effort is an experimental late-night music event called Pulse. The spring 2012 concert began with a DJ spinning electronic dance music. Then orchestra members performed traditional pieces by Mozart, Andrea Gabrieli, and Stravinsky, with electronic-infused musical interludes. Beneath vivid light displays projected on large screens, audience members were free to mill around the hall, drinks in hand, entering and exiting as they pleased. They could watch the musicians up close or mingle with other concertgoers. Preliminary findings are encouraging. For the Pulse program, which has had four sold-out performances over two seasons, 40 percent of ticket buyers are new to the New World Symphony database.

This book is replete with examples of effective marketing strategies developed by people thinking laterally. Taking a lateral marketing approach can help marketers think creatively to develop viable and valuable marketing innovations.

CHAPTER 7

CONDUCTING AND USING MARKETING RESEARCH

HISTORICALLY, MANAGERS HAVE DEVOTED MOST OF THEIR ATTENTION TO MANAGING their products, their money, and their people, while paying less attention to another of the organization's critical resources: information. The need for marketing research information is greater now than at any time in the past. Said economist Joseph Steiglitz, "What you measure affects what you do. If you measure the right thing, you do the right thing."[1] As marketing segmentation strategies become more sophisticated, segments become smaller, and people expect increasingly individualized service, organizations need to learn more about the needs and wants of their various target markets. Also, as consumers have become more selective and demanding in their buying behavior, sellers find it harder to predict buyers' responses to different features, benefits, packaging options, and other attributes unless they turn to marketing research. This need for information has grown over recent years with the emergence of sophisticated technologies that have revolutionized information handling and have made it accessible to and inexpensive for even the smallest organizations.

The amount of data in our world has been exploding, and analyzing large data sets—so-called big data—is a function that leaders in every sector will have to grapple with, not just a few data-oriented managers, according to research by MGI and McKinsey's Business Technology Office.[2] The increasing volume and detail of information captured by enterprises, the rise of multimedia, social media, and the Internet will fuel exponential growth in data for the foreseeable future. All this rich data has enabled and required businesses to move from managing relatively large market segments to treating each individual as his or her own segment. Says Ethan Roeder, data director of Obama for America, "In 2012 you didn't just have to be an African-American from Akron or a suburban married female age 45–54. More and more, the information age allows people to be complicated, contradictory, and unique. New technologies and an abundance of data may rattle the senses, but they are also bringing a fresh appreciation of the value of the individual."[3]

Modern databases offer arts marketers the power to capture detailed information about people's purchase behavior and leverage this information by reaching out to people with programs and offers that are likely to be highly appealing. Arts managers should use their data to effectively personalize their marketing communications; they should share the data across departments to enable a holistic approach to each patron. Data is one of the most powerful tools ever available to marketers. It should be used wisely and well.

Big data tells us what is happening now in great detail. *Long data* tells us what is happening over time—what changes and what doesn't. It is important to look at longitudinal information and learn what is a part of human nature, and discriminate that from what is trendy or environmentally produced, such as economic conditions.

Says *New York Times* columnist David Brooks, "Our lives are now mediated through data-collecting computers. In this world, data can be used to make sense of mind-bogglingly complex situations. But there are many things data does poorly. Computer-driven data analysis excels at measuring the quantity of social interaction but not the quality."[4] In other words, data analysis can't make decisions about social relationships. Nor can data put situations in context: Why do we want to see a particular show or performer? With whom? What resonates with us about certain music? Certain stories? Also, big data obscures values. Brooks concludes that big data is a great tool, but it's important to keep in mind that it's good at some things and not at others.[5]

Marketing research is the systematic design, collection, analysis, and reporting of data and findings relevant to a specific marketing situation facing the organization. It plays a critical role in understanding customer attitudes and behavior and in planning marketing strategies, for major decisions and for the ongoing functions of the organizations.

When Arena Stage was building a new venue, it moved its performances to temporary locations. In preparation for this move, Chad Bauman, former director of marketing and communications for Arena Stage, and current managing director of the Milwaukee Repertory Theater, undertook marketing research and developed strategies from the research results. Here is Chad's blog on the subject.[6]

Arena Restaged

In January 2008, Arena Stage moved from its SW DC home to two temporary locations, where we would remain for nearly three years while the Mead Center for American Theater was built. During that time, we had to minimize patron attrition caused by the move and work to grow our audience base, as the new building would require a significantly increased patron base. I turned to Shugoll Research to help map out a strategy. I wanted to know what barriers existed for our patrons in moving to our temporary locations. What would motivate them to stay with us through the construction years? What competitive advantages existed at our temporary locations that were good selling points? How could we make the move less onerous? We tested messaging, sales strategies and tactics. From that, I learned that if our patrons got lost on their first trip to our new theaters, they wouldn't return. I learned that we had to make sure that parking and public transportation were readily available. I learned that dining options were incredibly important. I spent months on signage plans. In collaboration with the Crystal City Business Improvement District, we installed more than 100 new directional signs within a two-mile radius of our temporary theater in Virginia. In coordination with the MidCity Business Association, we aggressively marketed the restaurants on U Street and offered valet parking for every performance, as the neighborhood had very few parking options. We sent out personalized websites to each of our subscribers, which, among other things, offered up step-by-step directions from their house to the new theaters. For these efforts, Arena Stage was recognized with the Box Office of the Year Award from INTIX and the Helen Hayes Washington Post Award for Innovative Leadership in the Theatre Community. More importantly, we were budgeted to experience seven percent attrition during the move and only realized 1.9 percent—and it all started with market research.

ORGANIZATIONAL RESISTANCE TO MARKETING RESEARCH

Although marketing research can provide much benefit for performing arts organizations, its use is not widespread. Some artistic directors express concern that patron research results may have negative implications for their artistic freedom. Marketing staff is often overworked and doesn't have the time or talent to conduct research or leverage information readily available in their own databases. Marketers' knowledge of the correct use of marketing research and its technical aspects is limited. Performing arts organizations have limited budgets and assign higher priority to other expenditures. In the following paragraphs I refute the reasons for resistance to marketing research.

Costs of marketing research. Marketing research need not be expensive. There are low-cost forms of marketing research, such as audience surveys, focus groups, and systematic observation. Much can be learned from just listening to audience comments during intermission or inviting audiences to stay after performances to discuss their reactions. Focus group discussions with eight to twelve participants can yield rich data, often for only the cost of offering complimentary tickets. Another low-cost method is the analysis of currently available internal information, such as ticket sales by production, performer, time of year, day of week, time of day, or by audience segment. Even when higher levels of sophistication in research methodology are needed, such as when careful field study projects are being planned, arts organizations can often get low-cost assistance from marketing students and professors. Also, arts organizations can invite marketing research professionals to sit on their boards of directors. Most importantly, web-based ticket sales and donation software programs, in addition to the organization's own website, Facebook page, and other digital sources, provide a wealth of cost-free, readily accessible information.

Inadequate technical knowledge. Arts marketers planning to undertake research programs should acquaint themselves with the various approaches to marketing research, the marketing research process, and the rudimentary principles of probability sampling, questionnaire design, and interpretation of results. But arts marketers need not be focus group leaders, survey designers, statisticians, or computer experts. They should be familiar enough with market research methods, terminology, and processes to facilitate communication with the researcher and to guarantee that the organization's research needs are fulfilled.

Resistance by artistic decision-makers. There is concern among some managers and artistic directors that applying marketing research results will compromise the organization's artistic mission and integrity. These artistic directors are missing an opportunity to use market research information to their advantage. For example, managers could identify opportunities to repackage the programs they are already providing. Based on consumer feedback, the Vancouver Symphony developed a "Great Composers" series and placed its more adventuresome programming in other concerts. This allowed for a more focused choice on the part of the patrons, and therefore, greater satisfaction among both the more traditional music lovers and those who prefer the new and different.

The Scope, Objectives, and Uses of Marketing Research

Marketing research aims to help managers make better decisions, whether the issue is, for example, to move a theater to a new location, to add a new series, to develop a marketing plan for a new target group, to adjust pricing structures for both revenue and audience size maximization, to determine the most effective communications, or to decide which customer service components should be improved. By using the various tools available to market researchers, arts organizations can carry on a range of studies that generally have one of three purposes: description, explanation, or prediction.

DESCRIPTION

Marketing research can be designed to tell management what the marketing environment is like, such as how many people from which demographic and psychographic profiles attend each performance, where they heard about the production, how frequently single ticket buyers attend, and what product features they prefer. Descriptive data usually serves management decisions in three ways: (1) by monitoring sales performance to indicate whether strategy changes are needed, (2) by describing consumers to inform segmentation decisions, and (3) by serving as the basis for more sophisticated analysis.

EXPLANATION

Usually a manager wants to understand the forces that lie behind the descriptive findings. The simplest level of explanation is to discover what seems to be associated with what. A manager may use research to help identify socioeconomic, demographic, or lifestyle characteristics of the people who allowed their subscriptions to lapse in the past year. The findings help management develop stronger appeals to certain people in the lapsing audience. The next level of explanation is causation. The ultimate level of explanation is to know not only that A caused B, but also why A caused B. Management must often dig for explanations if research is to lead to the right decisions and actions. Research found that some patrons of the New National Theater of Japan in Tokyo, who were seated in the rear of the most expensive section, were unhappy that they were paying top price for seats adjacent to a much less expensive section. Managers used this feedback to reprice the "sensitive" rows in the hall for the less popular programs. For high demand programs, people don't complain about the prices at all.

PREDICTION

Ideally, explanations will lead to predictions, making them more useful. Suppose management is willing to lower the price of subscriptions for seats that have poor visibility. It would be useful to predict how many more subscriptions will be purchased and/or renewed at different price points and how this will affect revenue and costs. The Joffrey Ballet found they could sell subscriptions for seats in an upper balcony that had not sold in the past by making them available at very low cost on Groupon. By utilizing this popular resource, the ballet was able to attract many people who had not attended the ballet before but who were attracted by the low prices and the Groupon "comfort zone," as this is the resource people commonly use to buy restaurant certificates and make other common nonarts purchases.

Approaches to Marketing Research: Sources of Information

Central to the marketing research process is data collection. Two broad categories of data may be gathered: *secondary data* and *primary data*. An organization may choose to collect one or the other, but most often it will use both to enrich the quality of the information.

Secondary Data

Secondary data consists of information that already exists, having been collected for another purpose. Researchers usually start their investigation by examining secondary data to see whether their problem can be partly or wholly solved without costly collection of primary data. A rich variety of secondary data sources is available, including external sources (government publications, trade magazines, other periodicals and books, foundation reports, and competitors' publications such as annual reports and season brochures) and internal sources (organizational financial statements, sales figures, information on attenders, and the information-rich data available continually regarding website usage, email open and click-through rates, and the like). Foundations, such as the Pew Charitable Trust; government agencies, such as the NEA; and trade organizations occasionally undertake basic research on behalf of a number of arts organizations or the field as a whole. The Theatre Communications Group (TCG) publishes an annual report titled *Theatre Facts*, which provides detailed information on current issues, trends, and comprehensive statistics in the theater industry. Some other secondary sources are cited in this book. Basic research might not have immediate application to specific management decisions, but can help lay the groundwork for better decisions in the future.

Researchers can use internal documents to track the popularity of various programs according to such variables as market segments attending, performance schedules, ticket prices in various seating areas, and the communications media used. They can also analyze how audience response has changed over time to an organization's various offerings. Secondary data provides a starting point for research and offers the advantages of lower cost and quicker findings.

Primary Data

Primary data consists of original information gathered for the specific purpose at hand by going out into the field. When planning a research project, the research manager needs to know the full array of methodologies that might be used to solve a particular problem in order to select the process that will best meet management's decision-making needs. Primary data can actually be collected in three broad ways: through *exploratory* research, which includes observation, individual in-depth interviews, and focus groups; through *descriptive* research, which includes surveys, conjoint research, and panel studies; and through *experimental* research.

QUALITATIVE RESEARCH METHODS

Qualitative research is exploratory in nature, seeking to identify and clarify issues. Sample sizes are usually small and findings are not generally appropriate for projections to larger populations.

OBSERVATIONAL RESEARCH

Marketers can learn a great deal about what customers are thinking and feeling during the ticket purchase process, while attending performances, and during postperformance discussions. Observation techniques are simple, informative, and inexpensive approaches for learning more about patrons and their reactions to offerings. Marketing personnel can simply listen to comments expressed by patrons in the lobby and can listen in on conversations over the phone or in person at the box office. Box office personnel and ushers should be asked to report their observations as well. More formalized techniques are mystery shopping and transactional surveys administered by box office personnel and customer service managers who receive email messages and calls from patrons.

MYSTERY SHOPPING

Organizations can utilize "mystery shoppers," people who pose as customers and rate the performance and service quality of personnel such as ticket sellers or ushers. This means standing in line at the box office, waiting on hold on the phone, searching the website for information, ordering tickets online, being seated by an usher—in other words, going through the entire patron experience at the key organization and for various competitors.

It is not necessary that this type of research always be done anonymously. In fact, it is revealing for marketing directors to observe encounters with patrons and listen in on phone conversations in the box office, during telemarketing sessions, usher training, and so on.

Employees should be educated on why and how mystery shopping will be used, and the research results should be used constructively—to reward personnel who receive outstanding scores and to help those with poor scores to improve. It is necessary to observe individual service providers multiple times to get a true picture of how they communicate with patrons and to minimize potential bias.[7] Researchers should systematically and comprehensively record their evaluations. Marketing directors can use the findings to detect weaknesses in their systems and offerings, and design and implement improvements.

TRANSACTIONAL SURVEYS

Transactional surveys are conducted when a person contacts the organization to purchase tickets or request information. This approach allows for feedback while the experience is still fresh and facilitates corrective actions with dissatisfied customers. When the contact is by phone or in person, patrons can simply be asked if they have been satisfied with the encounter, if their needs and preferences have been met, and if they have any further questions or concerns they would like to share. When the contact is via the Internet or email, a response mechanism should be built in, asking people to share their questions or concerns. Employees should share with a manager any comments that require a quick response. And whenever possible, the marketing director or even the executive director should respond to the customer, demonstrating that people at the highest level of the organization care about each patron.

Transaction surveys are also useful for tracking people's ticket purchase behavior. For example, knowing the amount of time between ticket purchase and the performance helps marketers select the most advantageous timing for their advertising efforts. Knowing how many calls came in just after an ad, promotion, public relations article, or other communication helps marketers determine the effectiveness of the strategy. Knowing whether a seating area that sold 100 percent of capacity could have sold 105 percent or 150 percent is important information for the marketing directors as they develop pricing structures and determine if a production has enough demand to be extended. It is also important to record whether or not people purchase tickets when they phone and if not, whether they were able to get the performance date or price they wanted.

A simple one-page survey can be composed for box office personnel to fill in for each patron who phones for tickets, except when the box office is so busy that customer service will suffer if extra time is taken to complete the surveys. Since demand varies by production throughout the season, it is best to conduct this study over a full season of performances and to repeat it several years later to test changes in consumer behavior.

This process can also be done for web-based orders, tracking each person's search history and eventual purchase. The rapid growth of online ticket sales makes it relatively easy to track the timing of purchase relative to the performance date, but creates new challenges for marketers to capture other information suggested here. When a person does not complete a transaction, a dialogue box can appear or the person can be sent an email message saying, "You stopped at step three of your ticket purchase transaction. Please tell us why." Several options can be offered for people's responses, such as "no seats available for my preferred date/seating area/ticket price," "I had to answer an important phone call just then," "I needed to check with companions about the date and time, or "other" with a blank space for completion by the patron."

FOCUS GROUP RESEARCH

Focus group research involves bringing together groups of eight to twelve persons for two to three hours one time only to respond to questions posed by a skilled moderator. It is typically a good practice to hold three or four focus group meetings, each consisting of different segments, such as frequent and infrequent attenders by age group, life cycle status, travel distance from home or work to the theater, or other special segments the organization would like to understand better. Focus groups are also a useful step in developing a brand campaign for the organization as participants can share their perspectives on the organization and help managers decide which characteristics should be emphasized and which repositioned.

Trained moderators have special techniques for getting beneath the surface of issues and for eliciting in-depth responses. The interaction among respondents may stimulate a richer response or newer and more valuable thoughts than subjects would have on their own. Sometimes the organization's director or marketing manager sits in on focus group sessions and responds to questions that may arise. Focus group sessions are often videotaped and discussed later by a management team. While focus groups are an important preliminary step in exploring a subject, their results should be used cautiously as they cannot always be projected to the general population. Marketers often use information gleaned from the groups in designing surveys to be administered to a larger sample. Focus groups are complementary to quantitative research tools such as surveys, but most of the time, they should not stand alone.

CUSTOMER ADVISORY PANELS

Customer advisory panels involve meetings, phone interviews, or mail questionnaires with sample groups to provide periodic feedback and advice. It is necessary to recruit highly cooperative respondents because of the membership nature of the group. This approach allows for in-depth questioning and for fast access to customer viewpoints when decisions affecting customers must be made quickly. The organization can create panels differentiated by loyalty status, such as long-term subscribers or infrequent attenders; by demographic characteristics, such as age, gender, or ethnicity; or by life cycle status or certain lifestyle characteristics, such as people with children at home or snowbirds who go to warmer climates for the winter. Panel members need to be rewarded in some way, such as with gift certificates or a special event.

Individual In-Depth Interviewing

Individual in-depth interviews involve lengthy questioning of a number of respondents, one at a time, often using disguised questions and minimal interviewer prompting so that the subjects will not be influenced by biased questions. A primary objective of this approach is to get beyond the parameters by which market researchers and the organization's marketing manager define the problem.

Individual interviews are used instead of focus groups when a greater depth of response per individual is desirable, when the subject matter is so sensitive that respondents might be unwilling to talk openly in a group, and when it is helpful to understand how attitudes and behaviors link together on an individual basis.

Gerald Zaltman, author of *How Customers Think: Essential Insights into the Mind of the Market*, claims that marketers misuse surveys and focus groups in an effort to get consumers to explain or even predict their responses to products. "Standard questioning," says Zaltman,

> can sometimes reveal consumers' thinking about familiar goods and services *if* those thoughts and feelings are readily accessible and easily articulated. Yet, these occasions occur infrequently. Fixed-response questions, in particular, won't get at consumers' most important thoughts and feelings if the manager or researcher has not first identified them by penetrating consumers' unconscious thoughts. Most fixed-response questions and focus group moderator questions address at a surface level *what consumers think about what managers think consumers are thinking about.*[8]

Thus, the importance of individual in-depth interviews.[9]

Quantitative Research Methods

Quantitative research is conducted in order to reliably profile markets, predict cause and effect, and project findings. Sample sizes are usually large, and surveys are conducted in a controlled and organized environment.

Marketing Experiments

The most scientific way to research customers is to present different offerings to matched customer groups and analyze differences in their responses. By controlling the variables, the organization can measure response variations. For example, an organization can test possible price points by offering different prices in a survey to different groups and evaluating the response rates.

CONJOINT RESEARCH TECHNIQUES

Traditional survey approaches, which will be discussed in depth later in this chapter, ask about choices one at a time and ignore the reality that most consumer choices involve trade-offs among desired benefits. Conjoint research techniques are specifically designed to capture this process. They permit analysts to estimate the importance to an individual or group of each quality or attribute and to project the likely success of new choices not included in the original set. For example, an organization can ask respondents to rank the importance of such factors as performer renown, seating priority, travel time to the venue (e.g., less than 15 minutes, less than 30 minutes, more than 30 minutes), and so on. By looking at the trade-offs consumers make among these factors, researchers can detect their relative importance and their interactions.

PANEL STUDIES

Researchers may wish to monitor the performance of a target market over time. For example, they may want to analyze patterns of single ticket buyers becoming subscribers or subscribers reverting to single ticket purchase. Changes in consumer behavior can be studied in the following ways:

1. *Retrospectively*, by asking a single sample of consumers what they are doing now and what they did at some past point in time.
2. *Cross-sectionally*, by comparing behaviors of consumers at different stages in a process (e.g., comparing theater attendance of 25- to 35-year-olds with that of 35- to 45-year-olds, or comparing previous theater attendance patterns of sudden subscribers (those who subscribe without previous attendance) with those of gradual subscribers (those who subscribe after building attendance frequency over a period of time).
3. *Cross-sectionally over time*, by asking about behaviors of different samples at two points in time (such as different age groups changing their attendance patterns over time).
4. *Longitudinally over time*, by taking behavioral measures of the same panel of consumers at different points over time. Government agencies or foundation funders may choose to do very long-term studies on people who are exposed to the arts at a young age and measure the impact of various factors on their future performing arts attendance patterns.

SURVEY QUESTIONNAIRES

The survey is the most popular and widely used device for investigating, describing, and measuring people's knowledge, beliefs, product and media preferences, satisfaction levels, demographics, competitive choices, and decision-making processes. To quickly and briefly survey current attenders, questionnaires can be stuffed into programs. Marketers seeking extensive information from their patrons or desiring to survey nonattenders should administer surveys by snail mail, email, or by a web-based survey program. It is important to keep in mind that people with different characteristics respond to different formats of surveys, such as those who prefer online surveys to mail surveys.

Total market surveys are more comprehensive in detail than transactional surveys as they seek to investigate an overall assessment of a company's service and offers. Also, they usually seek demographic customer information, which must be done in a situation that allows for anonymity. Staff and consultants should brainstorm to identify all aspects of the customer's experience that they would like to evaluate to best serve the decisions at hand. In addition to closed-ended questions, space should be provided for patrons to write comments about other issues they have in mind. Because experiences change over time, these surveys should be repeated every few years.

Several years ago, I consulted with a symphony orchestra that offered two concerts each of five different programs over the season. The concerts had been presented on Friday and Saturday evenings for about 30 years and I recognized an opportunity to attract more families and seniors with a Sunday matinee. This meant eliminating the Friday evening performance, as the orchestra would not be in the financial position to add performances until the audience had significantly grown. Some key board members, who liked to attend on Fridays, resisted the change. I surveyed the audience and asked them to rate their preference for performance day and time. A significant but small percentage of the respondents indicated that they would prefer attending a Sunday matinee. This is exactly what I anticipated would happen for two reasons: people are reluctant to change, especially after a long-standing habit; and, most people who would be attracted to a Sunday matinee were not current patrons, simply because we did not offer concerts at their preferred time, so we did not have the opportunity to survey them. Now, some years later, with some new board and staff members who have come to realize this change would provide them with an opportunity for growth, they plan to move the Friday concerts to Sunday afternoons.

Surveys may be administered, for example, with new, nonrenewing, and former customers. Asking new customers why they came can provide powerful information. The organization may find that poor service or other factors under the marketer's control are the cause of customer defections. When this is the case, the organization may be able to win people back and keep others from defecting.

Survey questions are typically designed to be codable and countable so as to yield a quantitative picture of customer opinions, attitudes, and behavior. By including personal questions, the researcher can correlate the answers with the different demographic and psychographic characteristics of the respondents. In using the findings, the organizations should be aware of possible biases resulting from a low response rate, poorly worded questions, or other flaws.

THE MARKETING RESEARCH PROCESS

Effective marketing research involves several steps. The process should begin with the end in mind. Says Alan Andreasen, "The secret here is to start with the decisions to be made and to make certain that the research helps management reach those decisions."[10] I have often seen marketers conduct surveys just because they think they should and copy survey instruments used by other organizations. This is usually a waste of time and money as it most likely will not address the organization's unique issues and pending decisions. Even when several organizations have similar issues, their patron bases and other factors are different enough to warrant customized questionnaires.

DEFINING THE PROBLEM AND SETTING OBJECTIVES

The first step calls for the managers and the marketing researcher to define the problem carefully and agree on the research objectives. Unless the problem is well defined, the findings are not likely to be valuable to the organization. The manager should make clear to the researcher what the decision alternatives are and then the two can work together to determine what information is required to make those decisions. Also, the manager should share with the researcher the financial, political, and other constraints under which the organization operates that may affect the way the research is approached and evaluated.

DEVELOPING THE PLAN

The second step of the process calls for developing the most efficient plan for gathering the needed information. Marketing managers must determine for whom the research is being conducted, to whom it will be presented, and what specific information is needed to help in the decision-making process. Who should the respondents be? From whom should the information be collected? Whose opinions matter?

Next, managers must decide which technique is the most effective and efficient way to gather this information. Designing a research plan calls for decisions about the research method to be used, the sampling plan, contact methods, and questionnaire design. How many respondents should be surveyed, given desired statistical confidence levels? From where can names and contact information of potential respondents be collected? How is the sample to be selected from this population to ensure that the data is representative of the target audience? How should the survey be designed to best avoid research biases?

Sample Size

Large samples give more reliable results than small samples. However, it is not necessary to sample the entire target population, or even a substantial portion, to achieve reliable results. Responses from a small percentage of a population can often provide good reliability, given a credible sampling procedure. Rather than thinking in terms of surveying a specific percentage of a large population, researchers will generally seek a specific number of responses.

Survey Research in San Francisco

Some years ago, I conducted a major survey for four arts organizations in San Francisco: the San Francisco Symphony, the San Francisco Ballet, the San Francisco Opera, and the American Conservatory Theater. The primary objective of the survey was to determine the pricing sensitivity of current patrons as the organizations participating in the study wanted to know how much of a ticket price increase the market would bear. Because ticket price is a function not only of the *ability* to pay but also of the *willingness* to pay—based on the person's interest level—we found it necessary to explore the attitudes, interests, and preferences of current attenders. What factors attract them to attend? What factors might serve to increase their frequency of attendance? What barriers were limiting or preventing their ticket purchase? Additionally, we asked specific pricing questions, about demographics, past attendance habits, and more. All these questions required that the survey be eight pages long.

Because there were so many variables in the responses, we needed a large sample size to guarantee a significant number of responses in each category. We decided to mail surveys to eight thousand households (two thousand per organization) from a random sampling of the audience members of the four participating arts organizations. The samples were chosen from an equal number of long-term subscribers (five years or more), short-term subscribers (one to two years), and single ticket buyers. The resulting list of subjects was merged and purged, so that no one would receive surveys from two or more of the organizations. Surveys were color-coded by organization to help us organize the responses as they were returned to us for analysis.

Sampling Procedure

No sample is likely to produce results that are precisely the same as those for the entire population from which the sample was drawn. There is always some chance that those who are included in the sample are not perfectly representative of the whole population. It is always possible to pick, strictly by chance, a particular group that happens to be different in some important attribute from the population as a whole. If such differences between the sample data and the population data result purely by random chance, this is known as *sampling error*. Smaller samples are more likely to differ from the population than larger ones; so smaller samples are subject to greater sampling error. The higher the sampling error, the lower the reliability; therefore, the smaller the sample, the lower the reliability of the data.

Since it is impractical to measure the entire population, researchers and statisticians usually think of sampling error in terms of comparisons between similar samples. Data from a sample is relatively free from sampling error and is reliable if another sample of the same size, taken from the same population with the same selection technique, is very likely to provide results that are the same or very similar.

To obtain a representative sample, a probability sample of the population should be drawn. Probability sampling allows the calculation of confidence limits for sampling error, so that one could conclude, for example, that there is a 95 percent chance of being correct that patrons will attend at a temporary venue or that they will be willing to absorb a 5 percent ticket price increase. There are three types of probability sampling:

1. *Simple random sample*: Every member of the population has a known and equal chance of selection.
2. *Stratified random sample*: The population is divided into mutually exclusive groups (such as age groups), and random samples are drawn from each group.
3. *Cluster (area) sample*: The population is divided into mutually exclusive groups (such as neighborhoods), and the researcher draws a sample of the groups to interview.

When the cost or time involved in probability sampling is too high, marketing researchers often take *nonprobability* samples. Some marketing researchers feel that nonprobability samples can be very useful in many circumstances, even though the sampling error cannot be measured. For example, if management is chiefly interested in frequency counts (say, of words used by consumers to describe the organization's offerings or of complaints voiced about its staff), sampling restrictions need not be so tight. Three examples of nonprobability samples are:

1. *Convenience sample*: The researcher selects the most accessible population members from whom to obtain information.
2. *Judgment sample*: The researcher uses his or her judgment to select population members who are good prospects for accurate information.
3. *Quota sample*: The researcher finds and interviews a prescribed number of people in each of several categories.

Research Biases

Research can be effective only if it avoids the two major sources of error: sampling error and systematic bias.[11]

Systematic bias refers to extraneous sampling factors that affect survey results and reduce the validity of the data. This bias in a survey usually results from one of the following sources:

- *Frame bias* is caused by drawing a probability sample from a poor representation of the sampling universe.
- *Selection bias* results when the procedure for drawing actual sample members always excludes or underrepresents certain types of universe (frame) members. This would occur if telephone interviewers called only during the day and at home, thereby underrepresenting people who work. It would also occur in field surveys if interviewers were allowed to pick and choose whom to interview. Suppose a survey was conducted to measure how satisfied or content people are. If the field-workers tended to only interview people who looked like they were very friendly, the data would be likely to reflect a bias in the direction of greater satisfaction.
- *Nonresponse bias* results when a particular group of those contacted declines to participate.
- *Interviewer bias* results when an interviewer deliberately or inadvertently leads the respondent to deviate from the truth. This can occur, for example, when interviewers read or "clarify" a question in a way that suggests that a particular answer is preferred.
- *Questionnaire bias* can arise due to poor or confusing wording, leading questions, identification of the research sponsor when this information is supposed to be withheld, or omission of important possible responses.
- *Respondent bias* occurs when respondents distort answers. Often this type of bias manifests itself when respondents "upgrade" their income or education level or their frequency of attendance.
- *Processing bias* would occur if the interviewer (or the respondent in a written study) wrote down the wrong answers, if the office entered the answers into the computer incorrectly, or if the computer analysis was programmed incorrectly.

Catching all these glitches is not easy. It takes careful attention to the design process, to interviewer training, to checking on data processing, and to reviewing reports for errors.

Designing the Questionnaire

When designing the questionnaire for the San Francisco study, I knew that it would be irrational for people to select the highest price when presented with multiple ticket prices, when asked how much they would pay for a ticket, so I created a strategy that avoided this problem. Two sets of surveys were developed for each organization, each with one price structure to which people would respond. One-half of the sample population (one thousand people for each organization) received version one of the survey, with lower prices for each offering, and the other half received version two, with higher prices. This strategy of offering each respondent one set of prices, differing by seating area rather than price options for each offer, was used to avoid a response bias, because when given a choice, people naturally select lower prices. In the data analysis, we were able to compare the responses of the two groups to get a much more realistic view of pricing sensitivity than we would have otherwise.

Another factor I took into consideration when designing this survey was how people respond to the question, "What is your gender?" Although it seems to be straightforward, consider that typically, although two people per household attend performances, they receive one questionnaire. More often than not, the woman fills in the survey, possibly consulting her spouse when deciding how to answer the questions. But when asked "what is your gender," the woman will respond "female." Therefore, the results are often skewed, claiming a much larger percentage of women attending than is actually the case. Therefore, in the survey, I asked: "What is your gender? Do you typically attend with another person? If so, what is that person's gender?" This way I was able to obtain a much more accurate count of the people of each gender attending.

THE QUESTIONS ASKED. A common type of error concerns the nature of the questions asked. The researcher must be careful to include all pertinent questions and to exclude those that cannot, would not, or need not be answered. Each question should be checked to determine whether it contributes to the research objectives. Questions that are merely interesting should be dropped because they increase the response time required and strain the respondents' patience.

Questionnaires can be a marketing tool in that the questions asked may lead the reader to learn about attractive programs and services at the organization. Top-level managers should be given the opportunity to review the questions in advance. Some managers express concern that certain seemingly innocent questions may set up unrealistic expectations among the people surveyed, resulting in disappointment. One symphony director requested that researchers exclude from their survey a question asking patrons if they would like a babysitting service, because he was not interested in providing such a benefit. Management participation in the design decision has other advantages. It serves to win managers' support of marketing research and deepens their understanding of research details. Participation will also sensitize management to the study's limitations. The researcher might ask the decision-maker, "If I came up with these numbers, what would you do?" The researcher may discover that management would take the same course of action no matter what the results. Such a conclusion would suggest a disagreement between the general manager and the marketing director that needs to be resolved.

When I consulted with the Chicago Opera Theater, some patrons had told the marketing director that they would not return the following season unless supertitles were offered for all opera performances, including English-language operas, because sung English is difficult to understand. Brian Dickie, the general director at the time, was opposed to offering supertitles for English-language operas, but he agreed that we could include questions about this subject in the survey we were about to administer. The results came back with a strong, clear preference for these supertitles, and Mr. Dickie relented. We then publicized "Supertitles for all operas!" on our season brochure. Not only did we retain audience members with this change, but we also attracted new patrons who were enticed by the increased accessibility.

THE FORM OF THE QUESTION. The form of the question can influence the response. Marketing researchers distinguish between closed-ended and open-ended questions. Closed-ended questions specify all the possible answers, and respondents make a choice among them. These questions provide answers that are relatively easy to interpret and tabulate. Open-ended questions are especially useful in the exploratory stage of research, when the researcher is looking for insight into how people think rather than measuring how many people think in a certain way. It is useful and often enlightening for organizations to encourage respondents to make comments about whatever interests or concerns them at the end of closed-ended questionnaires.

Exhibit 7.1 provides examples of some commonly employed forms of closed-ended and open-ended questions.

WORDING OF QUESTIONS. The researcher should use simple, direct, unbiased wording so that the questions do not stimulate the respondent to think about the organization in a certain way. The questions should be pretested with a sample of respondents before they are formally included in the survey.

SEQUENCING OF QUESTIONS. Whenever possible, the lead question should create interest. The questions should follow a logical order. Classificatory questions about the respondent are asked last because they are more personal and less interesting to the respondent.

Incentives

A typical response rate for a mail survey administered to arts patrons is 20–25 percent (a much higher response rate than surveys for most goods and services). But in the San Francisco study, we wanted an even higher response rate to provide the basis for a richer analysis, so we offered incentives to increase the response rate. Two free tickets to select performances were offered by the symphony, the ballet, and the theater. The opera offered a one-time 10 percent discount at the Opera Shop.

Exhibit 7.1 Types of survey questions

Closed-ended questions	Example
Dichotomous: A question offering two answer choices	Is this the first time you have attended this theater? Yes ( ) No ( )
Multiple choice: A question offering three or more answer choices	With whom are you attending this performance? No one ( ) Spouse ( ) Other relatives/friends ( ) Business associates ( ) An organized group ( )
Likert scale: A statement with which the respondent shows the amount of agreement/disagreement	Good critical reviews are an important factor for me in choosing to attend a performance. Strongly disagree 1 ( ) Disagree 2 ( ) Undecided 3 ( ) Agree 4 ( ) Strongly agree 5 ( )
Semantic differential: A scale is inscribed between two bipolar words, and the respondent selects the point that represents his or her opinion	High ticket prices ____ • ____ • ____ • ____ • ____ Low ticket prices Convenient location ____ • ____ • ____ • ____ • ____ Inconvenient location
Importance scale: A scale that rates the importance of some attribute	Ticket exchange privilege to me is… Extremely important 1____ Very important 2____ Somewhat important 3____ Not very important 4____ Not at all important 5____

continued

Exhibit 7.1 Continued

Closed-ended questions	**Example**
Rating scale: A scale that rates some attribute from "poor" to "excellent"	The quality of our productions is . . . Excellent 1_____ Very good 2_____ Good 3_____ Fair 4_____ Poor 5_____
Intention-to-buy scale: A scale that rates the respondent's intention to buy	If subscriptions were offered for packages of three productions, I would . . . Definitely buy 1 ( ) Probably buy 2 ( ) Not sure 3 ( ) Probably not buy 4 ( ) Definitely not buy 5 ( )

Open-ended questions	**Example**
Completely unstructured: A question that respondents can answer in an unlimited number of ways	What is your opinion of Centre East Theatre?
Word association: Words are presented, one at a time, and respondents mention the first word that comes to mind	What is the first word that comes to your mind when you hear the following? Theater_____ Chamber music_____ Centre East _____
Sentence completion: Incomplete sentences are presented, one at a time, and respondents complete the sentence	When I choose a performance to attend, the most important consideration in my decision is _____ _____

The value of a desirable incentive was clear. The response rates for the symphony, the ballet, and the theater averaged 46 percent, approximately double the usual rate. The response rate from opera patrons, who obviously did not value the weak incentive offered to them, was 23 percent. Incentives are an extremely effective and low-cost way to increase the quantity of responses since free tickets can be offered for programs that have anticipated excess capacity. The organizations also benefit by getting more people in the hall for programs that do not sell at capacity. The only real cost to the organization is the time required for handling the ticket vouchers.

Although it was important for us to offer survey respondents anonymity to encourage them to share personal information, we needed contact information in order to provide them the ticket vouchers. We managed this by having a voucher form separate from the survey, and we promised patrons that we would separate this voucher from their responses as soon as we opened their envelope and verified they had completed the survey.

Pretesting

Researchers should *pretest* the survey questionnaire or focus group discussion guide with various persons to make sure the questions are clear and lead to the desired information. People both inside and outside the organization who have various levels of involvement with the organization should be included in the pretest.

Contact Methods

Respondents to a survey may be contacted by email, by snail mail, or by a web-based program. Brief surveys may also be inserted in program books at performances.

Web-based survey instruments are the most cost-effective way of reaching a large number of people. They also garner the largest and quickest response. The response rate of snail mail surveys tends to be lower and people are slower to respond. However, when reaching out to people who are not fluent in computer use, the paper-and-pencil approach is likely to be welcome. An organization may consider using both methods for one survey, reaching out to older audiences with the snail mail survey and everyone else with an online survey.

For surveying current patrons, arts organizations are in the enviable position of having a "captive" audience. By placing questionnaires in programs and collecting them at intermission or at the end of the performance, researchers can benefit from the advantages of mail questionnaires while avoiding the problem of sparse or slow responses. Even then, patrons need encouragement from someone on stage or from ushers to complete and turn in the surveys.

Collecting and Analyzing Data

The next steps are to collect and analyze the data. The data collection phase is generally the most susceptible to error and research biases should be controlled for as best as possible. Once an adequate sample has been obtained, pertinent findings are extracted from the data. The researcher tabulates the data and develops frequency distributions, averages, and measures of dispersion. Based on this information, the researcher analyzes the results in terms of the managerial decisions to be made and forms a series of recommendations, in effect transforming research data into marketing insights.

Statistics Used by Researchers

It is helpful for marketing directors and general managers to familiarize themselves with some of the statistical terminology and procedures that are used to organize and analyze the data and determine its meaning. Here are brief descriptions of statistical terms.[12]

Terms Describing the Distribution of the Data

Mode: The response or score that occurs with the greatest frequency among findings.

Median: The value (score) halfway through the ordered data set, below and above which lie an equal number of values.

Mean: The simple average of a group of numbers. Managers must take care in how they use this number, however, as averages really tell us very little.

Range: Determined by subtracting the lowest score from the highest score.

Terms Describing Measures of Variability

Margin of error: A measure indicating how closely you can expect your sample results to represent the entire population (e.g., plus or minus 3.5 percent).

Confidence interval: A statistic plus or minus a margin of error (e.g., 40 percent plus or minus 3.5 percent).

Confidence level: The probability associated with a confidence interval. Expressed as a percentage, often 95 percent, it represents how often the true percentage of the population lies within the confidence interval.

Standard deviation: A measure of the spread of dispersion of a set of data. It indicates whether all the data scores are close to the average or whether the data are spread out over a wide range. The smaller the standard deviation, the more alike the scores.

TERMS DESCRIBING ANALYTICAL TECHNIQUES

Cross-tabs: Used to understand and compare subsets of survey respondents, providing two-way tables of data with rows and columns allowing you to see two variables at once (e.g., the number of people aged 55–74 who subscribe compared to subscribers aged 35–54).

Factor analysis: Used to help determine what variables (factors) contribute the most to results (scores). This analysis, for example, might be used to help determine the characteristics of people who have lapsed their subscriptions, who are first-time donors, or who attend once and do not return.

Cluster analysis: Used to help identify and describe homogeneous groups within a heterogeneous population relative to attitudes and behaviors. This analysis can be used to identify market segments.

Conjoint analysis: Used to explore how various combinations of options (prices, benefits, etc.) affect preferences and behavior intent.

Discriminant analysis: Used to find the variables that help differentiate between two or more groups.

CREATING REPORTS AND DISSEMINATING FINDINGS AND RECOMMENDATIONS

Once the data is analyzed, the researcher creates reports to submit to managers and board members. An executive summary of the findings is used to report and analyze the results and make recommendations in a few succinct pages. The executive summary should not include a multitude of numbers and descriptions of complex statistical techniques; it should present the major findings that are relevant to the marketing decisions facing management. Often, the researcher will make a presentation about the major findings at a meeting of the board of directors.

Generally, the researcher will also submit an extensive, detailed report of the findings and analyses of each part of the questionnaire. Such detailed information is useful to the marketing manager and sometimes to the executive director or a marketing professional on the board; these are possibly the only people who have the patience and the necessary skills to wade through all the data.

IMPLEMENTATION AND EVALUATION

The most important thing about planning is the process itself, not the plan. Says Philip Kotler, "Planning forces you to think deeply and futuristically in a more systematic manner."[13] Yet, too often, the final report is shelved before many of the recommendations are acted upon. One way this shortsighted approach can be avoided is by keeping the research consultant involved during the implementation phase. He or she understands the data and its nuances and has already thought hard about what the results should mean to the manager. Continued teamwork by the manager and the researcher will increase the likelihood of good decisions being made and followed through.

The research process just undertaken should be thoroughly evaluated to see if it could have been carried out more effectively, to see if further research is needed in certain subject areas or with other segments, and to decide if and how often similar research should be repeated. It is worthwhile to note that as extensive as it was, the San Francisco survey researched the pricing sensitivity of current attenders only and offered no way to determine what pricing strategies would be attractive to segments who have not attended in the past. Other research may be undertaken to explore how to attract new target segments.

The Power of Marketing Research

Marketing research provides managers with powerful information that dramatically increases their effectiveness in appealing to and satisfying customers. Customer expertise also provides marketers with the courage of conviction they need to promote their point of view to the leaders within the organization. Marketing research provides the insights and proof managers need to move forward with marketing strategies and tactics.

CHAPTER 8

Using Strategic Marketing to Define and Analyze the Product Offering

The works presented on the stages of performing arts organizations are their raison d'être. Yet the product consists not only of the performances themselves; it is the complete bundle of offerings and experiences provided by the institution to the public.

Consumers seek products, services, and experiences that meet their needs and desires. According to marketing strategist Theodore Levitt,

> People buy products... in order to solve problems. A product is, to the potential buyer, a complex cluster of value satisfactions. The generic "thing" or "essence" is not itself the product.... Customers attach value to products in proportion to the perceived ability of those products to help solve their problems. Only the buyer or user can assign value, because value can reside only in the benefits he wants or perceives.[1]

Shakespeare may have said "The play's the thing," but what is truly the "thing" for the audience is the entertainment or the aesthetic, intellectual, emotional, or social experience, or some combination of these experiences. In a study of Broadway theater patrons, Olson Zaltman Associates found that for aficionados, the show itself is of the greatest importance while other factors such as the hustle and bustle of Broadway, restaurants, and bars have minimal importance (or appeal). For those defined as frequent attenders (but not aficionados), these other factors rated equal in importance with the show; and among infrequents, the Broadway environment was by far the most important factor, followed by bars, clubs, restaurants, and last, the show itself.[2] Dominique Bourgeon, who surveyed nearly fifteen hundred people in the regions of Caen, Dijon, and Nancy, France, also found that, in general, the play itself is the main determinant of a frequent attender's feelings about the performance. But for occasional attenders, other intangible and atmospheric factors are more important.[3]

What this suggests for building frequency among occasional attenders and for building new audiences is that the total experience should appeal to the imagination, to emotions, and especially to pleasurable, joyful feelings. This does not mean that theaters must always present lighthearted plays; it means that the organization will attract a broader audience if it makes the total theatergoing experience a source of emotional satisfaction.

In 2012, during a discussion of the future of theater, producers and directors recognized that if Broadway (and other theater) is going to compete with the digital world, it has to give theatergoers a "bigger bang for their bucks" (more value for their money). Says Jordan Roth, the president of Jujamcyn Theaters, "Each artist will have a different reason for why this story's being told live. Some will make it more interactive, while others will make it more of a 360-degree experience with a set extended into the lobby and dancers mingling with the audience."[4] Others expect to look for material that goes deeper, something people can take home.

The desired and satisfying experience varies not only from segment to segment but also from individual to individual, which complicates the marketer's task. A first-time operagoer will respond far differently to a production of *La Boheme* than will someone who has seen it five times. Furthermore, each patron's experience will vary according to his or her knowledge, preferences, background, and mental state at the time of the performance. Also, people's interests, needs, and tastes change with repeated exposure and experience. Audiences change generation by generation. Therefore, organizations must periodically redefine what is attractive to any given segment of the population.

The Total Product Concept

The definition of the arts organization's product extends beyond the work presented on stage to include all the organization's offerings. One can describe a product as comprising two different levels.

The Core Product

The core product is that which is being offered to the target market for purchase or consumption. The core offering of a symphony orchestra may be a single piece of music to be performed (Tchaikovsky's *1812 Overture*), the program for an entire concert (a Tchaikovsky Spectacular), the collection of programs for an entire subscription series (a Great Composer Series), or the programs and series for an entire season. It may also be considered in terms of the specific orchestra, the conductor, and the guest soloist(s).

People place different values on various aspects of the core product. For example, consider a performance of the Tchaikovsky Violin Concerto by Itzhak Perlman with the Chicago Symphony Orchestra, conducted by James Conlon at the Ravinia Festival. There are several components to this product, each of which is valued differently by the audience members. Perlman's name is a magical enticement and many patrons are eager to attend his performances. Some patrons will base their attendance on the musical selections being performed, others on the orchestra's high quality. The ambience of picnicking at Ravinia on a warm summer evening is also a draw.

The core product choice is in the domain of the artistic director. Yet programming is only partially driven by the artistic decision-maker's vision. Selecting programming is a complex activity, requiring that the artistic director and the managing directors work together to solve a perpetual problem: how to create a series of programs that has artistic merit; is congruent with the organization's mission, competencies, and constraints; and serves the needs and interests of the community. Ideally, an organization's programming is both highly artistic and highly satisfying to the audience. That being said, an art organization's season is best designed to balance artistic exploration with the clear preferences of current and potential patrons.

Entertainment, which is market centered, has customer satisfaction as its core goal. Art by definition is provocative, challenging, and often unfamiliar and disturbing. If fine arts patrons were all satisfied, artistic directors would not be living up to their responsibility to challenge and provoke. The separation of arts from entertainment in much of the media is intimidating to some people and causes them to think they won't enjoy art and certainly won't understand it. But, those of us who love the arts find them highly entertaining.

A performance is essentially a communication between the artist or performer and the audience. This communication cannot take place if the audience does not relate to what is happening on the stage. Says Howard Shalwitz, director of Woolly Mammoth Theatre Company in Washington, DC, "We're in a state of evolution about how we think about our role in relation to our audience. We're still working from our historical emphasis on challenging our audience, but with an increasing emphasis on trying to do plays that connect with pressing conversations happening in our community." Woolly Mammoth staff members ask playwrights: "Who do you imagine is the audience you want to be talking to with this play?" This is what Shalwitz and his staff call "audience design." They are looking for audience members who are going to complete the story of the play, who are going to make the conversation that the play stimulates more meaningful by their presence in the theater.[5]

THE AUGMENTED PRODUCT

The augmented product consists of features and benefits created by the marketer to stimulate purchase and enhance consumption of the core product. Augmentation includes such offers as subscription packages, ticket exchange privileges, newsletters, pre- or postperformance lectures, blogs, videos, audio clips, educational programs, and special events.

The augmented product includes the customer's normal expectations regarding the purchase and consumption of the product. Patrons may expect to be able to choose their own seats when purchasing tickets online. Subscribers expect ticket exchange privileges. Customers may expect a well-lit parking lot, especially if the theater is in an unsafe neighborhood. They expect to be treated well by box office personnel and ushers. People expect to be able to readily find comprehensive information online about the organization, each production, the people involved, and in-depth interviews and analyses of the play or music to be performed. Many organizations mail or email newsletters and program notes to their patrons in advance of performances. Other product augmentations may include discounts at nearby restaurants, box meals during the intermissions of long shows, and pre- or postperformance lectures. A few short years ago, such augmented features as blogs, video clips, and the like were rare. Now they have become expected features, along with Twitter feeds, Facebook pages, and the like. The expectations of different audience members, of course, will vary. Some patrons may expect that the acoustical quality of the hall will be high; for others, adequate leg room may be more important. Elderly and disabled patrons may not attend unless there is ease of access, no matter how attractive the evening's performance may be.

Companies tend to think in terms of related *products*. Customers think in terms of related *activities*. The augmented product often consists of features and benefits that enhance the experience of current patrons. Typically, augmented product features do not attract people to attend if the core product doesn't interest them, but a lack of some of these features can serve as a barrier that keeps people away, even when the core product is highly desirable.

It is in the ways that the core product is packaged, priced, and promoted to the publics and the ways that information is shared, accessibility is engendered, and interest is built that is the domain of marketing.

Twenty-First-Century Marketing at Piccolo Teatro di Milano

Since 2001, Piccolo Teatro has effectively boosted its audience development mission by implementing a web-based marketing plan including an interactive website, a virtual community, online services, and multimedia archives. To meet the growing demand for more in-depth information on the Internet, in 2009, Piccolo Teatro started a special website (www.piccoloteatro.tv). It was the first theater web.tv in Italy and included interviews, unreleased backstage footage, and show previews with the artists seen on Piccolo's stages. As well as being screened in Piccolo's three theaters and in the theater's Cloister with restaurant and lobby areas, the videos and documentaries of web.tv are linked to the Piccolo site and posted as a means of promotion and service to journalists, members of the community, and Piccolo's Facebook fans. This product extension serves to build accessibility, promote the plays, build interest and enthusiasm, and to inform and educate the theater's various publics.

THE PRODUCT AS A SERVICE

Performing arts organizations are basically in the service business. A service is essentially intangible and does not result in the ownership of anything. Certain characteristics that are unique to services have special implications for marketers.

INTANGIBILITY

Services are intangible; they cannot be seen, heard, felt, tasted, or otherwise experienced before they are bought. A patient having a haircut cannot see the result before the hairdresser has completed his or her task; a theatergoer cannot experience the play before the performance. To reduce the uncertainty inherent in the purchase decision, the service buyer seeks signs or evidence of quality—thus, the importance of a big name, reputation, and good reviews. The service provider seeks to "manage the evidence," to "tangibilize the intangible."[6] Whereas tangible product marketers are challenged to add abstract ideas (McDonald's "Happy Meal" links entertainment with eating), service marketers are challenged to associate physical evidence and imagery with their abstract offers (Prudential Insurance says, "Get a piece of the rock").

Performing arts marketers can tangibilize their offerings in a number of ways. When promoting relatively unknown performers, marketers refer to concrete symbols of their quality such as awards or a past performance at Carnegie Hall. Such symbols function the way brand names do for physical products. When the artistic work or the artists do not have a track record, other factors should be presented that indicate the nature of the experience to expect. Special attention must be paid to "atmospherics."[7] The visual quality of the brochures and the character of the facilities can affect the customers' expectations. A no-frills auditorium sets the stage for the young, adventuresome theater company as well as a marble lobby does for the opera company.

People are also important conveyors of a performing arts organization's positioning. Publicity photos of formally dressed chamber musicians create different expectations than do photos of musicians in brightly colored silk shirts. Even the audience provides a tangible signal about the appeal of the performance. By showing which people go (e.g., demographic characteristics), how they dress, and the serious or animated expressions on their faces, marketers can convey an image of the theatergoing experience.

PERISHABILITY

Among an arts marketer's greatest challenges is the fact that services are perishable; they cannot be stored or preserved. A car or appliance can be kept in inventory until it is sold, but the revenue potential of an unoccupied theater seat is lost each time the curtain rises. As interest in a theatrical production builds over time, the empty seats from earlier performances cannot be filled. And an organization presenting a once-only concert has no opportunity to benefit from the interest generated afterward. Perishability is less of an issue when demand is steady, as is the case for heavily subscribed organizations. But when demand fluctuates widely, perishability is a serious problem. This is why arts organizations seek advance publicity for their performances, offer discount pricing for previews, and promote subscriptions.

INSEPARABILITY

Unlike physical goods that are manufactured and put into inventory, services are typically produced and consumed at the same time. A service is inseparable from the source that provides it. Consider the emotional impact on an audience expecting to hear Renee Fleming perform if an announcer tells them that Ms. Fleming is indisposed and that someone else will substitute.

This concept applies to the organization's customer service as well. An otherwise enjoyable evening can be sabotaged by unresponsive or unhelpful personnel. Some factors that affect a patron's satisfaction with the experience are beyond the organization's control, such as a flat tire on the way to the performance or an emergency at home in the middle of the evening. But there are external factors to which the organization can respond. When a bad snowstorm or a major traffic jam affects many patrons, the curtain can be delayed a few minutes or patrons can be seated late, against custom. Such efforts go beyond a customer's expectations and may actually serve to add to the satisfaction and enjoyment the patron would have experienced had the inconvenience not occurred at all. The effect may be somewhat offset, of course, by the dissatisfaction felt by those who arrived on time.

VARIABILITY

Since a service is so closely linked to its source, its quality can vary depending on who is providing it and when it is being provided. A performance of the Dvorak cello concerto by Yo Yo Ma is likely to be of higher quality and more exciting to watch than the same concerto played by a young musician. And the quality of the audience's experience with Ma performing can vary depending on factors such as the temperature in the hall or the quality of the orchestra and conductor with whom he is performing. Purchasers of services are aware of this high variability, and the more unknowns there are about a service provider (performer, playwright, composer, director, presenting organization), the more those purchasers will engage in risk-reducing behavior to learn whether the offering is worthwhile.

Service providers can manage variability within the organization. Consistency can be attained with good personnel selection and training and by routinizing as many parts of the service as possible. Also, organizations should develop adequate customer satisfaction monitoring systems, using suggestion and complaint systems, customer surveys, and comparison shopping with other arts organizations.

CUSTOMER INVOLVEMENT AND EXPECTATIONS

The nature of the customer's involvement is an integral aspect of the service exchange. A performance of the Brahms Violin Concerto will be "consumed" differently by the various audience members according to their knowledge, preferences, backgrounds, and mental states at the time of the performance. To help increase the likelihood of a highly positive experience and to minimize consumer disappointment, art marketers can help patrons to be better informed art consumers with preconcert lectures, postperformance discussions, informative newsletters, in addition to many other online and in-person educational opportunities.

Performing arts organizations should make their purpose and positioning clear to their target audience. When patrons are repeatedly unmoved by performances, they may either think that they are missing something or that they just do not enjoy the art form. A major source of dissatisfaction on the part of many service customers is not inferior service, but exaggerated expectations. A theatergoer should know in advance if a play will be depressing or contain violence. A bittersweet play with a few comic moments should not be billed as "hilarious."

The essence of services marketing is fine service. Whether the product is a performance by a pianist, the advertisers, or the box office personnel, nothing is more important than the quality of that performance.[8]

Determining the Product Offering

Selecting programming is a complex activity. It requires that the artistic director and the managing directors work together to solve their perennial riddle: how to create a series of programs that have artistic merit; are congruent with the organization's mission, competencies, and constraints; and serve the needs and interests of the community. There is a distinction and delicate balance that must be maintained between art for art's sake, art for society, and art for the survival of the organization.

Although the artistic product derives from a unique vision, creativity does not occur in a social vacuum. All artists are sensitive and responsive to the worlds in which they live. Shakespeare created Nick Bottom and other colorful characters after the "common folk" he observed at the Globe Theater. Haydn composed a forte portion in his "Surprise Symphony" at the point where he knew his patrons were likely to fall asleep. Playwrights through the ages have expressed strong political and social messages in their writings. Consider Arthur Miller's *Crucible*; David Mamet's *Race*; Bruce Norris's *Clybourne Park*; and Matt Stone, Trey Parker, and Robert Lopez's *Book of Mormon.*

Thematic Programming

Thematic programming is most commonly used for concerts, since a variety of music is played at each performance, but the concept is also used by theaters in programming their entire season.

One common orchestral approach is the musical theme: "Romantic Music," " The First and Last Works of Beethoven and Bartók," "Sounds from Norway," "From Classical to Jazz." Often these programs are unique and include rarely heard musical selections. However, this approach is typically a timid, product-centered effort to attract new listeners into the concert hall that does little to enhance the audiences' experience or to create new levels of understanding or appreciation.

Consider instead how the American Symphony Orchestra (ASO) in New York City takes a customer-centered approach to thematic programming to offer an accessible, meaningful, and enriching experience for its audience.

Thematic Programming at the American Symphony Orchestra

According to music director Leon Botstein, the ASO has a mission of taking bold steps "to reimagine the tradition of orchestral music and concerts within the larger culture—to link music to the visual arts, literary life, and politics as well as popular culture." One performance entitled "New York Avant Garde" tied together music and visual arts from the early twentieth century, a period studied and enjoyed by many arts lovers and history buffs. Says the ASO's website, for example, "From Strauss to Antheil to Copland, ASO presents a glimpse of New York's modernist musical culture in the years surrounding the 1913 Armory Show, in partnership with New York Historical Society's retrospective on the historic exhibit." The programs are often linked to current New York art exhibitions and cultural events and the preconcert and mid-concert talks open every possible avenue for education within the performance. According to *New York Times* critic Ed Rothstein, Leon Botstein has become "the dominant figure in the most important contemporary trend in concert programming during the last 50 years."[9]

PROGRAMMING FOR THE COMMUNITY

Some organizations find they can be most effective when programming specifically for the community of which they are a part. Specific recognition of the arts organization's interest in the community can develop not only a larger audience but one that is proud to have an orchestra or theater that cares. If the area has a dominant ethnic background, the orchestra can plan an appropriate concert or festival. If an important historical event has shaped local history, the orchestra can schedule a concert of music from the period, perhaps including a neglected American work of the time. If a corporation, hotel, or other business is promoting a major anniversary, it can show its appreciation for the community by commissioning a work for the local orchestra. Consider how the Helsinki Philharmonic Orchestra's Godchild Project reaches the very youngest members of the community.

Helsinki Philharmonic Orchestra's Godchild Project

Said the Helsinki Philharmonic Orchestra's website: "[Our] first godchild project (2000–2007) was a great success and aroused widespread interest both in Finland and abroad. In 2012, as the orchestra celebrates its 130th anniversary, the project is to have a sequel. The orchestra is accordingly inviting all Helsinki babies born in 2012 to be its godchildren."

For this program, godfamilies receive annual invitations to concerts geared to the child's age at the time. In the first year, little concerts are held in different parts of Helsinki; later, they are given in the Helsinki Philharmonic Orchestra's home hall at the Helsinki Music Centre.

The presenter at these concerts is Satu Sopanen, an expert on children's music who, in 2011, was awarded the State Prize for Merits in Children's Culture. Also contributing to the concerts are students at the Sibelius Academy, experts on early music education, theater directors, composers, and conductors.

Families who register for the godchild project receive the orchestra's CD called "Nallekarhu konsertissa/Teddy goes to a concert." The items on this disc are classics from the fairytale world, featuring butterflies, bumble bees, princesses, and goblins. During its first project (2000–2007) the Helsinki Philharmonic produced 15 concerts, each different, in more than 80 performances under the baton of eminent conductors. These were attended by 4,500 godfamilies, and the feedback was extremely positive. "We're extremely proud of our musical godparents for 'opening up' the world of orchestral music for our children. Many thanks!"

APPROACHES TO PRESENTATION

Since the early 1990s, orchestras and other performing arts organizations have been urged by their trade associations and other experts in their fields to rethink not only their programming, but also their approaches to program presentation. Harvey Lichtenstein, former long-standing president and executive producer of the Brooklyn Academy of Music, has been a leader in questioning every aspect of the concert experience. Says Lichtenstein, "The orchestra, more than most cultural instruments, needs a radical approach, because it's stuck in the past, and it has more of a problem gathering a young audience. And let's face it: the orchestra, visually, is fairly dull. You've got to keep questioning all the old traditions. Some of them may be valid. But lots of them just make no sense."[10] Arts organizations are taking up the challenge of making performances more inviting, exciting, relevant, and accessible by varying the performance rituals and environment.

THE PERFORMANCE RITUAL

Arts organizations have traditionally followed a ritualistic pattern of performance presentation. The symphony concert, in particular, is generally characterized by a short opening work, a concerto with soloist, an intermission, and then a longer symphonic piece, all performed by up to one hundred musicians identically dressed in formal attire. There are many ways this ritual can be altered to enliven the experience, including concert-related activities during intermission, small ensemble and solo performances during the orchestra concert, shorter concerts and concerts with different starting times to accommodate people with different schedules, more variety in the kinds of works presented, and unscheduled encores in the middle of the program. Harvey Lichtenstein envisioned concerts with dance, theatrical events, and opera. The use of color can add excitement and interest to a normally staid and formal art form. At one concert, when the women players arrived wearing brightly colored long gowns, the atmosphere immediately became more festive.

THE PERFORMANCE ENVIRONMENT

Ambience is a critical factor in the audience's performing arts experience. Creating an ambience that enriches the programming is a unique challenge for each organization—unique because it is dependent on the nature of each organization's product offering; its physical, financial, and human resources; and its audience's preferences, interests, and needs. Lingering at New York's Lincoln Center or London's Barbican Centre makes the whole concert-going experience more pleasant. Patrons can browse through music-related books and recordings. Meals provided before a performance or during intermission accommodate those who do not have time to eat before the show. Postconcert wine or dessert bars in the lobby, or even on the stage, stimulate socializing and discussion about the performance. Postconcert events also answer such questions as these: where should we go after the concert, where can we meet some new people, and how can we learn more about the performance/performers/composers?" A public "green room" allows audience members to meet the musicians, dancers, or actors after the concert and provides an opportunity for managers to hear audience comments and answer questions.

Using Multimedia

Arts organizations often capitalize on modern technology to enhance the audience's experience both within and outside the performance hall. In large halls, some organizations utilize onstage cameras to relay video images of the conductor and performers to a projection screen at the rear or sides of the stage. A visual component may be added by using film projected on scrims or screens, most often with live musical accompaniment, as was done when composer Philip Glass created instrumental and vocal music for filmmaker Jean Cocteau's *Beauty and the Beast*. Orchestra purists worry that this approach will distract from the music, but when done well, such enhancements add to everyone's experience, while attracting a younger crowd.

Outside the performance hall, arts organizations can use modern technology to increase exposure, familiarity, and comfort levels with the arts. Some organizations such as Lincoln Center in New York and Millennium Park in Chicago project select performances on large screens outside their halls. Video and audio clips of the performance—with commentary—can be posted on the website and YouTube to help audience members become better "tuned in" to what they are about to hear and see. Community access cable television, public television, and radio can provide forums for orchestras and theaters to introduce repertoire to potential audiences. More and more, arts organizations are capitalizing on the power and accessibility of the Internet, social media, and dedicated applications (apps) to communicate with audiences and to provide opportunities for interactive exposure and learning.

The Event

As the well-known Stratford, Ontario, Shakespeare Festival and the Bayreuth Festival demonstrate, an event may provide the basis for an organization's total programming. Some organizations offer festivals to enhance and extend their regular, ongoing programming, such as the NY Phil Biennial, initiated in 2014. Jazz presenters in Montreal capitalized on the event concept to create their own successful festival.

The Montreal Jazz Festival

In the late 1970s, people would not come out for a jazz concert and Montreal jazz clubs were folding. When legendary bassist Charles Mingus came to town, only six hundred seats were sold, and the concert's producers lost money. Yet, in 1980, the same producers presented Montreal's first jazz festival, launched as a weekend event with a budget of $250,000. It attracted 12,000 Montrealers. By the next year, 22,000 listeners appeared; by 1990 the festival had hit the one million attendance mark for a staggering schedule of three hundred concerts. "That's the difference a festival can make. It makes people pay attention," said Alain Simard, president and cofounder of the event. The festival offers a two-week combination of free outdoor performances, ticketed indoor concerts, special film screenings and premieres, themed concert series, cabaret performances, and more. Performances are staged inside and outside the city's spectacular Place des Arts complex and in nearby auditoriums as well.[11]

By 2012, the Montreal Jazz Festival had become the largest such festival in the world, attracting world-renowned performers and a wide variety of performing groups to 14 concert halls around the city. Music starts at noon and runs until midnight, but there are after-hours jam sessions that go until 3:00 or 4:00 a.m. The festival attracts more than two million visitors, about two hundred and fifty thousand of whom are tourists.

NY Phil Biennial

In 2013, the New York Philharmonic announced the creation of the NY Phil Biennial, a ten-day festival (every two years, as the name implies) that Philharmonic officials describe as a "veritable playground of new and recent music from around the world." Says music director Alan Gilbert, "For a contained period of time, audience members can get a glimpse of what we feel has been exciting in the recent past." The orchestra chose the word "biennial," said Gilbert, "to imply that the festival was permanent and to convey the feel of a broad international survey. We want to provide for music a nexus and rallying point that the great biennials in the arts world have become." The project extends efforts by the orchestra, which has sought in recent years to fend off critical perceptions that it is stodgy, to program more contemporary music and make an initiative in the search for more relevance and newer audiences.[12]

THE ORGANIZATION'S PERSPECTIVE ON PROGRAMMING

Each organization must take many concerns into account when determining its programs. Even if a particular program meets artistic-related criteria, other issues add to the complexity of the programming decision. For an artistic director, the process boils down to balancing personal vision, audience impact, and costs, among other factors.

THE ORGANIZATION'S MISSION AND OPPORTUNITIES

When evaluating current or potential product offerings, managers will want to ascertain that they are working to realize the organization's mission and are taking advantage of its best opportunities. To do so, managers may wish to use the criteria of *centrality*, *quality*, and *market viability*. Centrality is the degree to which a program or activity is central to the organization's mission. The quality of the program or activity should be evaluated relative to the organization's own standards and those of its competitors. Market viability is the degree to which the market for the program or activity is sufficient in size and growth potential.

These issues may not all carry equal weight for a performing arts organization. For example, an orchestra with a strong mission to educate its public may choose to program more contemporary and unfamiliar music despite the resistance of a large segment of its audience (high centrality, low anticipated market viability). On the other hand, consider a symphony orchestra that presents orchestral, chamber, and solo recital series of classical music. The director is considering adding a jazz series to its programming because there is sizable demand; it is likely to increase revenues, may stimulate some interest in classical programming, and is not in conflict with the orchestra's mission (high quality, high market viability, neutral centrality).

COST CONSIDERATIONS

Many organizations not only have to manage within the financial constraints of the current season, but carry deficits from previous seasons as well. Ardis Krainik, former, long-standing general director of the Lyric Opera of Chicago, attributed her organization's firm foundation to rigorous financial vigilance. Ms. Krainik always determined her season within the constraints of a balanced budget, which frequently meant "dusting off" an old production rather than creating a new one.

After suffering severe financial problems for many years and with little prospect for increasing contributed income, managers at the Vancouver Symphony Orchestra (VSO) concentrated on improving the returns on performances. To do so, they examined the economies of previous program offerings and estimated the costs of future programming options. A procedure called *marginal contribution analysis* was used to assist in the process. The steps in the procedure are:

1. Calculate revenue from ticket sales, fees, concessions, sponsorships, and restricted grants.
2. Less (subtract) variable costs (promotion, guest artists fees, production expense, materials).
3. Add to fixed overhead (musician and staff salaries, building costs, marketing expenses, etc.).
4. Divide by the number of orchestra services consumed by the activity (a rehearsal or a performance is one service).
5. Equals per-service contribution to overhead.

Managers should use the results from those calculations in the following way: If the result is a negative number, the activity should be discontinued (unless it has extraordinary redeeming artistic or community value); the overall financial result would be improved by not doing it. If the number is positive but low compared to other activities, and a superior alternative use exists, the services used by the activity should be reallocated to the alternative. Techniques such as marginal contribution analysis help managers keep their organization on track financially while making programming decisions.

MANAGING PRODUCT LIFE CYCLE STAGES

Over time, managers need to periodically adjust or reformulate their marketing and positioning strategies. There are ongoing changes in the environment, such as evolving audience preferences, growing competition, social media, and other high-tech advances. The organization may undergo a major internal change such as a move to a new venue or the arrival of a new artistic director. The organization is also subject to change according to its life cycle stage—from introduction to growth and maturity, and, one hopes, to staving off decline. The organization itself, its core products, and its augmented products are all subject to life cycle analysis.

INTRODUCTION STAGE

In the introduction stage the primary challenge is to build awareness of the organization and its product offerings. New organizations will rely heavily on public relations and other low-cost promotional approaches.

GROWTH STAGE

During the growth stage the organization capitalizes on the audience's strong response and seeks to develop patron loyalty. It may add new product features and benefits such as visible improvement in the performance hall, higher-quality production values, and a greater focus on customer service. Investing more heavily in production values may help to garner awards, which helps an organization earn the label "best" in some category. The organization may add new products such as matinee concerts or plays performed on a smaller, second stage to attract different audiences. It may consider new distribution channels, such as performances in neighborhood churches or on local cable channels, collaborations with area businesses, or tours.

MATURITY STAGE

When the growth rate slows, often to the point where managers are working harder just to maintain past years' audience levels, the organization has entered the stage of maturity. It can try to expand its number of users by converting nonusers, by encouraging more frequent use among current users, and by increasing the amount of use per occasion—for example, by encouraging patrons to bring friends or family members. The organization can also stimulate demand by modifying the product. This can take the form of quality improvement, usually by investing more heavily in production values; or feature improvement, such as adding multimedia technology to a performance. The marketing director should also consider what modifications could be made to each nonproduct element of the marketing mix to stimulate demand.

The organization must determine which tools would be most effective for meeting its needs. Each organization faces trade-offs and must weigh the relative benefits and costs of each opportunity. A SWOT analysis, as described in chapter 5, would be very useful in helping to make this analysis.

Piccolo Teatro: Capitalizing on New Technology to Build the Audience

Piccolo Teatro of Milan was founded in 1947 as the first public theater in Italy. From the 1940s to the 1980s, Piccolo grew steadily, to a point where it was offering more than 30 shows each season and had more than 10,000 ticket holders. The subsequent changes in the social, political, and economic scenarios led to a more fragmented public in the world of culture and theater in Italy. Within the context of these changes, the development of new technologies and the advent of the Internet represented a crucial step in Piccolo's marketing strategy. In particular, in response to the economic crisis and to take advantage of technological innovation, Piccolo chose to focus on (1) creating an offer specifically targeted to the public through increasingly flexible season ticket options, (2) developing new educational activities for the public in general and for students in particular, (3) introducing a customer relationship management (CRM) system fed by continuing off- and online surveys held among its audiences, and (4) developing its web system.

In 2012, Piccolo performed more than 60 shows in its three theaters (Teatro Strehler: 980 seats; Teatro Grassi: 500 seats; and Teatro Studio: 370 seats), staging classical and contemporary drama, dance, music, opera, and special performances for the young. In the 2012–2013 season, Piccolo sold 288,000 tickets, of which approximately 100,000 came from the sale of more than 22,000 season tickets, the highest number in the theater's history. Of these tickets, about half come from groups and students—preschoolers to postgraduates.

A wide range of educational activities is arranged every season, including various training programs for students and their teachers. During the workshops the shows being staged at Piccolo are used to illustrate the different theater disciplines and to bring the theater experience to students of all ages. The completeness of these in-depth study projects is one of the reasons why 45 percent of the theater's public is made up of people under age 26. Approximately 50 percent of patrons come from outside Milan.

In 2001, Piccolo opened its Community, an email list of contacts who receive a monthly newsletter with information and special offers for Piccolo performances. The "Community del Piccolo" has been a huge web success, generating growing sales from more than one hundred and twenty thousand registered members in just over ten years. The possibility of contacting people by email and the constant evolution of the Internet led to the creation of new services for registered members of the Community, such as digital tickets with a print @ home barcode, which eliminates the need to collect tickets from the box office; the create your own "virtual agenda" with automatically generated messages that remind you when to go to the theater; and the opportunity to receive an email with a brief satisfaction questionnaire the same evening as the show, which enables Piccolo to "listen" and to profile its audience more effectively.

Given the success of online sales through the Community offers, in 2008 it was decided to create a real online sales channel connected to the database of Piccolo's Customer CRM system.

The sales site was named after "PiccoloCard," the loyalty card that is issued free to all buyers and that allows them to accumulate points to use on future tickets and get discounts and special services in the cafeterias or library of the Piccolo Teatro. By July 2012, Piccolo had more than 50,000 PiccoloCard holders, all of whom were added to Piccolo's CRM database system. The sales site, together with the Community channel, grew to generate sales of 50,000 tickets per year.

Moving the sale of 4,000 tickets away from the Community toward independent sales through the PiccoloCard site brought about an increase in revenue as people were now buying full price tickets, as opposed to the discounts generally available through the Community site. From 2008, when the PiccoloCard site was begun, to 2011, the average price of tickets rose from 15.40 euros to 19.1 euros, an increase of 24 percent. During that period, total online sales rose 47.4 percent. The success of sales through the PiccoloCard site still generated an increase in members registered on the Community site however, as purchasers are required to register, for free.

In the 2012–2013 season, the PiccoloCard site was integrated inside the main website www.piccoteatro.org. With this change, the customer remains in the Piccolo website and can more easily find and purchase tickets, choosing his or her preferred seats in an interactive seat map. This simplification of the purchasing process increased web ticket sales by 20 percent in just one season.

Other features of Piccolo Teatro's initiatives to stay relevant in a changing economy, society, and technological environment will be discussed in the chapters on branding and high technology.

Piccolo Teatro is a mature organization that has thrived, grown dramatically, and served ever more diverse audiences by perpetually staying in touch with people's preferences and needs and by capitalizing on the technology that has made much of this growth possible.

Decline Stage

At some point many products outlive their value to consumers. If a product or organization has not sustained itself or been rejuvenated through the maturity stage, its sales will eventually decline. There is a great deal of concern that as more orchestras than ever are facing shrinking audiences, musician strikes, huge deficits, and bankruptcy, the whole field is at risk. As a result, many orchestra managers are reevaluating their organizations' roles in society, their programs, and their modes of presentation.

The most striking example of augmented product decline, pervasive in the performing arts industry as a whole, is the erosion in recent years of the full-season subscription. Some organizations have been able not only to maintain but also build their full-season subscriptions with their high-quality offerings and attractive benefits. But for most of the industry, alternative packages such as mini-plans and an emphasis on single ticket sales online have served to maintain or build audience size in the face of declining subscriptions. Creative arts managers and marketers who think broadly enough to imagine what changes will attract new attenders and build frequency among current attenders are likely to avoid decline. Adaptation must come from within the organization.

A state of decline does not necessarily portend a death knell for an arts organization. Those who think broadly enough to imagine how artistic integrity can be maintained within the context of sweeping changes are the most likely to avoid decline. For the rest, the future is at risk.

CHAPTER 9

MANAGING LOCATION, CAPACITY, AND TICKETING SYSTEMS

WHEN WE THINK ABOUT A PERFORMANCE, ITS SETTING TYPICALLY COMES TO MIND. Says Alan Brown, "Settings may be formal or informal, temporary or permanent, public or private, and physical or virtual. In the broadest sense, 'setting' is a sort of meeting ground between artist and audience—a place both parties occupy for a finite period of time to exchange ideas and create meaning."[1] The setting plays a significant role in that it influences both the art itself and the audience response.

A venue can serve to draw people in or to keep them away. A world-class venue will attract many people in and of itself. A lack of accessibility and comforts will serve as barriers to attendance. Is the location convenient or otherwise desirable? Does the venue provide comfortable seating, adequate bathroom facilities, nearby parking at a reasonable cost, and other significant features? Is it in a safe, well-lit neighborhood? Does it provide access for disabled people? In a study conducted with performing arts attenders in New Orleans, 70 percent of respondents said that venue was an important factor when deciding which events to attend.[2] In order to remain relevant, arts managers are recognizing the need to reconceptualize the relationships between their programs and their spaces in order to reach diverse audience segments.

Audiences not only identify organizations closely with their performance venues, but often depend on consistency of location. In the mid-1990s, the San Francisco Ballet and A.C.T. performed in alternative yet convenient venues while their usual performance halls were undergoing extensive repairs following damage due to an earthquake. During this transitional time they lost many audience members who eventually returned when the venues reopened. In chapter 7, I showed how Chad Bauman carefully implemented strategies to retain his audience during a temporary venue change.

Organizations that perform in a variety of halls, especially presenting organizations that offer a wide variety of performing groups, have difficulty building a strong brand identity and a loyal following for the organization itself. Says Alan Brown, "While multipurpose venues can expand access to the arts, important connections between arts and setting have been lost."[3]

Performing arts venues vary from school auditoriums, community houses, and store-front theater spaces, to grand halls renowned worldwide like Carnegie Hall and the acoustically outstanding and architecturally significant Walt Disney Concert Hall, designed by Frank Gehry for the LA Phil. At the Sydney Opera House, approximately 30 percent of the patrons buy tickets just to be in the hall, no matter what is being performed. Not only are the productions in such halls world-class in quality, but the venues themselves have tremendous symbolic, social, and political significance that serves to attract audiences.

The Metropolitan Opera House

The Metropolitan Opera House, according to music critic Alex Ross, is "an extravagant point in space"[4] with an oversized budgetary and artistic scale. Much of the repertory both before and after the nineteenth century's elaborate operas is ruled out there, as small-scale sets are virtually lost on the Met's gargantuan stage and in the house's cavernous interior. Having a four thousand–seat house also means the collective taste of a huge public must be satisfied with the productions. With a big house, the stakes are higher and fewer chances can be taken. Some critics say this necessarily leads to a least-common-denominator approach to both repertory choice and set design. An adventuresome late twentieth-century production of Benjamin Britten's *Death in Venice* presented in the 1990s drew a mere two-thirds capacity audience. Many an opera company would be thrilled with an audience of two thousand six hundred people. But for the Met, it was a near financial disaster, and plans to bring the work back the following season were scrapped. The musical worldview of music director James Levine dominates the Met's repertory selections, but there is no question that the place itself is a crucial factor in deciding what will be produced and how each opera will be presented.

In recent years, the Metropolitan Opera has spread far beyond the limits of its Lincoln Center venue to reach out to audiences worldwide, thanks to advances in high technology. Met performances are available on YouTube and on the Met's website, via On Demand, on the Met's iPad app, and on CDs and DVDs through the Met's Opera Shop online, and are offered in high definition at movie theaters internationally. Through these showings, people who live in small cities and towns where live opera is not available can now watch highest-quality opera on the big screen in their local theaters, with the added features of performer interviews and backstage tours during intermissions. These performances are also offered in big cities that have their own opera companies, possibly eroding attendance at the local venues. The effect of these showings and those offered by some other operas, symphonies, and theaters on live local productions remains to be seen.

Site-Specific Performance Venues

At the other extreme of performance venues are outdoor sites like parks and piers, which provide ideal backdrops for the performances of Dancing in the Streets. Since 1984, Dancing in the Streets has presented site-specific performances, pioneering the union of choreographers, public sites, and communities and making the performing arts a vital and enriching element of public life. Every July, Dancing in the Streets offers families its three-week Wave Hill performance series in the renowned public gardens above the Hudson River in the Bronx. And each May the organization presents the Young People's Performance Festival at the Beard Street Piers in Brooklyn, in which public school children collaborate with local arts organizations and established artists. From Grand Central Terminal to nineteenth-century warehouses on the piers of Brooklyn's Red Hook, Governors Island, and the Roosevelt Island ruins, Dancing in the Streets' productions engage communities directly in unexpected places and their productions challenge the boundaries of where performance belongs and whom it can involve.

The Long Beach Opera (LBO), which holds most of its performances in Long Beach's theater spaces, also features unique, unlikely "Out of Bounds" locations for some productions. In 2006, *The Diary of Anne Frank* by Grigori Frid was performed in underground parking garages as a metaphor for dislocation and abandonment. In 2008, artistic director Andreas Mitisek staged Ricky Ian Gordon's *Orpheus and Euridice* in an Olympic size swimming pool; and in 2009 he staged a double bill of Viktor Ullmann's *The Emperor of Atlantis* and Carl Orff's *Die Kluge* in the former engine rooms of the Queen Mary. In 2013, LBO's "Outer Limits" production of *King Gesar* by Peter Lieberson, which recounts the story of a Tibetan warrior king who rose from obscurity to battle the demons that enslave humankind, was performed as a "campfire opera" under the stars at Harry Bridges Memorial Park.

For the marketer, *location* implies three different meanings. First, it refers to managing the benefits and constraints of the organization's own performance venue in efforts to realize audience-building and customer-satisfaction objectives. Second, location refers to all the places an organization can consider performing or providing lectures and demonstrations. Third, location may refer to all the ticket distribution sites and methods the marketer may use to make the product offering available to the public. Decisions about facility size and features, performance location, and ticket delivery systems should be consciously related to the organization's overall marketing strategy and its specific marketing objectives.

THE PERFORMANCE VENUE

When organizations consider how to serve their markets most effectively, their thinking about distribution patterns and systems is usually colored by their existing investment in facilities. They consider how to attract people to their current facilities, selected at some point in the past for reasons that may or may not be relevant today. Grand halls for orchestras and opera companies were built in the beginning or middle of the twentieth century in the central locations most preferred by the social elite. These facilities tend to have major symbolic, social, and sentimental significance for those who have traditionally provided financial support and volunteer leadership for the organizations. But for many organizations, past decisions about location have created imbalances in current times. In recent years, some organizations have taken on the challenge of building new, expensive performance halls that place a huge financial strain on the organization and its supporters. As arts organizations try to broaden their audience base to include multicultural populations and younger people, they may have to reach out to those audiences in areas and venues that these people find comfortable and familiar.

In our modern society, consumers increasingly expect a great deal of choice in their leisure pursuits. While many people, especially traditional performing arts attenders, enjoy sitting back and quietly taking in a live performance, such static experiences are less desirable to younger generations who like to get up for a drink, talk with their companions, and text or tweet about their experiences during the performances. Alan Brown reports about an "imaginary tour" of a hypothetical jazz venue, developed in a focus group session of young adults. "During the day, the venue would be open as a coffee house and music lounge, where anyone can come to hear, share, and acquire music. At night, it would transition to a venue for live concerts where patrons can move fluidly between different spaces designed for intensive listening, 'partial-attention' listening, and socializing while watching the concert on a large screen. This need to offer consumers more opportunities to personalize their experiences has implications for the art itself, in terms of a diminishing audience for what some consider 'passive' experiences, and most likely foreshadows waning interest in the more restrictive setting in which professionalized art is offered."[5]

The organization's uniqueness and the nature of its product also affect the decision about location. People readily travel hundreds of miles to attend the Oregon Shakespeare Festival and thousands of miles for the Bayreuth Festival. But organizations that consider moving their venue from one part of town to another must take into account that they are likely to lose a significant portion of the audience members who prefer the "old" neighborhood. And organizations that offer select performances in an alternative location to attract new audiences usually find that most of these new patrons are not motivated to attend performances at the regular venue, no matter how much they liked the show. When an organization loses its performing venue and is forced to make a move, it is important to promote the benefits of the new space to patrons, rather than be apologetic. For example, the new space may be further from home and work for many patrons, but it may have superior acoustics, more convenient parking, and a better choice of restaurants nearby.

Whatever the nature of the venue, it is up to the marketer to ascertain that the facilities are responsive to audience needs with comfortable seating, adequate restroom facilities, and efficient box office lines. Audience-responsiveness also affects the design of circulation spaces, lobbies, education spaces, concession stands for refreshments and souvenirs, and spaces for special events. Organizations should capitalize on opportunities for community building with spaces that are more sociable, intimate, informal, and comfortable. Typically organizations focus on two primary factors when trying to grow and retain their audiences: programming and price. Clearly, facilities can and should play a more central role in the lives of their communities by engaging people in new and exciting ways.

MANAGING CAPACITY

The degree to which available seats are sold is critical to the artistic and financial success of an organization. Posting *Sold Out* or *Standing Room Only* signs indicates that the organization is successfully meeting its publics' needs and interests, but also means that it probably could sell more tickets if capacity allowed.

Capacity utilization can vary according to several factors. First is the degree of fluctuation in demand; a heavily subscribed organization is far less susceptible to fluctuating demand than one that relies primarily on single ticket sales. The popularity of a show is the key element in demand. An organization that offers a wide variety of programming will experience high fluctuations; an evening of contemporary chamber music will not attract the same size audience as Beethoven's best-loved symphonies do.

In most other industries, supply can be adjusted to meet demand. For example, grocery stores provide more shelf space for the more popular cheddar cheese than for blue cheese. But arts organizations, driven by their artistic vision to offer less-popular works, even though they will draw smaller crowds, must creatively manage the fact that their costs and capacity remain virtually the same no matter what the ticket demand. This is a major dilemma that the nonprofit arts have and will always face. Sometimes organizations are able to manage these issues by using a second stage at their own venue for certain productions, by offering performances at another location in the area, or by taking performances on tour.

Capacity utilization also varies according to the degree to which capacity is fixed. Fixed capacity is a function not only of the number of seats in the hall, but also of the feasibility of extra performances during and after the show's normal run. Just beyond the status of *sold out* is the point where patrons must be turned away. Once demand exceeds capacity, an organization is exposed to a potential loss of business. In addition to the negative financial implications, this impacts the organization's ability to expose its artistic product to the largest possible audience, a factor that may be central to the organization's mission.

Organizations that could sell more tickets than they have seats for can either offer additional performances or consider a move to a larger hall. Both of these choices involve serious cost considerations. Additional performances incur the variable costs of expenses on a production-by-production basis, such as for hall rental, artist fees, and marketing. A move to a larger hall, even if a good choice exists in the area, will incur the additional costs typically charged by larger facilities, in addition to the construction costs. Building a new facility or an addition to the organization's own hall requires serious, honest analyses of potential fund-raising and projected future demand. Too many organizations have taken on capital projects that couldn't be paid for and that exceeded the organization's needs into the future.

More commonly, organizations perform in halls where the capacity exceeds their usual demand. Too many empty seats in the house can have a devastating effect not only on financial return but also on the quality of the experience for both the audience members and the performers. A full house makes the experience more festive and compelling. People like to be a part of something highly sought after and tend to think there is something wrong when many seats do not fill. As Charles Dickens wrote in *Nicholas Nickleby*, it is "a remarkable fact in theatrical history, but one long since established beyond dispute, that it is a hopeless endeavour to attract people to a theatre unless they can be first brought to believe that they will never get into it."[6] Marketers must take care, though, not to overstate the scarcity of tickets. People may think they'll never be able to get tickets so they do not even try.

The choice of performance venue may be a function of the availability of a space that meets the organization's production needs for such things as acoustics, sets, lighting, and rehearsal space, rather than a function of the space's alignment with current and potential ticket sales. In one midsize city, the local opera company performed for several years in a hall with 2,500 seats but sold only about 700 tickets per performance. In such situations there are several strategies marketers can employ to improve the experience. If no smaller hall appropriate for opera productions is available in the area, managers can consider closing off balconies so the main floor, boxes, and mezzanine are relatively full, and keeping the balconies dark so people are unaware of them.

Marketers can also distribute free tickets (comps) for a sparsely sold performance to artists, colleagues, staff, and other special groups of people who are likely to serve as opinion leaders and spread the word about the organization. They can also offer deeply discounted tickets, preferably to select target groups such as students. Of course the best solution, if possible, is to perform in a venue that better matches audience size.

ALTERNATIVE VENUES

Many arts organizations worldwide are responding to the idea that they should bring art to the people before they can hope to bring people to the art. The purpose is to show people in compelling ways that the performing arts can make a difference in their lives. Some fine examples of bringing classical music to people in meaningful ways are the performances during memorial ceremonies for the victims of the September 11, 2001 tragedy; the performance by Yo Yo Ma at the opening ceremonies of the 2004 Olympics; and local orchestras performing at halftime during regional sporting events. Additionally, music groups may perform at area churches (then invite church members to a concert, often at a significant discount), present performances and discussions in the lunchrooms of corporate offices, offer free concerts in a park during the summer, and engage in other such projects that expose people to the arts in their own world.

Touring enables the organization to share its performances with different audiences and serves to establish a regional, national, or international reputation and raise the organization's public profile. It also provides enrichment for the performers as they are exposed to other arts professionals and new audiences. Touring is usually a very costly endeavor but is well worth the expense when the conditions are right. Orchestras and dance companies that have year-round contracts with their artists often depend on touring and summer venues, such as Tanglewood for the Boston Symphony, to provide them enough work throughout the year. Sometimes performing arts organizations can share the financial risks—and revenues—of a production on tour with the presenting organization. Each organization must carefully analyze all the variable costs of each tour in advance of committing so that there are no surprises of having costs exceed revenue. Some organizations undertake major fund-raising in advance of tours to support these projects. People often like to support touring financially as they enjoy how performances on tour build respect and visibility for the organization outside its hometown.

A recent phenomenon that has caught on in all kinds of unexpected places is the *flash mob.* A flash mob is a group of people who assemble suddenly and seemingly randomly in a public place, then begin a preplanned performance. A classical choir sang the *Hallelujah Chorus* in the food court of a shopping mall during the holiday shopping season; the Copenhagen Philharmonic performed Ravel's *Bolero* to an unsuspecting and delighted "audience" of people waiting for trains at Copenhagen's Central Station; members of the Longwood Symphony Orchestra "spontaneously" performed Bach's *Brandenburg Concerto #3* to surprised patients and staff at the Dana Farber Cancer Institute, providing music theapy and a joyous interlude during a period of difficult medical treatments. Similar events have taken place in a supermarket in London, in a plaza in Sabadell, Spain, and many other places, surprising, entertaining, and brightening the day of people young and old who just happened to be there.

TICKET DISTRIBUTION

In addition to determining the location and nature of performance spaces, managers must make strategic decisions about how and where tickets to performances will be made available. For some consumers, ease of access to tickets and information about availability is crucial if they are to actually make the ticket purchase. For regular attenders, satisfaction with the ticket purchasing process is increasingly important to their commitment and loyalty. In a study of performing arts attenders undertaken at the University of New Orleans, researchers found that the more effort it takes to attend an event, the harder it is to attract an audience. Ticketing procedures were cited as a deterrent by 44 percent of respondents.[7]

ONLINE TICKET SALES

Buying tickets online has become the most popular and widely used purchase method. By administering a survey to thousands of patrons from several of its client arts organizations around the United States in 2012, Patron Technology investigated how people prefer to transact their ticket purchases. The researchers found that 66 percent of patrons purchased tickets online, up 26 percent since 2007. Nineteen percent of the respondents called the box office to purchase tickets, 9 percent visited the box office in person, and 6 percent ordered tickets through the mail. According to Roger Tomlinson, this means that ticketing, which used to merely close the sale, can now also be the launching pad for patron communication and recognition—often mobile and personalized—and lead to greater customer loyalty and higher sales, when, for example, an online ticket buyer alerts her Facebook friends to her purchase. The ability to make purchases through many channels puts customers in control, buying what they want, when they want, and how they want.[8]

A 2012 study by Group of Minds.com found that 44 percent of arts emails were opened on smartphones. Analysis by theaters of online ticket sales has shown that as many as 36 percent come from mobile devices. So now, marketers must prioritize thinking about how their patrons access their emails, websites, Facebook updates, and Twitter feeds from their smartphones or tablets and use these devices to seek further information and to place orders.

Not only must marketers consider the user's devices in designing messages, but different messages should be customized and sent to various people depending on the stage of their relationship with the organization. For example, send different messages to people whose performance is coming up soon (Be sure to see this special show!) than to those who have already seen the show being promoted. (If you loved this show, share this link with your friends!) To engage in this aspect of CRM, art marketers should ask themselves two questions: (1) The customer knows and remembers what his relationship with us; do we remember what our relationship is with him? (2) If we contact the customer on his smartphone, we are reaching him personally; is our message relevant to him, specifically?

Recently, I purchased full-priced tickets online on a Wednesday for a theater performance that Saturday. On Thursday, I received an email from the theater saying: "It's not too late! Buy tickets for this weekend at a discount of 35 percent!" I emailed the managing director and strongly suggested that he use his databases's filtering system so that such offers are not sent to people who have already purchased tickets. He agreed, thanked me, and offered me a 35 percent refund. The promotional email I had received was not relevant to me, but the managing director was appropriately concerned about his relationship with me. I am certain that most patrons would not go to the trouble of contacting a manager with a complaint such as this and that they would be silently disgruntled with the organization about the situation.

With the customer in control, she can *pull* the content she wants and the organization can *push* content specific to her and make the experience personal. The user's login can go straight to the organization's database, making it possible to send back content that is relevant to her personally. Most systems also let customers see their booking history and manage their account for updating information. With ticketing systems this personal, marketers can add value to the ticket purchase with offers such as vouchers for program books, interval drinks, and parking. Marketers should also be sure to alert patrons to any anticipated traffic disruption on their way to the venue and make special offers at local restaurants before or after the show. Technology, used smartly and creatively, enables organizations to be more open and accessible to their customers in a way most customers appreciate.

In 2012, Capacity Interactive developed a digital media subscription offer for its client, New York City Center. The offer consisted of using retargeting technology, which means advertising online to people based on their previous Internet actions, in cases where these actions did not result in a sale. After one season of using retargeting to sell single tickets for *Encores!* performances, the organization had a pixel pool of *Encores!* single ticket buyers, who were considered to be the best prospects for new subscriptions. Instead of saving their addresses and sending costly direct mail, marketers served banner ads to these users, plus any users who had visited the *Encores!* subscription pages. Also, ads were served to the social connections of the *Encores!* single ticket buyers because theater fans usually have friends who also enjoy theater. No other efforts were made except one Playbill.com sponsored email. The strategy was to test if digital ad media alone could move new subscriptions. The result of the campaign was that subscriptions were sold to more than 230 new households. The ROI (return on investment) was 846 percent based on the combined media cost of these two efforts. If direct mail were included, it may have eroded this profitability since it is so expensive to execute.[9]

The development and adoption of new technologies is changing so quickly that it is up to marketers to stay current with technological advances and behavioral trends.

Ticketing Features

As online ordering has evolved over the years, people expect arts organizations to offer key features that were not possible under older systems. People want to be able to select their own seats from an online seating map. Special offers with discount codes are a welcome feature. Patrons also expect to be able to print out their own tickets at home. People would like to be able to make exchanges online, pay to upgrade their tickets to better seats, and forward an unusable ticket to a friend or put the value of that unused ticket into a credit account for the future. Subscriptions are available online at some organizations, but this has not yet become the norm.

Social Media and Ticketing

Word of mouth is what has historically happened *after* someone has seen a show and discussed it with friends. With Facebook, Twitter, and other social media, people can post what tickets they have bought and encourage friends to come to the same event, thereby building word of mouth *before* attending. Since word of mouth has always been the best resource for building sales, personal endorsements both before and after seeing a show can be powerful.[10] According to HubSpot, 90 percent of people believe brand recommendations from friends, 70 percent believe consumer opinions, and 71 percent of people are more likely to make a purchase if referred by friends. Importantly, only 7 percent of people make a purchase if not referred.[11]

By increasing the ease and frequency of connections, social media offers a big boon to marketers. Some international airlines are offering ticketed passengers the opportunity to link their Facebook page to that of others on the same flight.

With this feature, people who so choose can read about others' backgrounds and interests and see if there is someone they would like to sit near or otherwise meet. Such an offer from arts organizations could help singles meet others with similar interests, could help seniors match up with someone for transportation to and from the venue, and so on. It is up to marketers to do their best to stay up to date with evolving technologies such as this.

THE BOX OFFICE

The organization's own box office can be a key link with the customer. The well-informed ticket seller should be able to provide detailed information about the productions and venue and should be prepared to respond to a variety of customer questions and special needs. Box office personnel may use the call they receive as an opportunity to suggest additional programs to the caller. The phone works both ways. Box office personnel can mine their database to identify patrons who may be interested in an upcoming production or a special offer. Patrons can be called or emailed with messages that show the organization understands their interests and needs, from the types of performances they like to their preferred day of the week and seating location.

CENTRALIZED TICKET AGENCIES

Some arts organizations, especially those that rent a venue that comes with its own ticketing agency contract, use intermediary ticketing services, such as Ticketmaster. These services offer ease of setup and operation but are far from ideal from the marketing perspective. They typically charge customers high fees, which has negative implications for the organization's pricing strategy and customer satisfaction levels. Organizations strategize their own prices so carefully that what is considered the "right" price can easily be sabotaged by hefty ticketing fees. Also, centralized agencies usually do not share their data with the arts organizations, so the marketing department cannot access all-important patron information. Furthermore, ticket agency personnel, who serve multitudes of venues, cannot be expected to provide information about each organization and its offerings, something the organization's own box office personnel should be trained to do.

PARTNERSHIPS

Performing arts groups may find it advantageous to collaborate with other arts organizations in their area to form a centralized ticket agency. Hours of operation could be extended beyond those of each organization's box office if many organizations share the costs. Specially trained and well-informed salespeople could provide valuable information to callers about each organization's offerings. Furthermore, the organizations could share patrons' names and addresses for marketing purposes and, when appropriate, develop joint offers, such as a sampler series.

The Pittsburgh Music Alliance (PMA) is a data-sharing partnership among the Bach Choir of Pittsburgh, Chatham Baroque, Pittsburgh Camerata, Pittsburgh Chamber Music Society, and Renaissance and Baroque. Formed in 2011 with a three-year grant, the PMA strives to grow audiences for all five organizations through shared data. Not only have ticket sales increased across all five groups, but an artistic cross-fertilization has also yielded programs of intersecting themes to deepen the shared audience's relationship to the music. Competition is not an issue for organizations that partner in this way. Studies show that arts attenders who attend one performance tend to attend many and that satisfaction increases as does cultural attendance.[12]

Ticket Discounters

Some cities offer their local arts organizations a centralized place to make tickets available at a significant discount. New York has its TKTS booth in three locations; Chicago has its HotTix booths and online sales at half price. Typically these tickets are offered only the day of the show, or the previous day for matinees. Goldstar, which serves about 30 US cities, sells tickets online for performances, boat cruises, sporting events, circuses, and the like. Tickets are usually available on Goldstar as much as a week before the show. Seat availability is usually limited to rear sections only, but this depends on the organization and the available seating capacity for the programs. At Goldstar, some tickets are half price, some even lower. Organizations decide week by week if they will offer tickets on such sites and if so, how many. It is beneficial for the organizations to sell seats at discounts that would not otherwise sell, and in the case of such outlets as Goldstar, the arts groups are reaching people who are not strictly arts goers who might be interested in seeing a promoted show, especially at a reduced price. The major problem with repeated discount offers is that organizations are "training" their patrons to wait until the last minute to buy discounted seats, rather than buying in advance at the regular price.

Choosing a Ticketing System

The best ticketing systems are those that are not just for ticketing, but manage the donor database and virtually all of the customer data. These customer relationship management systems enable arts organizations to have all their data in one system so they can act on it for better marketing, customer service, and fund-raising.

There are literally dozens of ticketing systems available for arts organizations to explore. The marketer's first step should be to review the systems used by other similar-sized organizations and those "aspirational" systems used by organizations with larger budgets. Then the marketer should consult with peers in other organizations. Find out what they like and don't like about their systems. Investigate online the features offered by several of the systems and select about three to research in depth. Marketers should rate the importance of various features to the functioning of their organization and make their choice accordingly. Remember that an upgrade in ticketing and CRM software will be an investment for the organization—one that is most likely to pay off well over time, if not in the first year or two.

There is a wide range of costs, usually depending on the features the system offers. At the high end, Tessitura features ticketing, subscriptions, fundraising, memberships, marketing, comprehensive CRM, reporting, and web services. PatronManager CRM offers a full range of benefits similar to Tessitura, is more user friendly, is less costly to use, plus it features a Patron Mail system for designing and sending mass email blasts, at a price that is affordable for most organizations.

At the low end of the cost spectrum is Brown Paper Tickets, which charges each ticket buyer 99 cents plus 3.5 percent of the ticket price, but charges no fees to the organization. Brown Paper Tickets promotes its services for events, and is probably much more appropriate for single or infrequent events than for ongoing seasons of performances. It offers several ticketing-related benefits and shares customer information with the organization, but does not have a CRM plan that integrates ticket purchase with donations, email, or other customer-centered functions.

CHAPTER 10

Focusing on Value and Optimizing Revenue through Pricing Strategies

ARTS MARKETERS FOCUS A GREAT DEAL ON PRICE, THINKING THAT PRICE DRIVES ticket purchase decisions. What people care about more than price is *value.* Said investor Warren Buffet, "Price is what you pay; value is what you get."[1]

Marketing guru Seth Godin claims that when people say "I can't afford it," they aren't making a true statement. Says Godin, "At least it's not true almost all the time. Very few of your prospects literally can't afford it. What they are really trying to say is, 'it's not worth it.'"[2]

How Much?, a pricing research project at Sheffield Theatres in England, was undertaken to investigate the importance of price sensitivity among young people, but researchers soon realized that price was by no means the only, and often far from being the most important barrier to young peoples' attendance at arts events. Responses from participants showed that many young people have enough money to spend, but rarely choose to spend it on theater. The research findings were summarized as follows: "For attenders and non-attenders, price was not an absolute constraint to theatre attendance, but the uncertainty of what they would get in return for their money resulted in theatre being more susceptible to financial reasons for non-attendance than other leisure activities. In other words, like many people, they were using price to rationalize their lack of understanding of the value that the experience offered."[3]

One way to increase value is to lower the price. The other way to increase value is to increase perceived or actual benefits. Too often, organizations take the route of offering discounts instead of improving and/or communicating well their benefits. Price discounts offered broadly (not to specific segments such as students) are short-term tactics that are likely to have negative long-term implications for the organization.

In the performing arts, pricing is an especially complex issue. In most industries, prices are determined largely by costs so that profitability is ensured. In nonprofit arts organizations, income generation is a means of fulfilling the arts organization's mission, rather than an end in itself. Arts managers must continually work to balance their seemingly contradictory goals of audience maximization and revenue maximization.

Issues in Formulating Pricing Decisions

In formulating pricing decisions, arts managers must take several factors into consideration:

- The costs faced by their organization
- The costs faced by consumers relative to their perceptions of value
- The organization's pricing objectives, which are determined on the basis of its long- and short-term goals

Understanding these factors helps managers choose the most effective pricing strategies.

The Costs to the Arts Organization

Intangible Costs

In addition to the tangible monetary costs common to all businesses, arts organizations face certain intangible costs due to the nature of their productivity and their societal and aesthetic values. Arts organizations believe in the social value of their product, and on moral grounds, seek to make it widely available by maintaining affordable prices. Also, arts organizations are driven by the high professional standards of their artists, who often value innovation and risk, meaning narrower appeal and therefore higher cost per audience member.[4] In this sense, arts organizations face the "cost" of being true to their artistic mission rather than presenting works that are sure to have wide appeal.

Monetary Costs

Arts organizations incur two types of monetary costs: fixed and variable.

Fixed Costs. Fixed costs are incurred even if no performances are held. Institutional overhead comprises such costs as building rent or mortgage and salaries of administrators and artists under annual contract. An orchestra with a large hall and staff and up to a hundred musicians under contract has high fixed costs. Conversely, a theater company that has a small staff, rents a hall, and hires artists on a per production basis has relatively low fixed costs. Fixed costs are difficult to reduce.

VARIABLE COSTS. Variable costs are the expenses associated with each staged production or those that are more flexibly increased or reduced. They include wages of part-time actors, musicians, directors, and other temporary personnel; royalties paid; transportation; sets, costumes, and other production costs. These costs vary from show to show, depending on the number of people involved, the extent of special effects, performers' fees, and so on. Some administrative costs are also variable in nature. Marketing expenses may be adjusted according to the organization's needs and constraints, but keep in mind that as well-strategized marketing costs are reduced, revenue is likely to decrease.

The fixed production schedules of most nonprofit arts organizations exacerbate their variable costs. In any organization with a subscription series, repertory rotation, or other short-run scheduling constraints, a company cannot "milk" a successful production indefinitely by extending its run, as a Broadway theater might. Nor can a flop be closed before its appointed time to make room for a more popular show.[5]

When pricing levels are at least in part determined by a production's variable costs, managers generally employ a break-even analysis. This analysis helps to determine, for an anticipated level of demand, the break-even price or the price that must be charged to fully cover the production's costs. The break-even analysis also helps determine, for any proposed price, the break-even volume—how many tickets would have to be sold to fully cover the production's costs. Assume that the cost of staging a particular show is $275,000. Dividing $275,000 by the anticipated average ticket price (or anticipated demand) will indicate how many tickets need to be sold at each price (or what price should be charged) to break even—that is, to cover the costs of staging the show.

However, nonprofit arts organizations rarely, if ever, set their prices based on their costs. Typically, pricing strategies are set based on going-rate pricing and on customer segment pricing, rather than on the organization and its costs. Organizations depend on contributed income to compensate for the difference between the organization's expenses and its earned income, an amount that, in the United States, generally exceeds 50 percent of the budget.

Incremental costs are an aspect of variable costs. For an arts organization pricing its performances, incremental costs may be described as the additional costs involved in selling one more seat. Such costs tend to be extremely low. Box office personnel are already on hand, program booklets are preprinted, and most advertising is done through the mass media or very inexpensively online. The significance of low incremental costs to arts marketers is that the cost of selling an empty seat for a performance that is about to start is close to zero, so the incremental revenue of each empty seat sold is, in effect, the price of the seat. This is the economic justification for offering day-of-performance discounts, student rush tickets, and other promotions that sell otherwise unsold seats at a deep discount. Marketers must, however, carefully discriminate when, how, and for whom these discounted prices are offered.

Consumer Costs and Value

Today's arts marketers face a growing dilemma. As costs rise and attracting contributed income becomes more of a challenge, marketers face internal pressure to raise ticket prices. But as marketers try to expand their audiences and maintain satisfaction among current attenders, they fear that price increases will act as a barrier to their success. So, arts marketers need to ask: How important is price? And for whom is price important?

In a study of infrequent attenders at the Crucible Theatre Sheffield and Playhouse Theatre Leeds, 76 percent of respondents rated price in the second tier of issues, comparable with the importance of the performers, playwright, company, seat comfort, and parking, but well below the first-tier rating of quality of performance, entertainment value, and subject matter.[6]

For the purpose of analyzing the impact of price among nonattenders, P. Walshe drew a distinction between the *nonintenders,* those for whom pricing is not an issue since they do not intend to purchase a ticket for other reasons, and the *intender-rejectors,* those who would like to attend but become alienated as a result of pricing policy. For the vast majority of people, the nonintenders, rejection has set in before price becomes a consideration, since for these people there is probably an irreversible barrier: lack of interest. Therefore, one can conclude that there is little point in trying to use price cuts, even major discounts, to access the mass market. And generalized price-cutting is likely to work against the organization's financial best interest, since revenue would be greatly reduced among patrons willing to pay higher prices.

Arts consultant Chris Blamires suggests that pricing strategy research should be conducted on future or hypothetical events, in order to assess their appeal, rather than on events already attended. His rationale is that if one asks an individual if she or he would have paid more than the ticket price of, say, $40, to see each of the three plays attended in the last 12 months, the answer would be "of course" if the play was as good as the best play the person has ever seen and "of course not" if the play was as bad as the worst. Price, then, is a screening device used by those with a desire to attend a given event. It tests the value placed on the *promise* provided by the information received about the event.[7]

Importance of Ticket Price on Purchase in San Francisco

In the mid-1990s, I conducted an in-depth investigative market research study for four performing arts organizations in San Francisco: the San

Francisco Symphony, San Francisco Opera, San Francisco Ballet, and the American Conservatory Theater. As I described in chapter 7, our goal was to quantify and analyze the factors that influence ticket purchases and the effect specific ticket prices have on the purchase behavior of current audience segments.

The study found that 41–65 percent of respondents claimed that interest in specific productions is the most important factor in their purchase decision. Following programming in importance, 25–60 percent said that scheduling issues prevail. Only 8–29 percent said that ticket price drove their decision. The percentages varied by organization and by whether the respondents were subscribers or single ticket buyers. Although many of the single ticket buyers who responded to this study were former subscribers, price does not appear to have played a major role in this shift. Rather, the primary motives for their behavioral change are that many people prefer to select which programs to attend and many have difficulty scheduling in advance.

Among those who cited price as a primary factor, just as many people stated they were *unwilling* to pay a stated ticket price as said they were *unable* to pay. This distinction highlights the fact that for many people, *interest* is the barrier to attendance, not price.

In conjunction with the written audience survey, A.C.T.'s box office personnel recorded information as people phoned in for tickets. We were interested in knowing whether customers received their first-choice seat location, and if not, what their second choice was. For example, if people preferred mid-priced tickets and those were sold out, did they choose lower- or higher-priced seats? The survey found that on average, 77 percent of single ticket buyers were able to purchase tickets for the date and price they preferred. For the 23 percent who did not get their first choice of seat location, the great majority moved *up* in price, rather than down, indicating that a good seat location is far more important to them than a lower price. These responses suggest that theater attenders want the best experience possible and appreciate the value of good seat locations more than a lower-priced ticket.

The survey verified that most attenders—except those who purchase budget-priced tickets—have little sensitivity to price increases. A modest price increase has little or no effect on their ticket purchasing behavior. It is important to note that these studies surveyed current attenders only. Therefore they did not capture information about the price sensitivity of potential future attenders.[8]

In early 2013, the Metropolitan Opera announced that prices for the forthcoming 2013–2014 season would be reduced about 10 percent, back to the price levels for the 2011–2012 season. The 10 percent increase that had been put in place for the 2012–2013 season clearly exceeded a breaking point for many patrons, as there was a sharp drop in ticket sales.

In Spain, performing arts ticket prices skyrocketed in September 2012 when the government increased taxes on tickets (VAT) from 8 to 21 percent in an emergency measure to boost government coffers and reduce the burgeoning public deficit. In the four months following the tax hike, theaters reported that audiences diminished by about 1.8 million people, compared to the same period a year earlier, representing a ticket sales drop of 33 percent. As a result, even some of the largest and most successful halls had to drop planned performances and six hundred employees lost their jobs. Some small theaters devised ingenious ways of bypassing the sales tax to reattract audiences. A theater in Bescano, in the northeastern region of Catalonia, sold carrots for 13 euros (the typical ticket price) in lieu of tickets, because staple food products still have a 4 percent VAT charge. Experts concluded that the government plan to tax the arts would backfire and that its annual income from across the arts industry would actually fall by nearly 10 million euros as a result of the VAT increase.[9]

Actual and Perceived Costs

The price of a ticket is only one of the costs a consumer has to pay in order to attend a performance or to subscribe to a series. The consumer's actual and perceived costs of a proposed exchange can be defined as the sum of all expected negative outcomes.

There are significant psychological costs involved in attending a performance. Some people are concerned that they might feel ignorant about the art form, do not know if they will enjoy the experience, or feel they have to "dress up" or might not "fit in" with the crowd. Some people may not attend theater in what they consider to be an unsafe neighborhood. Others don't want to pay the "price" of dealing with heavy traffic.

Such costs may serve as greater barriers to attendance than the actual price of the ticket. Some of these factors are outside the marketer's control, such as inclement weather and traffic conditions. However, the arts organization can provide a guide to nearby parking facilities, along with discount coupons and pay-in-advance options where possible. Posting staffers and bright lighting in front of theaters might augment the perception of personal safety.

When selecting strategies to reduce perceived customer costs, the marketer must keep two questions in mind: (1) What is the marketer's cost of reducing a perceived customer cost? (2) What response can be expected from the customers to given levels of perceived cost reduction? Consumer responses are usually a reaction to a bundle of perceived costs and benefits. The problem in managing perceived costs is to figure out which of many costs to reduce and how much to spend to reduce them. For a given expenditure, the issue is which perceived costs should be targeted to yield the largest net gain in audience size and satisfaction.

Perceived Value

Perceived value is determined by the buyer. It represents the margin of difference, either positive or negative, between the producer value and what a consumer feels the offering is worth, regardless of its production costs. In the business sector, a good or service must be worth more to the consumer than its producer value, or it will not be produced. In most nonprofit arts organizations, the ticket price of a performance is significantly lower than its producer value, creating a gap that must be filled by other funding sources.

Arts consultant Andrew McIntyre says that "perceived value is established first by persuasive communications. . . . Arts organizations' communications are frequently not persuasive enough . . . and reducing the ticket price until it matches the lowered perceived value seems like a lazy option to me."[10]

Perceived value varies, as people differ in their perceptions of value for the money. At the Roundabout Theatre in New York City, people aged 18–35 are eligible for HipTix, a program that offers tickets for $20, while full-priced tickets range from $70 to $85 for similar seats in their nonprofit halls, and from $120 to $145 in their Broadway theater. The HipTix program has 40,000 members, 60 percent of whom purchase tickets for an average of one show per season. According to marketing director Tom O'Connor, a very small percentage of these people are purchasing full-priced tickets once they "graduate" from the program.[11] Is this a value issue for them? After being encouraged to see the shows for only $20 per ticket for many years, once they turn 36, do they think that those same tickets are not worth $70 or more? Roundabout does not know if those people are purchasing tickets through third-party discounters, which do not share customer information with the organization, or if they have stopped attending altogether.

To research the importance of subscription series discounts among the benefits offered to subscribers, Ryans and Weinberg surveyed subscription buyers for A.C.T. in San Francisco in the 1990s. Survey respondents reported that the main reason for buying a subscription series was not the savings, but to make sure they went to the theater more often and were assured of a good seat. Fewer than 25 percent of respondents even indicated discount as a valued benefit of subscribing. According to the report, "If a discount is a significant factor in converting occasional attenders into subscribers, then its use may be economically justified. On the other hand, if subscribers are primarily those who are the most enthusiastic theatre goers, then offering a discount may essentially be a price reduction for those who would attend in any case."[12] Validating the research results, A.C.T. abandoned its subscriber discount the next season with no significant impact on subscription sales. Thus, many organizations may be pricing their tickets lower than the market will bear, underestimating the perceived value of their product to many of their customers.

However, in recent years, as subscriptions have been declining dramatically, significant discounts are commonly being used by organizations to attract new subscribers and to keep current ones from lapsing their subscriptions. Arts marketers make such offers as six plays for the price of four; buy three, get one free; up to 30 percent off with your subscription; or "subscribe and guarantee the lowest price and best seats." Arts marketers must carefully consider whether big discounts are necessary, or if the *idea of a discount* is the motivator subscribers enjoy. If the latter is the case, a small discount will be more than adequate for satisfying customers and will increase revenue for the organization. Luxury retailers are among those companies offering free shipping on all orders. Luxury businesses do not like to discount their products, as this would compromise the value of their brands, but even high net worth individuals appreciate this small gesture of value.

Many people enjoy benefits that are not price related. Some like early booking opportunities to get the best seats or a special room in which to enjoy a drink before the show or during intermission. Marketers should carefully design benefits with the customers' interests in mind. For example, a complimentary wine reception on a separate day from a ticketed performance is unlikely to be taken advantage of by people who live and work quite far from the venue. A free seat upgrade for one performance in a subscription series is meaningless for people who already have the highest-priced seats.

Relative Value

People may evaluate prices differently depending on the situation. They may purchase tickets for a birthday or holiday celebration more readily than for an ordinary evening out. Someone looking for a family event will have a different framework for assessing price and value than a single person looking to impress someone on a date.

If arts organizations understand what needs people are trying to meet and what values they are seeking, they can begin to understand their competition and the context within which customers are evaluating the price and value that is offered.

SETTING PRICING OBJECTIVES

There are two basic objectives an arts organization may attempt to achieve through pricing its performances: *revenue maximization* and *audience size (*and *diversity) maximization*. These objectives are not mutually exclusive and arts managers often find they can meet both goals simultaneously.

Many arts organizations seek prices that will maximize revenue or at least recover a "reasonable" percentage of their costs. What is considered reasonable cost recovery for one organization may be quite different from that sought by another. In a study of 30 North American opera companies, earned income ranged from as high as 87 percent to as low as 18 percent of the annual budget. At the 1,782 theaters surveyed by TCG in 2012, total earned income, including that from ticket sales, touring, royalties, concessions, rentals, and so on, supported on average, 52 percent of expenses. Earned income fell by 5.7 percent over the five-year period 2008–2012 and supported 3.2 percent less of total expenses in 2012 than in 2008.[13]

Arts organizations also seek prices that will achieve the objective of attracting the largest possible audience. Some managers think this means setting a low, affordable price throughout the hall to stimulate demand.

PRICING PRINCIPLES AND STRATEGIES

Arts marketers have many different variables to consider when setting pricing strategies: how many prices, the price ranges, the difference between prices, the number of seats at each price, the location in the hall of different priced seats, promotional discounting, premium pricing with special benefits, revenue management, and surcharges. Managers should employ different pricing strategies, as appropriate, without creating confusion for the customers. Here we will examine competition-oriented pricing; image pricing; higher highs and lower lows; discriminatory pricing options; and dynamic pricing, a technique for changing prices according to demand.

COMPETITION-ORIENTED PRICING

An organization may choose to set its prices chiefly on the basis of what its competitors are charging, rather than on the basis of its own costs or level of demand. Most often, an organization will try to keep its prices at the average levels charged by its own industry group, such as regional theaters or professional orchestras.

Competition-oriented pricing is called going-rate or imitative pricing and is popular for several reasons. Where costs are difficult to recover, the going rate represents the collective wisdom of the industry concerning the price that would yield a fair return and elicit reasonable demand. The crucial operative factor is that people believe they are getting appropriate value for their money. The more differentiation there is between the arts offerings, the more latitude organizations have in their pricing decisions.

Price competition in the arts is rarely as direct as it is in the business sector. People may choose to buy Coke over Pepsi if it is on sale, but people rarely choose Theater A over Theater B on the basis of ticket price.

IMAGE PRICING

Organizations can capitalize on image differences in pricing their products. Patrons will pay a higher price for a particular play at an established, well-respected theater than for the same play at a new, start-up company. When a theater extends the run of a popular play, it may raise its prices because the production has gained the imprimatur of quality and broad acceptance.

HIGHER HIGHS AND LOWER LOWS

One strategy that helps achieve both revenue maximization and audience-size maximization is the principle of higher highs and lower lows: set prices as high as the market will bear for those eager and willing to have a premium experience, offer low-priced tickets for interested patrons who could not attend otherwise, and offer a range of prices in between, depending on the size and configuration of seating areas and on audience demand.

Many arts attenders want the best seats and are willing to pay for them. On the other hand, many people who enjoy attending the arts can only do so if inexpensive tickets are available.

Because patrons who buy high-priced seats are relatively *price inelastic*, meaning that price increases have relatively little effect on their willingness to buy a ticket, when raising prices, arts marketers can often increase the price of their most costly seats by a much greater percentage than that of the lower-priced seats.

Pricing Subscriptions at the Goodman Theatre

The Goodman Theatre is the largest theater in the Chicago area and one of the most respected regional theaters in the United States. The Goodman features five productions per season in its main hall, and three per season in its smaller venue. In the main hall, patrons see a range of great classics, such as artistic director Robert Falls's edgy interpretations of Shakespearean plays, recent Tony award winners, big musicals, and contemporary dramas. For the 2013–2014 season, the Goodman offers subscribers a wide range of ticket packages and pricing options. There are more than 15 price options for subscribing to the five plays. People who attend preview performances (before the official opening night) can pay as little as $115 (lower-priced B section; Sunday evenings). Those who purchase seats in the A section on Saturday evenings during the regular run of the show pay $390. And there is a wide variety of options in between these prices depending on day of the week and seat location. These price ranges are modified each season according to the anticipated popularity of the upcoming shows and according to the marketing director's past experience regarding audience demand.

DISCRIMINATORY PRICING

Discriminatory pricing occurs whenever an organization sells a product or service at two or more prices that do not reflect a proportional difference in costs. The use of various discriminatory pricing techniques can go a long way toward simultaneously maximizing audience size and revenue.

Product-Form Pricing

In product-form pricing, different versions of the product are priced differently. A presenting organization will charge more for a famous star than for a rising star. A theater may charge more for a full-scale musical production than for a two-actor drama with minimal sets. It makes sense for arts organizations to vary single ticket prices according to the program, but heavily subscribed organizations are likely to find this strategy impossible to implement as subscribers pay a fixed amount for the entire season in advance. Some organizations are starting to "back out" their subscription prices, meaning that they price their subscription packages based on what they consider to be appropriate single ticket prices for the next season's productions.

Seat Location, Time, and Special Event Pricing

Different seat locations and days of performances are often priced differently, even though the cost of offering each seat is the same. Scaling the house is the most commonly used price discrimination strategy, especially by organizations with halls that have more than three hundred seats. Theaters with small venues do not usually have the option of varying ticket price by seat location.

By dividing the hall into two or more seating areas, organizations can offer welcome choices to their patrons and realize higher ticket revenue. Many patrons will pay a premium for the best seats and for the more desirable weekend performances.

Arts managers may vary the number of seats in each price category by performance, so that when Renee Fleming is singing a recital, there may be many more higher-priced tickets than usual, and when an unknown conductor is performing twentieth-century repertoire, there may be more lower-priced seats.

People are also willing to pay significantly higher prices for special events, such as opening night, which may include a complimentary glass of champagne and an opportunity to mingle with the artists after the show. The operative word for this pricing strategy is *special*; people must feel they are receiving added value in terms of such features as exclusivity and excitement.

Location pricing, in addition to seat location, may also be a function of where the performance is taking place. Concerts at esteemed venues typically bring higher prices because of their higher perceived value.

Special low price strategies may be effective for motivating people to buy tickets for performances that are offered at unattractive times, such as performances of holiday shows offered between Christmas and New Year's Eve, as most people prefer to attend before Christmas.

Some cities, including London, New York, and Chicago, have ticket booths where half-price tickets to a variety of theaters can be purchased on the day of the performance. The half-price ticket booth discriminates in favor of people who are willing to wait in line to save half the price of a ticket for select shows, usually in the least desirable seat locations. The New Jersey Theatre Alliance offers discount tickets online for 43 participating companies. This program has the appeal of great convenience and has been successful in attracting new audiences.

Ticket scalpers, third-party providers who sell tickets at a premium, discriminate in favor of people for whom the convenience is worth more than face value. However, there are antiscalping laws in 16 of the 50 US states. Economist Richard H. Thahler says that some people think it's fair to make everyone who wants a discounted ticket stand in line, "but that forces everyone to engage in a totally unproductive activity, and it discriminates in favor of people who have the most free time. Scalping gives other people a chance, too. I can see no reason for outlawing it."[14]

Pricing Seats at Chicago Opera Theater

In 2002, Chicago Opera Theater (COT) began to plan for its 2004 move from the aging Athenaeum Theater, which was inconveniently located and had inadequate facilities and parking, to the new Harris Theater for Music and Dance, a state-of-the-art facility in downtown Chicago's new Millennium Park. Unquestionably, the market would bear higher ticket prices in the new venue, which features many amenities and adjacent underground parking. With its greatly increased costs for performing at the Harris Theater and many more seats to fill, COT was eager to maximize both audience size and ticket revenue.

As the marketing consultant to COT, I suggested that we first eliminate subscriber discounts, including an early bird renewal discount, which COT's managers had thought to be helpful in increasing renewals during a traditionally slow cash flow period. Instead of the early bird discount, we gave patrons an incentive to renew early: a chance to win restaurant coupons, which were donated to the organization in exchange for the marketing benefits to the businesses. This strategy proved to be highly effective as early renewals remained consistent with those of previous years. Most important, none of the COT patrons failed to renew in 2003 because of the lost subscriber discount. COT realized thousands of dollars of increased income from this strategy.

For the 2004 season, we had the opportunity to scale the new hall into seating sections according to anticipated capacity utilization and to price tickets according to what the organization's managers and I expected the market would bear. Given the extremely high quality of COT performances in recent years, the excitement over the new performance space, and the fact that a large percentage of COT patrons also attend Chicago's Lyric Opera, where ticket prices are significantly higher, we planned to institute moderate to significant price increases in various seating areas.

Previously, COT had offered lower-priced tickets for weekday evenings and Sunday matinees than for weekend evenings. However, a capacity utilization analysis of the three previous seasons showed that opening night performances on Wednesday evenings, Thursday evening performances, and Sunday matinees were popular and well attended, especially in the higher-priced seats, so we decided to eliminate day-of-week price differentials altogether.

We conducted a capacity utilization analysis of all seats sold for one opera production, Benjamin Britten's *The Turn of the Screw*, performed in 2003 at the Athenaeum, to help guide us in scaling and pricing the new house for 2004. In our analysis, we found that 69.5 percent of the available seats for all five performances of *The Turn of the Screw* were sold overall. In the three seating sections in the Athenaeum, 72 percent of Section A seats (the highest-priced section), 49 percent of Section B seats, and 60 percent of Section C seats were sold. Clearly, *the greatest number of patrons prefer the best and most expensive seats.* In evaluating the ticket purchasing behavior of subscribers and single ticket buyers, we found that the number of Section A tickets sold to subscribers and to single ticket buyers were nearly equivalent; however, the data showed that approximately twice as many single ticket buyers as subscribers purchased tickets in Sections B and C. A careful analysis of all this data helped us decide how to scale seating sections and devise pricing levels for the new hall.

Because the new hall has significantly more seats than the Athenaeum, we increased the number of seating areas. We raised ticket prices in Section C from $30 to $45, but in keeping with COT's mission of having some ticket prices that are affordable for virtually anyone who wishes to attend, we offered tickets in Section D, in the back half of the balcony, for $30. Furthermore, students are eligible to pay half-price for tickets in that section, so they were able to attend in Section D for only $15. It is interesting to note that significantly more students purchased seats in Section C than in Section D in 2004, showing that they wanted the best experience they could afford, not necessarily the cheapest price.

For the loyal subscribers and donors who were willing to pay high prices for the best seats, we created a new section, Section A+, in the center of the main floor. Section A ticket prices had been increased from $65 at the Athenaeum to $85 at the new Harris Theater with no subscriber discount—a significant increase—but I recognized the opportunity to offer A+ seats at an even higher price of $97. In early April, before renewals went out to all COT subscribers, I wrote a letter that was sent to current Sections A and B subscribers, saying in part: We understand that as a dedicated COT patron, you value an excellent seating location for our performances. We are pleased to announce that the new Harris Theater for Music and Dance has a seating section with such a premium view that we are able to offer a new seating category, A+ seats. The general renewal letter will go out in May to all of COT's patrons. Until that time, you, as a loyal subscriber, have priority to reserve your seats in the A+ section for next season and into the future. Once these seats are filled, patrons have the right to renew them as long as they wish. The quantity of A+ seats is limited and we may not be able to accommodate all the people who would like to reserve these seats.

COT began to receive orders for Section A+ subscriptions the day after this mailing was sent, and by the end of the season, 105 percent of A+ seats had been sold. We were able to sell more than 100 percent capacity because as some A+ subscribers exchanged their tickets for another performance or turned tickets back as a donation, single ticket buyers eagerly purchased them.[15]

Capacity Utilization Pricing

Typically, marketing managers set new prices based on a percentage increase from the previous year, or they make an intuitive decision loosely based on past and anticipated demand. As we have seen in the Chicago Opera Theater example, capacity utilization analysis is a more scientific approach that will generate far better results. When marketing managers express concern that their ticket prices are too high, I suggest they review their capacity utilization data over a period of time; high-priced tickets are rarely the problem they are assumed to be.

Some large-budget organizations contract with pricing consultants to harness the power of sophisticated data analysis. Such data can inform pricing decisions by individual seat, day of week, timing in the show's run of performances, and much more. As compelling as this data may be, it is important that marketers do not rely strictly on data but use it to help inform their decisions, which need thoughtful, creative human input as well.

Focusing on Hard-to-Sell Seats

Marketers often find that the most expensive seats sell best and that the mid-priced tickets are the hardest to sell. Highly targeted group sales and special promotions are among the tactics that can be used to sell more mid-priced seats. Marketers should consider rescaling the house, if possible, to add more A-level or C-level seats. Also, some marketers modify prices based on demand in real time.

William Poundstone, author of *Priceless: The Myth of Fair Value (and How to Take Advantage of It)*, recounts that the Williams Sonoma chain, known for high-quality and high-priced housewares, once offered a fancy bread maker for $279. The company later added a somewhat bigger model, pricing it at $429. The $429 model hardly sold at all, but sales of the $279 model nearly doubled. This lower-priced machine, which seemed expensive before, now seemed like a bargain. It was no longer extravagant and buyers rationalized that it does practically everything the larger one does.[16]

This exemplifies that high-priced items can sometimes serve to promote lower-priced items. Typically in arts organizations, the middle-priced seats are hardest to sell, so arts marketers should keep this concept in mind when scaling the house.

Customer Segment Pricing

With customer segment pricing, different customer groups are charged different prices to acknowledge differences in their willingness or ability to pay. Arts managers can maximize their audience, and therefore their revenue, by responding to these differences. For the sake of developing pricing strategies, the audience can be segmented by loyalty to the organization, such as subscribers, multiple single ticket buyers, and first-time buyers; by age and life cycle stage, such as students, adults with children, and seniors; by groups; and by myriad other segments that can be identified by managers as significant, viable, and which can be offered special pricing (either high or low) without other segments feeling they are being discriminated against.

Some people tend to book in price ranges—low price bargain hunters, mid-price value-seekers, or top price quality seekers—almost irrespective of what those prices are. It is important, therefore, to offer a range of different prices to meet the needs of those different segments. There are certain special events for which a one-price policy makes sense, such as general admission, first-come first-served events. These are the exception rather than the rule, however.

For senior citizens, a theater may offer discounted tickets for mid-priced and lower-priced seats. It is unnecessary to make this offer available for the best seats, since many seniors are able and willing to pay full price to have the best experience. A symphony may offer half-price tickets or special, flat-rate lower prices to students the week of, the day of, or beginning two hours before the concert, filling otherwise empty seats. Some theaters make offers on their Twitter feed for a limited number of seats for that evening's performance at a low price. Group sales are another form of customer segment pricing, as discounts are given to customer groups purchasing a block of tickets for a single performance. Gift certificates and gift subscriptions may also be offered at a special price to encourage current patrons to bring friends and family.

Discount Pricing

What devalues the performance-going experience more: deeply discounted tickets or a half-empty hall? The answer is that it's not one or the other. Unquestionably, everyone, from the performers to the audience, has a better experience if most of the seats are filled. Although some people need to feel they are getting a special benefit or opportunity with a low-priced ticket, if discounts become the norm, people will assume that this is all the ticket is worth and will never pay more.

Rather than reaching automatically for the price discount to make their programs more attractive to recalcitrant audiences, arts organizations need to develop a better understanding of what people value, and better skills in creating and communicating that value.

Jim McCarthy, chief executive officer of Goldstar, an online ticket discounter, addresses consumers' psychology regarding low-priced concert tickets. "The person who gets the best of those seats loves that," says McCarthy. "And [for the best seats, the organization] is definitely leaving money on the table. I think... [it] is probably negative overall for sales, maybe not in the short run but in the long run, because people stop knowing how to value the product."[17] As a better alternative, McCarthy suggests arts marketers "think about pricing your house as a service to the consumer: how can I give the consumer a *range* of opportunities that make sense to them and that give them a different way to experience the thing? If you scale the house right, it means people are paying what they value the ticket at."[18] McCarthy acknowledges that some segment of the audience is cash-constrained and may just be waiting for a low enough price to purchase a ticket. These are people who know and like what arts organizations do but cannot afford the regular ticket price. "But," says McCarthy, "the person who you should look at as the target of your discount efforts is a person who will potentially become a regular patron. Your job is not to sell them a ticket; your job is to bring them into the theater so that they'll get interested in what you're doing."[19]

Says Andrew McIntyre, "The practice of regular tactical discounting must surely erode price trust, devalue the product, and encourage late booking. We might call this *de-marketing* (italics mine). There are lots of short term income generating tactics here, but where's the strategy?"[20] Discount pricing can be a dangerous race to the bottom.

Subscriber Discounts

Why do arts organizations habitually offer discounts to their subscribers, those who prove by committing to a season of performances months in advance that they highly value the experiences they will have with an organization? Says Penn Trevella, former marketing coordinator of the Royal New Zealand Ballet,

> Our research showed that discounting tickets was not the key motivator for subscribers, but rather, people subscribed because they wanted to secure the best possible seats and they also wanted to feel as though they were involved and contributing to our organization. As a result of the findings, we overhauled our subscription offering and shifted the emphasis from price discounts to securing the best seats and to other benefits that they would not receive if they just purchased tickets on a show by show basis, such as complimentary programs, opportunities to be more involved with the company, complete flexibility to change their tickets should the need arise, and so on. The end result was that we doubled our numbers in the first year. I think the offer of discounted prices is a motivator to first time subscribers but it becomes less important over time as individuals' level of commitment changes and they begin to seek out other benefits and to develop their knowledge of the art form.[21]

Many marketers have found that enticing first-time subscribers with very low one-time prices will result in a lower renewal rate. The more price breaks and gimmicks that are used to attract people, the less likely they are to renew. Are arts managers offering deep discounts because this is necessary for selling tickets, or are the organizations jumping on the discount bandwagon because they are afraid not to?

Reduced Price Ticket Strategies

When price acts as a barrier, whether for first-time attenders or for those who would like to attend more frequently than they are financially able, there are a variety of strategies that can be employed. Typically, these policies appeal to those who are willing to take a chance on getting seats, are indifferent as to where they sit in the hall, and for whom price concerns are the prime determinant of their attendance. Among the most common of these strategies, in addition to third-party ticket sellers previously addressed, are rush tickets and stand-by tickets. Special promotional pricing strategies are frequently used also.

Rush Tickets. Rush tickets are deeply discounted tickets, typically offered the day of the performance, sometimes just a couple of hours before curtain time, according to the preference of the organization. Rush tickets are not advisable, of course, for organizations that enjoy a fair amount of "day of" ticket sales. It is in the organization's and the patron's interest to fill the best, most visible seats that are open at this time.

Stand-By Tickets. The difference between rush and stand-by is that with rush tickets, patrons are given a specific seat location at the time of purchase. Stand-by tickets are offered for the seats of "no-show" patrons at sold-out shows. Some organizations report that often 5–10 percent of patrons do not show up for a performance, leaving seats empty that could be resold. Typically, with stand-by, tickets are sold for half-price starting two hours before the performance. Patrons wait in line until show time or are given numbers that reflect the order in which they arrived. A few minutes before curtain time, stand-by patrons are invited to take any empty seat in the hall, with the understanding that if the people who purchased those seats show up to claim their seats at intermission, the stand-by ticket holder will have to move. It rarely happens that no other seats in the hall are available for them. (If this were to happen, the organization would refund the ticket price.)

Special Price Promotions. Performing arts organizations will often offer a discount to people who respond to a certain advertisement, and time-limited offers to people who purchase tickets by a certain date or for selected performances. A small discount for responding to a marketing communication is an inexpensive way for the marketing manager to track the ad's effectiveness. Time-limited price offers may stimulate ticket purchase, especially for those who want to attend but have not gotten around to buying tickets for one reason or another.

Says Dan Ariely, author of *Predictably Irrational*, "Getting something for FREE! feels very good. Zero is not just another price, it turns out. Zero is an emotional hot button—a source of irrational excitement.... Most transactions have an upside and a downside, but when something is FREE! we forget the downside."[22] Obviously arts organizations cannot afford to give away tickets to their performances on any regular basis, nor does it make sense to do this. I have seen many cases where an arts organization offers a "free day" once a year; free tickets to certain targeted segments, thanks to a donation that compensates for the lost ticket revenue; free giveaways through a popular radio host; and the like. When I follow up and try to learn if these "freebie" recipients actually paid for a ticket for a subsequent production, the responses I receive are either depressingly low or totally evasive.

But there are ways for arts organizations to use free offers productively. For example, they can offer each subscriber the benefit of one complimentary guest ticket. This serves as a gift to the subscribers, who then invite friends to be their guest at a performance. Often, the guest will offer to pay for dinner before the show, creating a mutually giving and enjoyable experience. And, importantly, the organization benefits by collecting the names and contact information of the guests, which provides the organization important value from this offer.

Arts organizations have been making offers of twofers (two tickets for the price of one) for decades. Why not use the magic of the word *free* and say instead: buy one ticket; get one *free*! Marketers should keep in mind that the concept of "buy one get one free" ignores the people who would like to attend alone or in odd-numbered groups. They should ascertain that they offer half-price seats to people buying an odd number of tickets and publicize this offer along with the "buy one get one free" offer.

Dan Ariely also addresses the fact that although we know rationally that, for example, $10 is always $10, a savings of $10 on a $20 ticket is more motivating than a savings of $10 on a $100 ticket.[23] The implication of this is that small percentage discounts are meaningless to people. However, many arts marketers report that their patrons like to see that they are getting a "deal," even if the savings are relatively small. It is important for marketers to understand their own audiences' behaviors and motivations when establishing such policies.

Controlling Other Direct Costs

Marketers should take into account direct costs other than the ticket price that the consumer faces.

Handling Fees

Marketers are wisely cautious about each dollar increase in the ticket price, and customers are frequently subjected to handling fees that they consider exorbitant. Such fees should be in a "reasonable" range depending on the ticket price and the number of tickets ordered. Processing fees must make sense to the customer as a cost of doing business, not as add-ons to boost the price. Arts managers should explain to their customers the reasons for their service fees during the ticketing process and explain when service fees are charged by third-party ticketing services. Ticketmaster's high "convenience fees" have irked concertgoers for years; the company was runner-up for the Consumerist.com award for the worst company of 2013.

A per order fee may make more sense than a per ticket fee, especially when a large number of tickets is being purchased at one time. Yet the fee should be appropriate for the order. Marketers may also find consumers are often annoyed when their purchase includes both a per ticket fee and a per order fee. If potential ticket buyers are hanging up the phone or leaving the website without making a purchase, how much is this due to what may be considered exorbitant surcharges? It is worthwhile for managers to investigate the answer to this question.

Conditions for Price Discrimination

If price discrimination is to work, certain conditions must exist. First, the market must be segmentable, and the segments must show different intensities of demand. For example, students tend to have stronger than usual demand for day-of-performance tickets. Second, members of the lower price segment must not be able to resell the product to the higher price segment. So, for example, a deeply discounted student ticket must be clearly identifiable as such. Importantly, the practice must not breed customer resentment and ill will, meaning that customer segments not eligible for a particular price advantage should not feel discriminated against. Typically, mature audience members are delighted to see young people among them, at any ticket price.

Revenue Management

Revenue management refers to strategies that include yield management and dynamic pricing. Robert Cross, the pioneer of airline revenue management, says, "Revenue management ensures that companies will sell the right product to the right customer at the right time for the right price."[24]

According to pricing specialist Tim Baker, "The key principle of revenue management is to adjust price differentials in response to customer demand in order to maximize both occupancy and income."[25] It includes reducing prices when demand is low to stimulate sales as well as increasing prices to exploit high demand.

Managers should first review their capacity utilization over several productions to determine how many seats have sold in each price category and which seats are selling. As a result of this analysis, managers will have data to help them optimize the price to charge for different seating locations in the hall, determine how many different pricing sections to offer, and where the section breaks should be. One approach to maximizing revenue is changing the number of seats in each price section. If the higher-priced seats are selling out and many of the lower-priced seats are empty, management can add some seating rows to the higher-priced category. Another alternative is to raise the price of the higher-priced seats. If the situation is reversed and many of the higher-priced seats are left empty while lower-priced seats are consistently filled, management may choose to decrease the number of higher-priced seats offered. This strategy is used effectively at the New National Theater of Tokyo where patrons clamor for top-priced tickets for highly popular productions but resist high prices for all but the very best seats for the less popular shows.

Managers may evaluate the efficiency of a pricing system by computing the yield ratio, in which the denominator consists of the sum total of available seats at each price level multiplied by their prices. The numerator consists of the sum of the number of seats sold multiplied by their prices. Assume that a theater with 400 seats offers 150 seats at $45 and 250 seats at $35. If the theater sells all of its $45 seats and 200 of its $35 seats, the efficiency ratio is 89 percent. If the theater sells 100 of its $45 seats and all 250 of the $35 seats, the efficiency ratio is 85 percent. If 25 more people would have been willing to purchase $35 tickets, making some unsold $45 seats available at $35 would raise the efficiency ratio to 91 percent.

Rescaling the House at the Mark Taper Forum

In advance of the 2010 season, at the Center Theatre Group (CTG) in Los Angeles, marketing director Jim Royce changed the Mark Taper Forum, the 732-seat venue, from a two-price house to a three-price house to provide more price points and increase income from the best seats. Using the Pricing Institute's HotSeatIndex™, the CTG marketing team divided the house into 28 seating zones within these three prices. Each zone was assigned business rules governing which zones could be discounted or used for complimentary seats; most prime location seats now could not be comped (given away as complimentary seats). Prices of particular zones could be changed on a performance-by-performance basis, enabling CTG to maximize revenue and optimize inventory throughout the run of each production.

Marketing managers teamed up with development personnel over the management of subscriber upgrades to the most desired sections, so that an upgrade to certain sections would now require the subscriber to also become a higher-level donor, and to continue at that level for subsequent seasons. When prime seat locations were not renewed or used to accommodate an upgrade, these seats became full-price, never discounted or comped (under the business rules for those sections), and were held until development found a suitable donor. The zones least in demand were then identified as available for special promotions such as HotTix discounted offers.

The combination of rescaling, managing flexible zones for each performance, and requiring a donation for upgrades to the best sections yielded significant results, including a 9 percent increase in average prices across the board, an 83 percent subscription renewal rate, and a 28 percent increase in subscription revenue.[26]

Dynamic Pricing

Dynamic pricing is the act and art of progressive pricing, or changing ticket prices over time based on the anticipated or actual supply and demand. Says pricing expert Tim Baker, "Whereas revenue management need only mean charging different prices for different seats at different performances, dynamic pricing can also mean charging different amounts for the same seat at different time in the sales cycle."[27] The perception of scarcity is necessary to the success of increasing prices with dynamic pricing and it works best when demand exceeds supply.

Until recent years, ticket prices were established long before a season began and marketers were constrained by the prices they published in their printed brochures. Special discount offers were made available to certain customer groups and for weakly selling shows, but premium pricing did not exist in the nonprofit performing arts world until the advent of (1) sophisticated databases that provide up-to-the minute sales information, in addition to complex analyses of previous sales, and (2) the dynamic pricing concept, whereby organizations update their prices on a weekly or daily basis, and in the case of some organizations and special shows, even more often.

People are familiar with dynamic pricing, which is relatively new in the arts, as it has been employed by the airlines for decades. When pricing dynamically, typically arts organizations change prices according to a formula or plan. For example, when 75 percent of the seats are sold, prices increase $5; when 80 percent are sold, prices increase another $5; when 90 percent are sold, prices may increase another $10. The right answer for changing prices is different for each organization and for each discrete situation.

Dynamic pricing applies to single ticket sales only; subscribers are guaranteed a stated price and usually pay for their season of tickets long before single tickets go on sale. Organizations that enjoy a large subscriber base have fewer seats available to be priced dynamically.

Some organizations adjust their prices within the parameters of their stated price structure, rather than increasing the top price, by rescaling their hall. So, if the top-priced seats are selling the best, the marketing director may include some of the middle-priced seats in the top price category, in effect shrinking the middle price section. Without dynamic pricing, a hit show that sells out will have a lower average ticket price than a show that sells moderately well because more low-priced seats will be sold than for weaker-selling shows. Conversely, if there is high demand for the lowest prices, organizations can move more seats into the low-priced category from the back of the middle section. Since it is the mid-priced seats that are typically the hardest to sell, this strategy can be quite effective in maximizing sales.

In keeping with the principle of higher highs and lower lows, dynamically priced tickets should be increased more in the highest price sections; lower-priced sections should have smaller increases or retain their former lower prices to accommodate people who cannot or will not pay a higher price.

Dynamic Pricing at the Metropolitan Opera

The Metropolitan Opera explains its dynamic pricing policies on its website. To paraphrase: Prices for all performances are subject to change (increase or decrease) based on sales. The prices are adjusted in real time on the website and reflect the most current price for each performance. Sales are analyzed on a per-performance and per-section basis and prices are adjusted based on changes in sales. If prices decrease after a purchase is made, there is no adjustment, as all sales are final. Operas purchased as part of a subscription are not subject to dynamic pricing; subscription prices are locked in for the entire season at the time of purchase. Additional single tickets purchased after the subscription package transaction are subject to dynamic pricing.

RATIONALE FOR DYNAMIC PRICING

In 2011, when *God of Carnage*, starring the late James Gandolfini, broke box office records at CTG in Los Angeles, the theater raised its top ticket price 66 percent to $200 as demand for tickets grew. "The people buying these tickets are underwriting our discounts. Premium prices pay for the mission," said artistic director Michael Ritchie.[28]

Not all organizations with hit shows agree with the principle of premium pricing. In spring 2013, when Next Theatre of Evanston, Illinois, extended its production of *Everything Is Illuminated* and continued to enjoy high demand for tickets throughout the run of the show, ticket prices did not increase, in keeping with Next's community-minded mission. The top ticket for Lincoln Center Theater's superbly popular *War Horse* remained at $135, even in the wake of winning the Tony Award for best play. Administrators discussed introducing premium pricing, but decided against it for "philosophical reasons," according to a Lincoln Center spokesman.

Pros of Raising Prices

Raising prices for high-demand shows gives organizations the opportunity to maximize their revenue. The more demand there is for a show, the less important price becomes as a driving factor in the ticket purchase decision. If people are eager to see a show, many will pay a premium price. Some arts marketers say that the very scarcity of tickets adds value.

In the 2008–2009 season, the Arts Club Theatre Company in Vancouver, Canada, earned $430,000 more than the previous season, a 10 percent increase in sales, attributed to dynamic pricing strategies. This was accomplished by making price adjustment decisions on a daily basis, section by section. Similarly, the Yale Repertory Theater brought in an additional $50,000 of revenue for a production using dynamic pricing, a 37 percent increase.[29]

Some marketing managers think that dynamic pricing may help fuel subscriptions. Organizations often motivate people to subscribe by locking in a guaranteed low price that is not subject to single ticket price changes. It is important to mention one caveat here: if an organization realizes high demand for its forthcoming season of shows, then it is true that single ticket buyers will pay higher prices than subscribers. But dynamic pricing works both ways; prices will be lowered if demand is weak. If enough shows in the season have weak demand and single ticket buyers can purchase discounted tickets, then this is possibly less costly than a subscription on a per show basis.

Cons of Raising Prices

If prices are rising as the performance date nears, then it is most rational for people to buy tickets in advance. However, many people have busy schedules or are otherwise unable to plan in advance. Should people be "punished" for buying tickets close to the performance time? For other luxury, discretionary experiences, such as dining in a fine restaurant, if someone is fortunate enough to get a reservation at the last minute, she will not pay more than if she had reserved a month or two in advance. People should be motivated to purchase tickets by their attraction to the show and the scarcity of available seats, not by looming price increases. When arts marketers do charge a premium price for an event, they should ascertain that they are offering benefits of premium value to the customer.

Arts consultant Andrew McIntyre is fundamentally opposed to dynamic pricing. He believes that marketers should be trying to build a community around their organizations with long-term emotional interaction rather than trying to simply maximize income with short-term financial transactions. McIntyre argues that dynamic pricing "undermines trust, diminishes the sense of belonging, and hemorrhages emotional brand equity." He says that organizations should pursue solvency and resilience by inspiring the audience to form rich relationships that generate the necessary financial support over the long term.[30]

Pros of Lowering Prices

If people aren't interested in seeing a show, they won't come at any price. But if people are undecided, a lower price may reduce their perceived risks. By lowering ticket prices, organizations can fulfill their mission of exposing more people to their art and create a better experience for both the artists and the audience with a fuller house, while garnering some additional revenue.

Cons of Lowering Prices

Organizations must take at least as much care when lowering their prices to sell more of their unsold capacity as they do when deciding when and how much to increase prices in the case of high demand. Arts managers may be "leaving money on the table" when they discount tickets that they could have sold at regular prices. This is especially the case when people choose to wait until a few days (or less) before a performance to purchase tickets, by which time prices have already been lowered. Years ago, I used to recommend that organizations offer "rush tickets," half-price tickets the day of the show, to general audiences, not just to special segments like students. Nowadays, when many people like to be as spontaneous about purchasing theater tickets as they are about movie tickets, a discount may be an unnecessary lure. Furthermore, people may become accustomed to "last minute" discounts and wait for them, working against the organization's best interest.

In this age of many third-party discount providers, such as Groupon and Goldstar, discounting has become all too prevalent. Marketing managers must stop training people to wait until the last minute for discount pricing.

Importantly, deep discounts may reduce the performance's value in people's minds. Managers know that price is an indicator of value, and if people repeatedly see low prices for events, they will assume that is all the events are worth. Then, when "normal" prices are listed, people will not want to buy. It is not that they can't afford a higher price; it is not that they do not want to attend; but they do not want to pay a price that differs significantly from their perceived value of the experience.

Cons of Dynamic Pricing

Complex pricing schemes shift the burden of the pricing decision and activity from the organization to the patron, so that patrons are tasked to determine what is the *right* price for a ticket and when is the *right* time to buy. Marketing managers know that people like having the freedom to choose, but having too many choices can paralyze a person's decision-making and possibly leave the person dissatisfied. If a person hesitates in buying tickets one day, and then the next week sees that prices have risen, he may (1) be deterred from making the purchase, or (2) buy the tickets at the higher price and set higher expectations than he held previously about the nature of the experience. Similarly, if someone buys tickets, then sees that the following week they are being offered at a discount, he will be dissatisfied that he paid the higher price. So organizations must be sure to carefully discriminate who is offered their special prices and how these offers are distributed. Most important, people must clearly understand how to value the offering.

DYNAMIC PRICING: CLAIMS BY INDUSTRY PROFESSIONALS AND MY RESPONSES

Many performing arts pricing specialists have expressed their views about the value of dynamic pricing. In this section, I will present some of their claims, many of which I consider to be fallacies, and my responses to them.

Subscribers feel rewarded when single ticket prices go up close to the time of performance. How much do subscribers care that they have paid a lower price? Subscribers feel rewarded by having the best seats, by planning in advance, by ascertaining that they will see all the shows, by being a part of the organization's extended family, and by having an emotional connection to the organization. These factors are much more important to them than price.

It is likely that last minute single ticket buyers at high prices will become subscribers to pay a lower price. Single ticket buyers like to choose which performances to attend and prefer spontaneity to subscribing. It is highly doubtful that single ticket buyers will subscribe in meaningful numbers for the sake of saving money. The exception is when the theater is offering an attractive roster of productions in the upcoming season. Then people may subscribe in order to guarantee good seats for those shows. But once that season is over, many of those new subscribers are likely to lapse. When single ticket buyers become subscribers, and stay subscribers, it is largely because they have developed a strong interest in and loyalty to the organization and because they want to see all the shows, not because they want to save some money. Actually, if they don't want to see all the shows, subscriptions cost them more because now they have tickets to shows they don't care to see.

Raising prices incentivizes people to buy early when they will get the best seat selection and best price. This helps the organization breathe and plan better. Buying early is helpful to the *organization* because it allows for better planning of ongoing strategies to attract more patrons. Earlier cash flow is very compelling to managers, but helping the organization is *organization-centered*, not *customer-centered*, as marketing should be.

You can reward people for buying early by charging a lower price. People buying early are eager to see the show and care more about guaranteeing good seats than about price. Typically, early ticket buyers are willing to pay full price; there is no reason to offer them a discount. Most organizations promoting early purchase discount about 10 percent off the regular price. Price-sensitive people are more likely to wait for deep discounts that may be available during the week of the performance than to buy at a small discount well in advance.

You need to charge a high price for perception of quality. The organization should set its prices according to many factors, including perception of quality. The "regular" price for the top-priced section of the hall should indicate the company's quality positioning. The organization should also ascertain that raising prices does not mean raising the perception of risk among potential ticket buyers.

Everyone is used to the airline model of perpetually changing and rising prices. The comparison many dynamic pricing aficionados make with the airlines is not supportable because taking a flight is a means to an end—a business meeting, a vacation, a family event—not an end in and of itself, as is attending an arts event. People are more likely to choose not to attend a performance because of the cost than to choose not to fly in order to experience the destination. Besides, the airlines are disliked for their pricing structure. Why do arts organizations want to imitate them?

Advocates of dynamic pricing say such things as: No one will notice the difference. It's just a few extra dollars. If they can pay, they should pay. Critics of dynamic pricing say that for arts organizations, revenue is a means to an end.

And that end is creating, nurturing, and sustaining a meaningful relationship between the art and its publics. Pricing too often takes the center stage away from the art itself.

PRICING: THE BOTTOM LINE

When determining pricing tactics, managers must consider the broad issues: How do their organization's pricing strategies affect accessibility? What message are they sending to their publics with their prices? How does this affect their relationship with their supporters? Will people want to make a donation if they see that tickets are frequently discounted for whoever will buy them? Will people want to consider making a donation if they have paid a premium ticket price? Will people donate if they have paid a very low, bargain price? Price impacts loyalty; fairness is a sign of success and stability. Are the organization's pricing policies seen as fair?

What really matters most to the patron is not price, but *value*. Say Baker and Roth, "Arts marketers must recognize that patrons won't realize the value they are expecting if they don't know about it. And often we don't communicate the true value (i.e. the art, the experience, the uniqueness). We talk in jargon, which means something to us, but nothing to the patron. Or we focus the conversation on price."[31]

It is important to remember that price is just one variable in the marketing mix; all the variables must be considered jointly to develop one coherent marketing strategy for each target segment. Also, the organization must recognize that it exists in a constantly changing environment. Pricing strategies must be regularly revised in light of the organization's current and potential markets, its competition, its funding sources, and the economic climate.

CHAPTER 11

IDENTIFYING AND CAPITALIZING ON BRAND IDENTITY

THE BRAND IS THE PRIMARY DRIVER OF PEOPLE'S INTEREST IN AND LOYALTY TO AN organization and its offerings. It is what the audiences, artists, staff, board, and entire community think of the organization. It is the company's reputation, the perceived quality of the works presented, the promise of its products, the value it brings to the participants, and the sum of every experience one has with the organization.

Arts marketing consultants Tim Baker and Steven Roth say,

> Brands enhance value by communicating it more effectively. Branding should therefore be an ideal means of communicating (and thus creating) perceived value for the arts.... Value not communicated is valueless. The corollary is even more powerful: The better you communicate the true value you offer—value as defined by the consumer—the more you increase that value, the more likely the customer will buy.[1]

Organizations must create their brand value proposition carefully, strategically, and analytically. Then the brand value proposition must be used as the key driver of the company's strategy, operations, services, and product development. Says Starbucks CEO Howard Schultz, "In this ever-changing society, the most powerful and enduring brands are built from the heart. They are real and sustainable. Their foundations are stronger because they are built with the strength of the human spirit, not an ad campaign."[2]

Branding is an organizing principle so broad and so defining that it can shape and direct just about anything an organization does. During the brand-building process, every aspect of the organization should be investigated, including all its business practices. This does not mean that branding will interfere with the organization's mission or artistic vision. On the contrary, a thorough brand development and implementation process guarantees that the organization's mission and artistic vision will remain in clear focus and stay central to all activities.

WHAT IS A BRAND?

The brand is an indication of quality and features to expect and services that will be rendered. The key aspects of a brand are *identity*, *integrity*, and *image*. The brand should be vibrant and engaging, should differentiate the organization from the competition, convey the value it offers its audiences, and provide a compelling case to experience its art. The organization's uniqueness is its greatest strength. First and foremost, a great brand requires a great product.

Says Philip Kotler, "Don't advertise a brand; live it."[3] The brand should be lived through every method with which the organization communicates with its publics, from its advertising to the style and quality of its direct-mail pieces and website, to the way ushers and box office personnel relate to the customers, and of course, to the quality and consistency of the artistic work presented on the stage. Marketers should target consumers' minds and spirits simultaneously to touch their hearts. An organization's positioning should trigger the consumer's *mind* to consider a buying decision. A brand requires an authentic differentiation for the human *spirit* to confirm the decision. Finally, says Kotler, "The heart will lead a consumer to act and make the buying decision."[4]

A brand carries meaning and associations. It is a symbol of trust between the organization and the customer. A brand is not a logo; it is not a label. The logo is developed to be a visual expression of a brand, but it is not and cannot be the meaning of the brand itself.

EMOTIONAL ASPECTS OF BRANDS

A brand taps into emotions. Jerome Kathman, an authority on the role of design in brand building, says, "Brands have the ability to influence and enhance people's lives. They provide a means of personal association, of internal reflection, and outward projection of self-image. Emotional brands not only support who we are, but also provide a tangible means of transformation into what we aspire to be."[5]

According to Scott Bedbury, "The near-universal desire for greater personal freedom, and the more particular American quest for rugged individuality, are what we might call *cultural emotions*. . . . Effective brand building requires making relevant and compelling connections to deeply rooted human emotions or profound cultural forces."[6]

Brands are powerful because they work psychologically. They can enhance people's sense of self-identity; articulate and confirm beliefs; change or endorse attitudes and values; influence perceptions, associations, and opinions; and when brands act as the deciding factor in purchase choices, they influence behavior. Brands also influence the use of language—how managers articulate their attitudes and opinions about the organization, its offerings, and other related topics.

SOCIAL ASPECTS OF BRANDS

Brands are social as they represent ideas that people have in common. Marketing expert Mohan Sawhney says that brands are conversations; they need a credible social voice. The power of a brand is demonstrated in the extent to which it brings people together for a common purpose—to share in an experience or to buy the same thing—while being personally relevant. The performing arts are ideal resources for bringing people together, for stimulating conversation, and for developing personal ties.

The organization must know what influences perception and what factors affect influencers through the market funnel. The organization should build brands around community, rather than the other way around. Sawhney says that branding is about *listening*. Listening is brand-centric, issue-centric, customer-centric, competitor-centric, insight-centric, and complaint-centric. In developing a brand strategy we must listen for trends and behavior. In addition to listening in person, we can listen through third-party sites, owned sites, social networks, micro blogs, and video networks. Our objective is to help customers achieve their goals and pursue their interests. Ask how you can help, not what you can sell. It's not about you. Don't drive sales; influence them. Influence them by telling your story.[7]

STORY TELLING IS CENTRAL TO BRANDING

Branding expert Paul Valerio agrees that as brands aspire to create deeper connections with an endlessly distracted consumer, storytelling has become ever more crucial. Valerio offers a few guidelines for creating an effective overture for any brand experience, for designing the experience *before* the experience:

- *Show, don't tell.* Your audience will learn more about your story by experiencing it directly, not by being told about it.
- *Know what your story is about, not just what happens.* This is the most important thing to figure out, and also the most difficult. The product is what your company sells, but the brand is what your company is about. You must know this inside and out in order to encapsulate the brand and communicate it effectively from the beginning.
- *Empathize with your audience.* Involve people on their terms, not yours. Put yourself in your target audience's shoes, and develop the best understanding you can about what they do and do not understand about your product and category.

- *Be honest.* Acknowledge what's difficult for the customer in the experience you provide. If it's a necessary evil, present it as evidence of what makes it worth the trouble.
- *Educate the newcomers while respecting the regulars.* Not everyone starts with the same level of understanding about your category or your product.[8]

A brand story has no meaning when consumers are not talking about it. Conversation is the new advertising. Positive word of mouth, the most effective advertising, originates from engaged, delighted consumers. Stories about a brand can create loyalty in consumers who see the brand as an icon. The brand's identity is rated by accumulation of experience within the community. However, one bad experience can spoil an organization's brand integrity and destroy its brand image.

The Customer Is at the Center of the Brand

The brand is no longer at the center of the universe; the user is. The brand, like beauty, lies in the proverbial eye of the beholder. Says marketer Lisa Baxter, "*Our product is not the art itself, but people's experience of it.*"[9] Baxter says marketers should create a visitor experience blueprint: a roadmap for designing differentiated experiences for intended and current patrons, from long-standing loyal subscribers to young, first-time attenders. The process is one of connecting, understanding, and empathizing with audience members about their passions, experiences, aspirations, needs, and what drives them. This then becomes the roadmap for *brand*. Therefore, the audience experience is ultimately the brand experience. Aligning audience and brand experience at the strategic level creates potential for positive, exciting change at an operational level. Then, brand becomes relevant for everyone within the organization: for artists, fund-raisers, volunteers, and all audience engagement personnel.

> Rather than messaging the brand as *it's all about me*, the focus will change to shaping brand experiences that deliver genuine value for audiences: *it's all about you.* This brings the audience into the heart of the organization because everyone is on "audience red alert" as caretakers of their experiences. This means that all staff members become creative producers of the audience experience and are responsible for delivering on an intended patron experience that delivers on the brand promise.[10]

Customer empowerment is the key to making a difference.

Brand Strength

Brand awareness may be the traditional measure of brand strength, but brand relevance and brand resonance are far more valuable as aspects of the branding package. Says Scott Bedbury, author of *A New Brand World*, "Top-of-mind awareness and other surface-level viewpoints of a brand reveal little about a brand's real strength or weakness. To fully understand a brand you have to look much deeper. You have to strip everything away and get to its core and understand how it is viewed and *felt* by people inside the company and the world outside."[11] The overall goal is to create a brand image that is compelling in its creativity, relevance, and dynamism in relation to various market segments. To build the brand, marketers must develop rich associations and promises for the brand name and manage all the customers' brand contacts so that they meet or exceed customers' expectations associated with the brand.[12]

According to British consulting firm Morris Hargreaves McIntyre, a brand's identity is supported by several component parts: brand value, brand personality, brand attributes, brand benefits, and finally, brand perception. Says Gerri Morris, "The battleground for the customer takes place in the sphere of brand perception: will it fulfill my needs or not?"[13]

Tag Lines for Chicago Opera Theater

Before COT began to perform in the new state-of-the-art Harris Theater for Music and Dance, board and staff members and I engaged in a long and thorough brand analysis and development process. At the end of the process we worked to develop a tag line that would encapsulate the company's personality, benefits, value, and would influence people's perceptions of the brand. We wanted to emphasize how COT is different from the larger and seemingly more traditional Lyric Opera of Chicago, which most COT patrons also attend, and also emphasize the theatricality of COT productions. Some of the words we came up with that define the COT experience were: exciting, intimate, engaging, theatrical, and adventurous. We consulted with some of the artistic directors who had worked with COT to help us distill the experience into a tag line. The result was: "Thrilling music, stunning theater." This message emphasized both the musical and the theatrical aspects of the performance experience. We knew that not all the music was "thrilling," but had other attractive characteristics. Yet we felt that the message was close enough to reality that we were being authentic while creating an emotional resonance.

Shortly after I left COT, a new board president, who had an interest in marketing, and a new marketing director came on board. They had not been a part of our extensive branding process and were not aware of the analysis we had done. (It is unfortunate and too typical that consulting work gets put on a shelf and not reviewed by future managers and directors.) They chose a new tag line: "Opera Less Ordinary." These three words appeared on their website and in all their promotional materials. The word "less" was emphasized in a bold color. I was not particularly tied to our former tag line, but felt that this new one was far off the mark. It contained three words, only one of which was appropriate, "opera." "Less" and "ordinary" are both negative words that, in my estimation, demean and devalue the brand. "Less ordinary" than what?, I asked myself. If it is meant to be less ordinary than performances at the great Lyric Opera, that message will not resonate well with COT's patrons, 87 percent of whom attend Lyric Opera productions. Why would COT aspire to being "ordinary," even if it is "less ordinary?" Instead of presenting a negative image, what can managers say about COT in a positive way?

In 2012, with the arrival of Andreas Mitisek as the new general director, the tag line was quickly changed to "Timeless Opera, Modern Attitude," which encapsulates the essence of COT.

DEFINING THE BRAND ELEMENTS

Several major brand elements can be identified.[14]

The *product* consists of the attributes associated with the ability of the organization to fulfill its mission through its programs, services, and funding. How the performance of those responsibilities is perceived is a key brand driver.

Experience refers to the quality of a user's interactions with the organization. This can encompass patrons' ability to access information and services and their interaction with staff, whether donors feel properly acknowledged, how volunteers are managed, the physical characteristics of an event site, and so on.

Identity encompasses the message conveyed by an organization's name and representations of its image, including logos, publications, the website, merchandise, and collateral materials, as well as ads and other forms of public communication.

Associations are other organizations with which the organization chooses to cobrand or otherwise link to, or which users perceive to be somehow affiliated with the property. Examples include sponsors, media partners, and related organizations.

Competition, which is a key factor in branding in the business sector, is not as strong a driver in the arts sector, as people commonly attend multiple organizations and report increased satisfaction with a wide variety of arts-going experiences.

BRAND IMAGE

In 1955, advertising guru David Ogilvy introduced the concepts of brand image and personality. Said Ogilvy, "Products, like people, have personalities, and they [the personalities] can make or break them in the marketplace.... Every advertisement should be thought of as a contribution to the complex symbol which is the brand image."[15] *Brand image* is an umbrella term for emotional added values.

According to public relations strategist William Rudman, "Image leads to our survival and growth, or to our failure. The way our institutions are perceived has much to do with how many tickets we sell, and to whom, and how many contributions we receive, and from whom."[16]

Historian Daniel Boorstin describes an image as "a visible public 'personality,' distinguished from an inward private 'character.'... By our very use of the term we imply that something can be done to it: the image can always be more or less successfully synthesized, doctored, repaired, refurbished, and improved, quite apart from (though not entirely independent of) the spontaneous original of which the image is a public portrait."[17]

In creating an image, public relations managers and journalists work with facts to develop stories, perspectives, or opinions; they always look for a fresh and exciting angle; and often prefer human interest stories over "substance." The image's "truth" comes from its congruence with the ideas, perceptual processes, beliefs, and self-interests of the audience.[18]

Naive managers and marketers often assume that good ideas, good productions, good composers (or playwrights or choreographers), and good performers have a persuasive power all their own ("See how many Grammys we have won!"). They believe that if presented effectively, this "truth" will be recognized and accepted and the source of what is "good" (the arts organization) will be rewarded. A more realistic view is that images gain power from their appropriateness to the audience to which they are presented.[19]

CHARACTERISTICS OF AN IMAGE

Boorstin describes six characteristics of an image: it is synthetic, believable, passive, vivid, simplified, and ambiguous.

An image is synthetic. It is planned, created especially to serve a purpose—to make a certain kind of impression. This characteristic is exemplified by the trademark or brand name: "The Met" (The Metropolitan Opera) or A.R.T. (American Repertory Theatre). A trademark is a legally protected set of letters, a picture, or a design identifying a particular product. Unlike other standards, trademarks can be owned. One task of PR is to disseminate, reinforce, and exploit them.

An image is believable. It serves no purpose if people do not believe in the image that the organization is trying to convey. One of the best paths to believability is understatement. One prudent public relations director takes advantage of the increasing use of superlatives to make his own hyperbole seem a conservative truth: "This orchestra is not very good... simply the best there is."[20] Of course, the organization must deliver on this promise that sets such high expectations or people will not return for a second visit.

An image is passive. Because the image is already supposed to be congruent with reality, the producer of the image (namely, the organization) is expected to fit the image—rather than strive toward it. The consumer of the image (e.g., a potential customer) is also supposed somehow to fit into it. Once an image is there, it becomes a more important reality than the organization itself. In the beginning the image is a likeness of the organization; finally the organization becomes a likeness of the image, and its conduct seems mere evidence. Sometimes image building is concentrated in a key individual rather than in the organization itself. This is especially true in the case of a strong, charismatic leader.

An image is vivid, concrete, and simplified. The image should be simple and graspable. One or a few of the products, the persons, or the organization's qualities must be selected for vivid portrayal. The most effective image is simple and distinctive enough to be remembered.

An image is ambiguous. In order to suit unpredictable future purposes and unpredictable changes in taste, an image should, according to Boorstin, "float somewhere between expectation and reality, between the imagination and the senses."[21] The fuzzier the image, the more people can clarify it by making it what they want it to be. Whether it is fact or fantasy, the purpose of the image is to overshadow reality.

These characteristics apply whether the organization is describing one of its productions, a performer, or itself. Consider the examples in the following sections.

A PRODUCTION'S IMAGE: The following is a blurb by publicist Susan Bloch for a production of *Playboy of the Western World.*

> A moonstruck dreamer starving for his place in the sun finds it, by chance, in a town hungry for heroes at any cost. Synge's folk tale for all time slyly captures the fire and joy of the Irish mystique as it compassionately satirizes the contradictions of a world gone topsy-turvy. Primitive, lyrical, romantic, and cynical, this masterful play of great wisdom and gorgeous language will be staged by John Hirsute.[22]

Examples of Boorstin's image characteristics abound in this description:

- *Believable*: slyly, compassionately, masterful.
- *Passive*: finds it, by chance.

- *Vivid and concrete*: moonstruck dreamer, fire and joy, primitive, lyrical, romantic, cynical.
- *Simplified*: masterful play of great wisdom and gorgeous language.
- *Ambiguous*: Irish mystique, contradictions, topsy-turvy.

A PERSON'S IMAGE: The Russian pianist Evgeny Kissin was once named Instrumentalist of the Year by *Musical America*. Aspects of Kissin's image were clearly presented in a feature article in *Musical America*'s annual directory. Some excerpts follow:

- *Vivid, concrete*: "In highly Romantic repertoire he demonstrates an instinctual flair and natural gracefulness—in addition to jaw-dropping technical control."
- *Simplified*: Describing the "young prodigy," music critic Harold Schonberg said, "Suddenly I was in the presence of greatness.... The boy had everything."
- *Believable*: "The effect on audiences is heightened by the shy, even awkward stage presence he projects. At 23, his movements to and fro the piano—a bit mechanical but clearly earnest—convey a youthful vulnerability that is instantly endearing. He is equally shy and equally endearing in person."
- *Ambiguous*: "What does the future look like from here? All expectations on his behalf are of a deepening musicality and an even greater range of repertoire.... I am curious," said the interviewer, "to hear how you will put your personal stamp on Viennese classics." Responded Kissin with a chuckle, "I am curious too!"[23]

AN ORGANIZATION'S IMAGE

When positioning an organization and creating its image, the marketer should take into account the perspectives of prospective customers by showing the consumer what's in it for him or her. Image marketers must keep in mind that their goal is to remain honest while being creative and convincing, never deceptive. This is important not only for the obvious ethical reasons, but also for the sake of customer satisfaction. If the organization does not deliver what the image promises, the customer will be far less satisfied than if expectations had never been set so high.

DEVELOPING A BRANDING CAMPAIGN

A brand campaign is not a plan to be implemented with the next season's marketing materials; it is a long-term investment in the organization. Each individual event, program, communication strategy, and customer service encounter is a crucial element of the branding process. The brand is a symbol of trust between the organization and its publics, and trust is something that must be developed and nurtured over time. A brand develops strength through consistency, yet because a brand must be relevant to its customers, aspects of the organization's brand identity must be regularly reviewed. Like the mission, brand identity has long-range implications, but it must be ascertained that the brand continually resonates with audiences in a dynamic environment replete with changing tastes and values.

AUDIENCE SURVEY

The first step of a branding analysis is to conduct an audience survey. Brand building is more intuitive than analytical, but research tools such as audience surveys can be invaluable in learning how patrons view the organization, its offerings, and those of its direct competitors. A well-designed survey can help the organization's managers determine which features of the organization's offerings the patrons value, indicating effective positioning strategies, and what concerns patrons have, indicating possible changes and modifications. Understanding how patrons view the organization helps managers determine which characteristics of the current image to retain and build and which to dispel. In order for a survey to be well designed, the managers and marketing researcher must carefully define the issues and agree on the research objectives.

BOARD OF DIRECTORS AND STAFF BRAINSTORMING SESSION

Once the results of the audience survey are tabulated, analyzed, and prepared for presentation, key members of the board and staff should participate in an intensive brainstorming session about the organization's image and identity. The agenda should include addressing perceptions of the organization, of its direct competitors, and of the field in general (i.e., opera, ballet); identifying the ways the session participants view the organization's audiences; and reviewing the audiences' views of the organization and its offerings, as revealed in the survey results. This process can expose discrepancies between the ways key stakeholders view the organization, the ways they think the audience views the organization, and the actual responses from audience members.

The analysis of these factors leads the group to determine how the organization is currently perceived, what image it desires to project, and to identify current and potential market segments and opportunities for building the image and identity.

DEVELOPING STRATEGY AND TACTICS

In conjunction with key staff members, the branding consultant or marketing director further analyzes the rich material gleaned from the audience survey and the brainstorming session. Results are used to select strategies and tactics and are translated into engaging stories and other copy that reinforce the brand for season brochures, postcards, the website, social media, print and radio advertisements, and press releases. Positioning statements are selected for emphasis with targeted segments. If the organization has its own hall, branding materials should be placed throughout the facility.

Branding Piccolo Teatro di Milano

When Piccolo Teatro of Milan was founded in 1947, everyone was hoping for a complete revival, particularly a cultural one, that would put the horrors of World War II behind them.

Piccolo's mission "Theatre of art for all" represented exactly that spirit. This provided the basis for creating a strong brand identity aimed at building and maintaining a close relationship with its audience. The objective was clear: to educate and to foster a relationship of trust and a sense of belonging, creating not an indifferent spectator, but one who was active and "ready." In order for this to happen, Piccolo decided to direct its artistic and organizational choices toward multigenre and international content for each theater season (classical and contemporary drama, music, and dance); and develop pricing policies that would be attractive to all audience segments.

This helped Piccolo form its public and create a brand synonymous with reliability and a sense of belonging, deep-rooted in the Milanese cultural scene.

From the 1940s to the 1980s Piccolo grew steadily, to a point where it was offering more than 30 shows each season to more than 10,000 season ticket holders. However, the economic crisis of 1992–1993 led to a fall in demand and the number of season ticket holders dropped to 5,000. The subsequent changes in social, political, and economic conditions led to a more fragmented public in the world of culture in Italy.

Within the context of these changes, the development of new technologies and the advent of the Internet represented a crucial step in Piccolo's marketing strategy. In the mid-1990s it switched from the idea of "Theater of art for all" to "Theater of art for each and everyone," to satisfy different needs and create offers and services for varying targets: season ticket holders, youths, groups, and more.

In 1991, the Union of Theatres of Europe (an alliance of the main European theaters) was founded, and since then, "Teatro d'Europa" (or Theater of Europe) has been an integral part of Piccolo's name and logo: "Piccolo Teatro di Milano—Teatro d'Europa." This expression emphasizes the international nature of the "Piccolo Teatro" brand. In fact, Piccolo has been taking its performances worldwide, usually performing in Italian with translated supertitles, to sizable and appreciative audiences.

CHAPTER 12

FORMULATING COMMUNICATIONS STRATEGIES

MARKETING IS A PHILOSOPHY, A PROCESS, AND A SET OF STRATEGIES AND TACTICS for influencing behavior—either changing behavior (e.g., encouraging attendance at certain performances) or preventing it from changing (e.g., encouraging patrons to renew their subscriptions). In previous chapters, we considered the offer, price, and place components of the marketing mix and saw how these components influence behaviors directly by providing incentives for action or for reducing disincentives. Everything about an arts organization—its programs, packages, employees, facilities, and actions—communicates something. But influencing behavior is largely a matter of communication.

ASPECTS OF COMMUNICATION

Communication is a matter of *informing, persuading*, and *educating* target audiences about the alternatives for action, the positive consequences of choosing a particular course of action, and the motivations for acting (and often continuing to act) in a particular way.[1]

INFORMING

In order to make a decision on whether to attend, a patron needs basic information on the event itself: what will be performed and who will be performing, as well as the date, location, time, cost of tickets, and how tickets may be purchased. This may sound obvious, but some organizations do not prioritize this information according to what is important or most useful to potential ticket buyers.

Information is a necessary ingredient of most communication efforts but can stand alone only for the enthusiasts—the patrons with such a strong interest in a particular art form that they will actually seek out information on future performances without the benefit of extensive promotion.

PERSUADING

There are legendary accounts that when the late, renowned pianist Vladimir Horowitz performed recitals in Moscow, one bulletin posted outside the box office was all that was needed to sell out the event in a single day. In most cases, however, prospective patrons need much more than simple information to encourage them to attend. All sales promotion techniques, public relations, personal selling, and any advertising that goes beyond basic information is intended to persuade. Persuasion is central to marketing communication.

Performing arts organizations should make their purpose and positioning clear to their target audiences. Furthermore, observes music critic and commentator Greg Sandow,

> What you'll almost never read is anything that might make an intelligent person want to go to a concert—something about how the music is going to be played, what it will feel like to hear the music, or why this concert might be different from any other. Mostly we brag that we offer "acclaimed" musicians, predictably playing—no surprise here!—"great music." Every concert, if you believe our advertising, seems more or less the same—uniformly "great" and uplifting. Why should anybody care?[2]

Similarly, elitist attitudes among those who produce and present the arts work against what are otherwise good efforts to attract new audiences. *New Yorker* music critic Alex Ross articulates the problem this way:

> For at least a century, music has been captive to a cult of mediocre elitism that tries to manufacture self-esteem by clutching at empty formulas of intellectual superiority.... They are making little headway with the unconverted because they have forgotten to define the music as something worth loving. If it is worth loving, it must be great; no more need be said.[3]

Tacoma Symphony Orchestra music director Harvey Felder sets himself apart from the long-standing stereotype of the authoritarian conductor whose primary concern is performing the music he prefers. Felder writes about how to attract members of Generations X and Y by establishing trust and showing respect with acknowledgment of the customers' attitudes and preferences. The key, he says, is *relevance*. "Any approach that implies even the slightest 'this is good for you' subtext needs to be abandoned," says Felder. "Any attempt to position musicians, classical music, or the concert hall itself as representing something above the ordinary will likely not be received well." Felder delves into the marketing function by discussing promotional materials, which, he says, "should be conceived and reviewed carefully, with particular scrutiny given to graphic-design elements of advertisements, brochures, and websites that convey an implied superiority. . . . Those of us who have worked in this field for years are often blind to these subtle yet powerful embedded messages. Something as standard as an image of an individual in white tie and tails can send a subliminal message of smugness. Fresh, outside eyes are needed to help us see what we are saying to people."

In addition to *what* we say, *how* we get our messages across is a crucial element in the communication process. For Generations X and Y, suggests Felder, marketers can "humanize the musicians of the orchestra via daily blogs, YouTube posts, and Twitter updates. An interview with the principal trombonist in Mahler's Symphony No. 3, or a discussion with the principal oboist in Strauss's *Don Juan*, could offer great personal insight. To be privy to their excitement, psychological preparation, and even their nervousness would be fascinating."[4]

Screenwriter Robert McKee believes that there are two distinct ways to convince people. The first one is to base your ideas on a set of facts and numbers and engage people in intellectual arguments. An alternative, which he thinks is much more effective, is to write compelling stories around the ideas and engage with people's emotions instead.[5]

EDUCATING

For most people, an appreciation of the performing arts is learned or acquired over time. But education is a difficult task. A great deal of information has to be conveyed, much time and effort are required, and often people's long-held attitudes and beliefs must change. This is a primary reason why arts organizations have traditionally focused their efforts on informing and persuading current cultural patrons rather than on the much more daunting task of educating nonattenders. Yet creative arts managers can and do offer a wide variety of educational programs and resources, in person and online. The Internet makes it possible for arts organizations to educate their publics, no matter what their level of sophistication, efficiently and cost-effectively.

FUNDAMENTALS OF COMMUNICATION

COMMUNICATION IS PERCEPTION

This means that it is the *recipient* of the message, not the sender, who is central to communication. The message transmitter can only make it possible for a recipient, or *percipient*, to perceive. Perception is based on experience, not on logic, so one can perceive only what one is capable of perceiving. Therefore, in order to make communication possible, one must first know the recipient's language and experience.

The situation is further complicated by the problem of selective attention—the recipient will not notice all of the stimuli. People are bombarded by numerous commercial messages from a wide variety of media. Jay Walker-Smith, president of the marketing firm Yankelovich, says we've gone from being exposed to about five hundred ads a day back in the 1970s to as many as five thousand a day today. "Everywhere we turn we're saturated with advertising messages trying to get our attention. It seems like the goal of most marketers and advertisers nowadays is to cover every blank space with some kind of brand logo or a promotion or an advertisement. It's all an assault on the senses," says Walker-Smith.[6]

The American Association of Advertising Agencies reports that of all the commercial messages a day directed at the average individual, 80 are consciously noticed by the individual and 12 provoke some reaction. Apparently we have become skilled in screening out thousands of unwanted messages a day and at selecting the desired or unavoidable 12.[7]

COMMUNICATION IS EXPECTATION

We perceive, as a rule, what we *expect* to perceive. This is the process of selective distortion; people twist a message to hear what they want to hear. Receivers have set attitudes and beliefs, which lead to expectations about what they will hear or see. As a result, receivers often add things to the message that are not there and do not notice things that are there.

COMMUNICATION MAKES DEMANDS

If the communication fits in with the aspirations, values, and purposes of the recipient, it is powerful. If it goes against his or her aspirations, values, or motivations, it is likely not to be received at all, or, at best, to be resisted. Therefore, there is no communication unless the message can tap into the recipient's own values, at least to some degree. Communication can produce the most effective shifts with regard to lightly felt, peripheral issues that do not lie at the center of the recipient's value system.

COMMUNICATION VERSUS INFORMATION

Communication and information are different and largely opposite, yet interdependent. Whereas communication is perception, information is data. Information is impersonal, rather than interpersonal. The more it can be freed of the human component—of emotions, values, expectations, and perceptions—the more valid, reliable, and informative it becomes. Yet, information's effectiveness depends on the prior establishment of communication. So the most perfect communications may be purely "shared experiences" without any information whatever. Perception, then, has primacy over information. True communication exists when the recipient perceives what the transmitter intends.[8]

Consider how the Columbus (Georgia) Symphony Orchestra utilized these principles in developing an advertising campaign.

Columbus Symphony Creates a New Image

Michael Burks, former executive director of the Columbus (Georgia) Symphony Orchestra understood that in his community, it is not the classical music or the price of a ticket that is keeping people away from the concert hall, but their lack of familiarity with the experience itself. So Burks created the following ad to be aired on local TV stations during the week before a concert:

The camera scans the audience and focuses on an African American couple in their late thirties or early forties. She is in a sequined dress; he is wearing a tux. Both are fidgeting uncomfortably while looking at other audience members. His eyes focus on a college-aged couple dressed casually. The audience can hear the tuxedoed man think: "I know I could have dressed more casually. I sure am uncomfortable." Meanwhile, the younger man's eyes are also darting uncomfortably around the hall, and we can hear him think: "I know I should have worn a jacket and tie."

The announcer then says: "You don't have to feel uncomfortable to enjoy a concert."

This ad functions on two levels. First, it dispels the symphony's highbrow, stuffy image to those unfamiliar with the experience by showing that one doesn't have to dress up or feel the need to "fit in." It's not what you wear; it's being there that counts. Second, by incorporating wit and humor, it shows that the symphony experience is not dull and boring.

After the ad was aired, ticket sales jumped by almost 40 percent for that weekend's concert. And several longtime attenders were heard to say how nice it was to see so many young people there.

This example underscores the key factors in effective communication. Senders must know what audiences they want to reach and what approaches and media will reach them best. Their first goal is to gain the attention of the recipient; the next is to modify the recipient's behavior, sometimes requiring a modification in beliefs and attitudes. The more the sender's field of experience overlaps with that of the receiver, the more effective the message is likely to be.[9] Much of what is called persuasion is self-persuasion by the recipient.[10]

Steps in Developing Effective Communications

There are several major steps involved in developing a total communication and promotion program. The marketing communicator must

- identify the target audience;
- determine the communication objectives;
- design the message;
- select the communication channels;
- allocate the total promotion budget;
- decide on the promotion mix;
- measure the promotion's results; and
- manage and coordinate the total marketing communication process.

Identifying the Target Audience

A marketing communicator must start with a clear target audience in mind. For example, the audience could be potential buyers, current users, deciders, or influencers. It could be composed of individuals, families, or lifestyle groups. The organization's communications responsibilities also include gaining the support and goodwill of groups such as the press, government agencies, foundations, and the corporate community. The target audience will critically influence the communicator's decisions about what to say, how, when, where, and to whom. The better the marketer understands the target audience and that audience's image of and attitude and beliefs about the organization, the more effective the communications will be.

Determining Communication Objectives

The more carefully the specific objective of a communication is defined, the more effective the communication is likely to be. Possible objectives include making target consumers aware of a product or service, educating consumers about the offer or changes in the offer, or changing beliefs about the consequences of taking a particular action. The organization may also want to build consumer preference for its offerings over other options, develop conviction that buying a ticket or a subscription is a compelling thing to do, prevent discontinuation of behaviors, combat injurious rumors, and enlist the support of collaborators such as nearby restaurants or other arts organizations.

The ultimate objectives, of course, are purchase and high satisfaction. But purchase behavior is the end result of a long process of consumer decision-making.

BUYER READINESS

The marketing communicator needs to know how to move the target audience to higher states of readiness to buy. We can distinguish six buyer-readiness states.

Awareness. If most of the target audience is unaware of the organization or its offerings, as in the case of a dance company touring in a new city, the communicator's first task is to build awareness—perhaps just name recognition. This generally takes considerable time.

Knowledge. The target audience might be aware of the organization but not know much about it. The dance company may want its young target audience to know that it presents modern, upbeat, energetic programming, some of which is performed to rock music.

Liking. Assume the dance company is returning to a city for the second time. If target audience members know the organization and its offerings, how do they feel about it? What is the nature of the word of mouth that was spread by audience members after the first visit? If positive feelings prevail, they can be used to advantage by the communicator. Testimonials by audience members in other cities can also be included.

Preference. The target audience might like the organization and its programs but not prefer it to other available options. In this case, the communicator must try to build consumer preference. Arts organizations combating the growing preference of many people to, for example, stay home and watch movies, must find ways to stimulate the preference for a live performance.

Conviction. Perhaps the target audience knows that this is a terrific dance company and prefers it to other dance companies, but is not sure it wants to attend a dance performance. The communicator's job is to build conviction that attending this dance performance is the (fun, exciting, best) thing to do. The organization might invite community leaders to attend a rehearsal or special event, or might offer additional benefits. An orchestra trying to attract subscribers may invite people to a preview concert that provides an overview of the forthcoming season.

Purchase. Finally, some members of the target audience might have conviction but will not quite get around to making the purchase. They may wait for more information or plan to act later. The communicator must lead these consumers to take the final step. The touring dance company can try to develop a sense of urgency: "Only in town for three performances!" Or they may offer something special such as an exclusive backstage tour after the performance by the production manager.

Communication objectives depend heavily on how many people already know about an offering and may have already tried it. Determining the buyer-readiness states of the target consumers is critical in developing a communication program that will be cost-effective while inducing the desired response.[11]

Designing the Message

After defining the desired audience response, the communicator moves to developing effective messages. Ideally, each message should gain *attention*, hold *interest*, arouse *desire*, and elicit *action*. This is known as the AIDA model. In practice, few messages take the consumer all the way from awareness through purchase, but the AIDA framework suggests the desirable qualities. Formulating the message will require solving four problems: what to say (message content), how to say it logically (message structure), how to say it symbolically (message format), and who should say it (message source).

Formulating the Appeal

The communicator has to develop an appeal, theme, idea, or unique selling proposition (USP). This amounts to formulating some kind of benefit, motivation, identification, or reason why the audience should consider or purchase the offering. The USP is what distinguishes an offer or organization from others. Consider the following appeals:

- "Celebrate the majesty of music."
- "Transcend the ordinary."
- "Every time you take your seat you will be moved to an experience beyond words."

Each of these appeals is so general that it could apply to a multitude of organizations and experiences other than the one for which the ad was produced. Now consider the uniqueness of the following appeals:

- "Climbers dream of Mount Everest; divers dream of the Great Coral Reef; music lovers dream of Carnegie Hall."
- "Who in the world has won 51 Grammys?" (The Chicago Symphony Orchestra).
- The Court Theatre is the place for "Classics that bristle with a new energy."
- Chicago Opera Theater: "Timeless Opera, Modern Attitude."

Each of these appeals identifies a characteristic unique to the organization, communicating an experience that cannot be found elsewhere.

The more direct competition an organization has, the more it is necessary to develop a USP specific to the organization itself. When a symphony orchestra is the only serious music venue in town, a byline such as "Celebrate the majesty of music" may be appropriate, if not compelling. However, messages such as "Transcend the ordinary" and "Every time you take your seat you will be moved to an experience beyond words" could apply as well to a rock concert or a sporting event.

Types of Appeals

Three different types of appeals can be distinguished.

Rational appeals engage the audience's self-interest. They show that the offering will produce the claimed benefits, and are exemplified by messages that demonstrate an offering's quality or value. For example, the Steppenwolf Theatre once made a strong rational appeal by calling itself "The nation's most important, most consistently successful resident professional theatre company."

Emotional appeals attempt to stir up some emotions—positive or negative—that will motivate purchase. The Milwaukee Chamber Theatre's season brochure asks: "Wanna Play?...So do we....In the Broadway Theatre Center." The New York Philharmonic uses a lighthearted, humorous approach when it states: "Speeding Tickets....Don't get caught without seats for the New York Philharmonic's fall season."

An emotion that has kept many people from attending the performing arts is a sense of inadequacy. This derives largely from the elitism that has been rampant in the fine arts world, keeping the less sophisticated from feeling they are able to "appreciate" certain art forms. Some arts organizations, in trying to attract new patrons, are dealing with this feeling directly. For example, a quote on the back cover of a Chicago Shakespeare Repertory brochure states in part: "(Artistic Director) Barbara Gaines does for Shakespeare what Julia Child does for French cuisine: She demystifies it, popularizes it and turns it into great entertainment."

Whatever the theme, focusing on an emotional appeal remains central. Says Kevin Copps, general manager of Elektra International Classics, "It's a very basic marketing concept. You have a product. It doesn't matter whether it's a deodorant, a breakfast cereal or a (classical) CD. The question is, how do you get the consumer emotionally attached to your product?"[12]

Moral appeals are directed to the audience's sense of what is right and proper. They are often used to exhort people to support social causes. In the arts, moral appeals are most commonly used to encourage school boards to support more educational arts programs and legislators to provide more financial support for arts organizations as well as for arts education. Moral appeals are probably the least effective for audience development purposes, because a "good for you" approach is far less compelling than a dynamic emotional appeal. It is certainly more effective to appeal to what people *want* to do than to what they *ought* to do.

Strength of Appeal

Some advertisers believe that messages are maximally persuasive when they are moderately discrepant with what the audience believes. Messages that only state what the audience believes attract less attention and at best only reinforce audience beliefs. And if the messages are too discrepant with the audience's beliefs, they will be counterargued in the audience's mind and be disbelieved. The challenge is to design a message that avoids the two extremes.

The marketer should select the best message from a set of alternatives developed. It has been suggested that messages be rated on three scales: *desirability*, *exclusiveness*, and *believability*.[13] The message must first say something desirable or interesting about the product or behavior. This is not enough, however, because many competitors may be making the same claim. Therefore the message must also say something exclusive or distinctive that does not apply to every alternative. Finally, the message must be believable or provable. The desirable feature or features will vary by event and by market segment. By asking consumers to rate different messages on desirability, exclusiveness, and believability, marketers can evaluate them for their communication potency.

When someone registers online to purchase tickets through the website of the exquisite Palais Garnier, home of the Paris Opera Ballet, the patron receives an email message that provides him or her with an account number. The message says in part, "This number is *precious.* Keep it in a safe place." What a great way to communicate how special it is to see a performance at this renowned opera house.

Breadth of Appeal

Some messages draw conclusions about the product's benefits or advantages to the consumer. However, drawing too explicit a conclusion can limit a product's acceptance. If the issue is simple and/or the audience intelligent, they might be annoyed at the attempt to explain the obvious. Or, if the issue is highly personal, the audience might resent the communicator's attempt to draw a conclusion. The Pittsburgh Symphony once created a "romance" brochure, with soft-focus photographs of two lovers nuzzling, complemented by overheated text. The symphony changed its approach the next season, as this message offended some of the older subscribers.

Research indicates that stimulus ambiguity can lead to broader market acceptance. The best ads ask questions and allow readers and viewers to come to their own conclusions.[14] When a message is ambiguous, it can mean whatever the recipient wants it to mean. Coca-Cola advertisers knew this when they developed the message "Coke—it's the real thing." (What is a *thing?* What is *real* about it?) Similarly, Nike's "Just do it!" campaign is as wonderfully ambiguous as a message can be. (Just do *what*?)

Message Structure: One- and Two-Sided Arguments

A message's effectiveness depends on its structure as well as its content. The choice between one- and two-sided arguments raises the question of whether the communicator should only praise the product or also mention some of its shortcomings. One would think that one-sided presentations would be the most effective. Yet, it has been found that:

- one-sided messages work best with audiences who are initially predisposed to the communicator's position, and two-sided arguments work best with audiences who are opposed;
- two-sided messages tend to be more effective with better-educated audiences; and
- two-sided messages tend to be more effective with audiences that are likely to be exposed to counterpropaganda.[15]

Order of presentation raises the question of whether a communicator should present the strongest arguments first or last. In the case of a one-sided message, presenting the strongest argument first has the advantage of establishing attention and interest. This is important in newspapers, email blasts, and other media where the audience does not attend to the whole message. In a medium with a more captive audience, such as an announcement by the house manager before a performance, a climactic presentation might be more effective. In the case of a two-sided message, the issue is whether to present the positive argument first, relying on the primacy effect, or last, hoping to benefit from the recency effect. If the audience is initially opposed, the communicator might start with the other side's argument. This approach will disarm the audience and also has the advantage of concluding with the strongest argument. In high-involvement situations, the target audience will engage in extensive internal cognitive activity, including consideration of costs and alternatives, and an external search that will make available to them the "other side" of the argument.

The Chicago Shakespeare Theater put this principle to use for one season brochure. Recognizing many people's reluctance to attend Shakespearean productions, the company's brochure cover shows a photograph of an actor dressed as Shakespeare, wearing a signboard that says: "Shakespeare is for repressed, uptight, highbrow, pompous bookworms." Inside the cover is a back view of the signboard reading: "*You'd be surprised.* Shakespeare Repertory turns stereotypes upside down and Shakespeare right side up!"

Two-sided arguments may also recognize other alternatives available to the target audience. For example, the marketer may want to confront the fact that going to the theater or symphony means not staying home to watch TV, and then explain persuasively why the live theater experience is well worth whatever hassle is involved.

Format Elements

Format elements can also make a difference in a message's impact. Advertisers use such attention-getting devices as novelty, contrast, arresting pictures, and video. Humor is often an effective attention-getter for any type of message and even rewards the listener for paying attention. It also distracts people from counterarguing. When humor is used, it should be directly related to the message and neither the speaker nor the message's recipient should be at the receiving end of the joke. Special care must be taken when using humor in advertising: multiple exposures are usually necessary for an ad to be effective, and humor tends to wear out quickly.

An economical way to convey a message is to take advantage of common visual or verbal associations. Potential patrons often wonder if people "like them" attend these performances, so the marketer might want to show people of various age groups and ethnicities in the audience, depending on the target market for the ad. Colors have symbolism; in the United States, white is pure, gold is rich, blue is soothing, red is exciting, and so on.[16] Symbols can help or hurt communicators. The problem, of course, is to choose the right symbols and to be assured that the audience receives them as they are intended.

Message Source

Messages delivered by attractive sources receive more attention and are more reliably recalled. In the arts, a well-known and well-liked figure can help bridge the gap to accessibility. Transference is likely to occur so that positive feelings for the ad and spokesperson become positive feelings for the product. Tony and Emmy award–winning actor Alec Baldwin hosts weekly radio programs featuring the New York Philharmonic. His knowledge of and passion for classical music allow him to talk as comfortably about orchestral music as about movie shoots. "What classical music does is a little removed from the mainstream," said the orchestra's music director, Alan Gilbert. "Frankly, anybody who expresses real interest and devotion to what we do is welcome. All the better if it happens to be a famous person who is really hot."[17] Baldwin made a major statement about his devotion to the New York Philharmonic in 2012 with a donation of $1 million.

Messages delivered by highly credible sources are the most persuasive. The three factors most often identified with source credibility are expertise, trustworthiness, and likability.[18] *Expertise* is the specialized knowledge the communicator appears to possess that backs the claim. *Trustworthiness* is related to how objective and honest the source is perceived to be. Friends are trusted more than strangers or salespeople. The spokesperson does not need to possess special knowledge, but should be trusted to report accurately. *Likability* describes the source's attractiveness to the audience. Such qualities as candor, humor, and naturalness make a source more likable. The most highly credible source would be a person who scored high on all three dimensions, as Leonard Bernstein did during his leadership of the New York Philharmonic Orchestra and as Yo Yo Ma does today.

Choosing the ideal spokesperson is not easy. Sometimes, unknown or lesser-known faces can seem more approachable, which could help influence consumers to buy products. However, celebrity endorsements tend to reach out to an aspirational audience that admires them, while simultaneously reaching an already affluent market.[19] "In terms of whether it is better for a brand to be represented by a celebrity, model or unknown person depends upon the product," said Rex Whisman, principal at branded Consultants Group, Denver. "In all cases, the spokesperson should be someone who aligns with the brand's core values and is believable in the mind of the consumer."[20]

The effectiveness of a famous spokesperson is limited by his or her popularity, which is likely to wax and wane. One possible alternative is to hedge one's bets by using multiple spokespeople. Also, people like things to fit; they are uncomfortable with cognitive dissonance. If people like a particular spokesperson but dislike the particular offering, they will seek to reduce the discrepancy in their feelings toward the message and the source. They may tend to favor the spokesperson a little less or the offering a little more. The principle of congruity says that spokespersons can use their positive image to reduce some negative feelings toward an offering but in the process might lose some esteem with the audience.

COMMUNICATION CHANNELS

Messages are transmitted to the target consumer through some medium. The medium may be nonpersonal, like an organization's brochures, website, or newspaper ads; or personal, like the organization's own spokespersons or a patron's family or friends. Email messages are usually nonpersonal, but can effectively be mass personalized with different messages to subsegments.

A further distinction can be drawn between *advocate*, *social*, and *expert* channels of communication. Advocate channels consist of an organization's efforts to influence members of the target market through both impersonal media (i.e., direct marketing) and personal resources (i.e., telemarketing). Social channels by definition are personal, consisting of friends, family members, business associates, and other people who influence target buyers. Expert channels may be independent experts such as music, dance, and theater critics; advocates, such as a conductor acting as spokesperson for the orchestra's upcoming special event; or social, such as a patron's musician friend who is touting a young pianist.

Nonpersonal Communication Channels

Nonpersonal communication channels carry messages without personal contact or interaction. They include media, atmospheres, and events. Media consist of print media (newspapers, magazines, direct mail), broadcast media (radio, television), electronic media (email, the Internet, social media), and display media (billboards, signs, posters). Most nonpersonal messages come through paid media.

Atmospheres are "packaged environments" that create or reinforce the buyer's leanings toward product purchase and satisfaction with the experience. Opera houses are often designed to create an atmosphere of elegance and classicism. Experimental plays are often produced in "raw" settings, not just for their low cost but also because they provide atmospheric versatility, spontaneity, and non-elitist comfort.

Public relations departments arrange news conferences, grand openings, and other events to achieve specific communication effects with a target audience. One opera company gained high conversation value when it performed the Triumphal March from *Aida* during halftime at the local university's football game. Similarly, star performers generate conversation when they do a well-publicized CD or book signing.

Personal Communication Channels

Personal communication channels involve two or more people communicating directly with each other, either face to face, person to audience, over the telephone, via email, social media, or through the postal service. Personal communication channels are effective for individualizing the presentation and feedback. The most persuasive form of personal communication for attracting attendance at performing arts events is word of mouth, conducted most often through social channels. Audience surveys repeatedly find that word of mouth is the most frequently used information source.

Building Word-of-Mouth Marketing

Given all the time and expense involved in creating and disseminating marketing communications—email blasts, tweets, Facebook pages, YouTube videos, advertisements, direct-mail pieces, press releases, and so on—marketers know that word of mouth is the most influential factor in most people's ticket purchasing decisions, often more influential than all other marketing efforts combined. Although arts marketers are well aware of the significance of word-of-mouth marketing, it is not a common enough practice to implement strategies that encourage patrons to spread the word. Says Emanuel Rosen in his book *The Anatomy of Buzz: How to Create Word of Mouth Marketing*,

> Most of today's marketing still focuses on how to use advertising and other tools to influence *each customer individually*, ignoring the fact that purchasing many types of products is part of a *social process*. It involves not only a one-to-one interaction between the company and the customer but also many exchanges of information and influence among the people who surround that customer.[21]

Among the benefits of word-of-mouth marketing is that it empowers the customer; it puts the customer in the position of being the marketer. In fact, claims Rosen, people will talk more about what they discover themselves. People enjoy their own credibility, and when they back a product, they know that they're putting their reputation on the line. Buzz capitalizes on the fact that people are connected—connected in their personal relationships, through blogs, and through email. Buzz travels most smoothly through channels built on trust.

Before they attempt to stimulate buzz, Rosen suggests marketers and managers first ask themselves the following questions: Is something about my offering "fresh and different"? Is there something I can do to make my product or its users more noticeable in a positive way?[22]

All too often, arts organizations send an email to people who have recently seen a show, offering a $5 discount to any friends to whom they forward the message. What people really need is not a small discount to share, but help in describing a show they loved—stories about the performers, or the set, or the fascinating way the story evolves (without giving away the ending, of course). Many people who enjoy a play, a dance performance, or a concert, and want to recommend it to others, feel inadequate about putting into words what they loved about it and why they think their friends should see it too. The organization can help patrons motivate others to attend by providing compelling descriptions of a production and selections from good reviews.

Marketers should regularly talk with customers on an informal basis. Subscribe to blog services. Know what people are saying about the organization and its competitors. Identify the hubs to which customers belong—both the expert hubs and the social hubs—and try to distinguish who are the connectors, bridging the distance between clusters. Use all available techniques to find even more hubs, and be sure to keep good records of the contacts as they are found. Develop a profile for each person including his or her contact information, how much influence, and what useful connections he or she has.

Finally, marketers must regularly provide the hubs with updated information and offers, and acknowledge referrals, rewarding customer loyalty. Organizations can use a sneak preview to capture the interest of a select group of customers, and take their customers behind the scenes to share the excitement of a production coming together. People love hearing about any human drama that takes place in the creation of a show, not only the drama unfolding on stage. If an organization has a charismatic leader, he or she should be brought before its customers as often as possible. Also, organizations should make it possible for customers to talk to leaders, and be receptive and responsive to customer feedback and input. Generating buzz about an organization's offering calls for an attitude different from the perspective marketers are used to. Rosen also encourages marketers to operate in a spirit of truth, honesty, and directness, as openness and candor are keys to developing strong, long-term, grassroots support.

Viral marketing, so named because it is intended to spread widely, is most efficiently undertaken through an organization's website, email messages, and social media. Arts marketers must make sure that their website and email messages have a simple-to-use pass-it-on mechanism that is presented in a friendly way. Although the social contacts are usually their own reward, sometimes it is helpful if the messages include special, valued benefits that come from successfully spreading the news.

The most important policy, says Rosen, is to keep customers involved. "If you involve them, engage them, make it interesting for them, they will talk. Involvement translates to action, which in turn translates to buzz."[23]

To create viable conditions for buzz, marketers must invest energy behind the product, in terms of both time and money. The key is to build networking by reaching people who can spread the news to many others. In his book *The Tipping Point: How Little Things Can Make a Big Difference*, Malcolm Gladwell describes the power of connectors—acquaintances who give us access to opportunities and worlds to which we don't belong. Gladwell claims that the *opinion leaders* among those acquaintances are the ones who garner the most respect and therefore have the most power to influence others.[24] Arts organizations can capitalize on the influence of opinion leaders by offering tickets for performances early in the run of a show to people who are likely to be interested and who will talk about it to others.

Thus, the Saint Louis Symphony offers free tickets and membership in the hospitality room to African American ministers whose churches participate in the symphony's community partnership program. At many organizations, corporate managers are given special perks if they purchase blocks of seats or stimulate sales among their associates and employees. An arts organization can work with influential community members such as radio announcers, class presidents, and presidents of membership organizations. Patrons can be offered free guest passes with each subscription and can be encouraged to give gift certificates for performances to friends, associates, and employees.

ESTABLISHING THE PROMOTION BUDGET

Several factors affect how promotion budgets are determined. Some art forms and some specific events have greater drawing power than others, due to a well-known performer, a "war-horse" production, a great review, or a high level of buzz. It is up to the communications manager to determine the most effective level of promotional spending for each event. A highly popular program may garner a strong response to advertising messages, while no amount of promotional activity will attract an audience to an undesirable program. Therefore, the promotion budget should be varied from event to event, according to an estimated response to different levels of promotion. Of course, a certain amount of flexibility should be built into the budget planning so that if a production gets rave reviews and/or an unexpected groundswell of word-of-mouth support, the organization can take full advantage and adjust its advertising accordingly.

Arts organizations have several opportunities to help control their promotion budgets. First, low-cost or free opportunities for advertising are available to some organizations. Some media will provide a certain amount of free advertising as part of their public service efforts. In some cities, the local trade organization or arts council makes bulk media buys at a discount and then provides the space or time to arts organizations at an attractive rate. Community newsletters and publications are available to some arts organizations. Local businesses often collaborate with arts organizations and exchange free publicity.

METHODS FOR ESTABLISHING THE PROMOTION BUDGET

Four common methods are used by organizations to set their promotion budget.

The Affordable Method

Many organizations set the promotion budget at what they think they can afford. However, making the decision according to currently available funding completely ignores the role of promotion as an investment and the immediate impact of promotion on sales volume.

The Percentage-of-Sales Method

Many organizations set their promotion expenditures at a specified percentage of sales (either current or anticipated). This satisfies the financial managers, who feel that expenses should bear a close relationship to the movement of ticket sales, and encourages management to think in terms of the link between promotion cost and selling price. Despite these advantages, the percentage-of-sales method has little to justify it. It uses circular reasoning in viewing past or predicted sales rather than market opportunities as the basis for promotion budgeting. It discourages experimenting with countercyclical promotion or aggressive spending. The promotion budget's dependence on year-to-year sales fluctuations interferes with long-range planning and may cause an organization to spend a lower amount than is ideal at just the time when it may be in the organization's best interest to increase promotional activity.

The Competitive-Parity Method

Some organizations set their promotion budget to achieve share-of-voice parity with their competitors. This thinking is illustrated by an arts manager who was overheard saying, "If I could afford weekly half-page ads like some of the wealthier theaters in this town, I would have no problem filling my hall." She assumes that what the bigger theaters are doing represents the collective wisdom of the industry. But there are no grounds for believing that what the competition spends on promotion and how it chooses to spend its allocation is right for other organizations. Organizations' reputations, resources, opportunities, and objectives differ so much that others' promotion budgets are hardly a guide.

The Objective-and-Task Method

The objective-and-task method calls upon marketers to develop their promotion budgets by defining their specific objectives, determining the tasks that must be performed to achieve those objectives, and estimating the costs of performing those tasks. The sum of these costs is the proposed promotion budget. This method has the advantage of requiring management to spell out its assumptions about the relationship between dollars spent, exposure levels, trial rates, and regular attendance.

After implementing the promotional plan, the communicator should measure its impact on the target audience as best as possible. Marketers should keep a log of when and where all marketing promotions are implemented and track response rates accordingly. Promotions that carry a special benefit are easy to track. Patrons may be asked about where they are most likely to get their information about the performances they attend, but typically, they are exposed to several media, whose effect may be cumulative.

INTEGRATED MARKETING MANAGEMENT

To promote an organization, a season, or an individual production most effectively, marketing managers should plan and implement an integrated marketing approach. *Integrated* marketing means using a multiple-vehicle, multiple-stage campaign. The multiple vehicles might include a combination of paid ads in such media as newspapers, radio, billboards, TV, and the Internet; direct mail, brochures, or postcards; telemarketing; email; and public relations. Multiple stage refers to the time line developed to best leverage each of these media.

For example, a subscription campaign might start by mailing brochures to current subscribers, followed up with another mailing about one month later to people who have not renewed. About two weeks after the second mailing, begin telemarketing to reattract subscribers who have not renewed by mail. Concurrent with these efforts, the organization should send email notices about the forthcoming season and reminders at appropriate intervals. (Subscribe by May 1 to retain your seats!) Before this deadline approaches, begin the campaign to attract new subscribers. Send postcards and emails to prospects, with a response mechanism for people who want more information. Direct people to the website and clearly offer opportunities for people to request more information via email or snail mail. Try to capture the email addresses of anyone who comes in contact with the organization and maintain ongoing communication, even if the person is not ready to buy.

An integrated marketing plan to attract single ticket buyers making their buying decisions within a week or two of a performance should guarantee intense visibility in multiple media all in the same key time frame. Multiple media are used because different individuals respond to different media and because people in general are more likely to respond when they have had multiple exposures to a message. To accomplish this, organizations can use a strategy known as *roadblocking*: air a radio ad at the same time on several stations frequented by the target markets. For example, organizations can plan that consumers will hear the ad on the rush-hour commute from work and again in the early morning rush hour, when a radio personality will pitch the show and an opportunity to win free tickets. Send postcards so they are waiting in the consumer's mailbox when he or she arrives home, and send email messages with links to the organization's website where in-depth information can be found. Place ads in the newspaper during this time frame as well, but do not assume that the ads necessarily belong in the arts or entertainment section. Try to reach out to people where they are, where they may be expected to be. This use of *response compression*, whereby multiple media are deployed within a tightly defined time frame, increases message awareness and impact.

Consider Many Marketing Costs as Investments, Not Expenditures

Many marketing costs, such as those dedicated to developing a branding campaign or building young audiences, are undertaken with the understanding that they will not provide immediate returns, but are investments in the arts organization's future. When budgets are tight, these expenses are often eliminated or severely cut back. Similarly, costs for advertising, direct mail, and other current audience development activities are often viewed as expenditures that can be cut when revenues fall below expectations. They are easy targets for budget-minded managers because their costs are variable, whereas many of an organization's other expenses are fixed. Arts managers must realize, however, that slashes in the marketing budget not only compromise the marketing department's ability to meet capacity goals, but also lead to cutbacks in communication and service and compromise the organization's ability to provide value and satisfaction to its target customers. Marketing needs to be seen as an investment center whose expenditures create long-lasting customers and revenue flows.

Managing and Coordinating the Communication Process

It is imperative that communication tools and messages be coordinated. Otherwise, the message might be ill timed, lack consistency, or not be cost-effective. For example, direct mail and telemarketing campaigns for subscription renewals should be carefully timed so that together they reap the greatest possible response in the most cost-effective manner. The style of newsletters should imitate the organization's image as put forth in brochures and advertisements, and the newsletters' content should enhance other new subscriber and subscriber-retention efforts.

In striving to achieve integrated marketing communications, the executive director should appoint a marketing communications director who has overall responsibility for the organization's persuasive communications efforts; works out a philosophy of the role of different promotional tools and the extent to which they are to be used; keeps track of all promotional expenditures by product or service, promotional tool by tool over time (e.g., first, second, and third year of a new mini-plan); observes effects as a basis for improving further use of these tools; and coordinates the promotional activities and their timing.

Integration of marketing communications will produce more consistency in the organization's messages to its audiences and other publics. It firmly imposes the responsibility to portray a unified image throughout each of the organization's many activities. And it leads to a total marketing communication strategy aimed at showing how the organization and its offerings can meet the customers' needs and desires.

CHAPTER 13

Delivering the Message: Advertising, Personal Selling, Sales Promotion, Public Relations, and Crisis Management

Current marketing practice is simultaneously exemplified by the seemingly paradoxical extreme goals of mass branding and one-to-one relationship marketing. The marketing communications mix, also called the promotion mix, consists of four major tools: advertising, personal selling, sales promotion, and public relations. Each tool has its own unique characteristics and costs. Crisis management is also a critical function and managers must be adept at handling issues as they arise, or better yet, anticipate them before they become problematic.

In this chapter, we will consider these promotional approaches primarily from the perspective of traditional media; digital media will be discussed in depth in the following chapter. Although digital media has superseded traditional media to some extent, traditional media is here to stay for the foreseeable future, and arts marketers need to be knowledgeable about how and when to employ each medium and how to develop plans to integrate their use.

Advertising

Advertising is any paid form of nonpersonal presentation and promotion of ideas, goods, or services by an identified sponsor. Advertising has many purposes: long-term buildup of the organization's image (institutional advertising); dissemination of information about a season, production, package offer, event, or special program (product advertising); or announcement of a special price offer (promotional advertising) to trigger quick sales.

CHARACTERISTICS OF ADVERTISING

It is difficult to make generalizations about the distinctive qualities of advertising as a component of the promotional mix because it has so many forms and uses. However, the following characteristics can be noted:

- *Public presentation*: Advertising is a highly public mode of communication. Its public nature confers a kind of legitimacy on the product and also suggests a standardized offering. Because many people receive the same message, buyers know that their motives for purchasing the offering will be publicly understood.
- *Pervasiveness*: Advertising is a pervasive medium that permits the seller to repeat a message many times. It also allows the buyer to receive and compare the messages of various competitors. Large-scale advertising by a seller says something positive about the seller's size, popularity, and success.
- *Amplified expressiveness*: Advertising provides opportunities for dramatizing the organization and its offerings through the artful use of print, sound, image, and color.
- *Impersonality*: The audience does not feel obligated to pay attention or respond to advertising. Unlike personal selling, advertising is able to carry on only a monologue, not a dialogue, with the audience.[1]

Paid advertising permits total control over encoded message content and over the nature of the medium, plus substantial control of the scheduling of the message. On the other hand, paid advertising permits no control over message decoding by the audience and typically results in little or delayed feedback to the received message.

DEVELOPING THE ADVERTISING PROGRAM

Developing an effective advertising program involves the following steps: (1) setting the advertising objectives, (2) deciding on the budget, (3) designing the message, (4) deciding on the media mix, (5) deciding on media timing, and (6) evaluating advertising effectiveness.

Setting the Advertising Objectives

The first step in developing an advertising program is to set the advertising objectives. These objectives must flow from prior decisions about the target market, market positioning, and the marketing mix. The positioning and marketing-mix strategies define the job that advertising must do. A complete statement of objectives includes four components:

- Target: Who is to be reached? Which segments meet the ad's objectives?
- Position: What are the offering's merits? What makes it compelling to the target audiences?
- Response desired: What audience response is being sought (e.g., awareness, interest, purchase)?
- Time horizon: Within what time period should the objectives be achieved?

TYPES OF ADVERTISING OBJECTIVES. Many specific communication and sales objectives can be assigned to advertising. These objectives can be classified by their purpose: to inform, persuade, or remind.

Informative advertising aims to create audience awareness and knowledge of the organization's programs. Informational objectives include telling the market about a new season or production; suggesting new occasions for attendance; informing the market about a program extension; explaining how an offering works, such as a flex plan or coupon offer; describing available services such as parking, nearby restaurants, or educational programming; disseminating rave reviews; reducing consumers' fears; or building an organization's image.

Persuasive advertising aims to create audience preference and stimulate purchase of the organization's offerings. Persuasive objectives include changing target audiences' perceptions of the organization and its offerings, persuading customers to purchase *now,* and attracting donations.

Reminder advertising aims to keep consumers thinking about the organization and its offerings. Thus, even a sold-out production is advertised to highlight the organization's success. Similarly, reinforcement advertising seeks to assure current patrons that they have made the right choice. In a newsletter or special mailing to patrons or even in a newspaper ad, the organization may reprint good reviews or high attendance records, serving to build goodwill that will carry over to future seasons.

Deciding on the Advertising Budget

The proper way to set the advertising budget is to use the *objective-and-task* method. This method calls upon marketers to develop their promotion budgets by defining their specific objectives, determining the tasks that must be performed to achieve those objectives, and estimating the costs of performing those tasks. The sum of these costs is the proposed promotion budget. This method has the advantage of requiring management to spell out its assumptions about the relationship between dollars spent, exposure levels, trial rates, and regular attendance.

If the objective is to fill the hall for a highly popular entertainer, a few large ads in the Sunday paper may be adequate. If the objective is to fill the hall for a relatively unknown performer and a less popular program, the cost may be prohibitively high and the organization should reevaluate its objective.

Measuring the Impact

After implementing the promotion plan, the communicator must measure its impact on the target audience. After running a special promotion, the box office personnel may ask callers how they heard about the production or marketers may administer an audience survey in which patrons are asked about which media influence them. Since increasingly more people purchase their tickets online, it is harder for marketers to capture information about their sources for learning about the show. Of course, marketers can often judge by the timing of various ads what most likely influences a surge of ticket purchases.

Because of the difficulty of measuring the actual effects of advertising, organizations are never sure they are spending the right amount. If the organization is spending too little, the effect is insignificant, and therefore the organization is actually spending too much. But if the organization is spending too much on advertising, then some of the money could be put to better use. In practice, advertising budgets are allocated to market segments according to their respective sales levels or sales potential. It is common to spend twice as much advertising money in segment B as in segment A if segment B has twice the level of some indicator of sales or sales potential. A budget is well allocated when it is not possible to shift dollars from one segment to another and increase total sales.

Organizations must carefully allocate their advertising expenditures to different market segments, geographical areas, and time periods. For example, should a theater place a full-page ad in newspaper sections going to two target neighborhoods or a half-page ad in four target neighborhoods' papers? Should a major advertising campaign be launched before the show opens or delayed in the hope of capitalizing on good reviews? How much should the organization allocate to attracting current versus new patrons?

One theater that appeals primarily to an older, Jewish audience produced Neil Simon's *The Sunshine Boys* as its "safe, sure" attraction for the season's final play. But the production was failing to draw nonsubscribers. So, in addition to newspaper ads, management bought radio advertising to be read several times during a one-week period by a popular talk show host. To evaluate the radio ad's effectiveness, each person who phoned the box office during the two weeks following the ad's first airing was asked where he or she had heard about the show. Of the 225 callers, 64 percent said they had heard about it by word of mouth from friends and family, 34 percent from newspaper ads, and only 2 percent from the radio ad. The minimal patronage gained by the ad failed to cover its cost, and the ad was dropped.

DESIGNING THE MESSAGE

All effective ads contain certain essentials. Bob Schulberg suggests beginning with Winston Churchill's five rules for successful speechwriting: (1) begin strongly, (2) have one theme, (3) use simple language, (4) leave a picture in the listener's mind, and (5) end dramatically.[2] Creativity is especially important. An ad needs an effective attention-getting device; it also should have built-in associations with factors or qualities familiar to the target audience so that the message will be more easily processed.

Execution Styles

A message can be presented in different execution styles, for example:

- Show people experiencing the offering. A summer festival may show a family picnicking on the lawn as the concert is about to begin.
- Emphasize how the offering fits in with a lifestyle. An ad for a singles subscription series may show young adults standing in small groups and sipping wine during a pre- or postperformance reception.
- Create a fantasy around an offering or the occasion of its use. The Dallas Opera's season brochure one year was headlined: "Passion in the evening; no regrets in the morning."
- Tie the offering to an evocative mood or image, such as beauty, love, serenity, or excitement. No claim is made about the offering except through suggestion. "When my parents take me to the ballet, it makes me feel loved," says one child at the Atlanta Ballet.
- Show the expertise of the organization or its performers, directors, or composers in creating the offering or in mounting productions. The Chicago Symphony Orchestra announces that it has won 62 Grammy Awards; theaters advertise their awards and nominations and regularly excerpt phrases from critical reviews that help promote their plays; "Sold Out" banners are placed over performance announcements, testifying to their popularity.
- Feature a highly credible, likable, or expert source touting the offering. It may be a celebrity, such as Alec Baldwin endorsing the New York Philharmonic, or an ordinary person praising the offering.
- Most important, marketers must keep in mind the target audience's perspective (What's in this for me?), not what people in the organization consider most salient. For example, one proud artistic director who got the rights to a show that had just left Broadway advertised "Midwest premiere!" This feature was meaningful to the director, but not to potential patrons.

Designing a Print Ad

A number of researchers into print advertisements report that the picture, headline, and copy are important in that order. All of these elements must deliver a coherent image and message. Since few people read body copy, the picture and headline must be strong enough to draw attention to the ad and must summarize the selling proposition. The headline must be effective in propelling the person to read the copy. The copy itself must be well composed. Format elements such as ad size, color, and illustration will make a difference in an ad's impact. A minor rearrangement of mechanical elements within the ad can improve its attention-getting power. Larger-size ads gain more attention, though not necessarily in proportion to their higher cost. Color illustrations are more effective than black-and-white, although some graphic designers have found that two-color ads can be just as effective as four-color ads and far less costly.

One direct-mail brochure for the San Francisco Ballet's *Nutcracker* was a cleverly conceived two-fold piece. On the cover it said: *A Dream, the Way It Was Meant to Be Danced.* After the first fold was opened, the copy changed to: *A Dance, the Way It Was Meant to Be Dreamed.*

Designing Radio Ads

Well-written radio ads can lead listeners to imagine themselves at an event, enjoying its ambience. Although subtle suggestion may be a virtue in print advertising, a more direct approach should be taken in radio advertising, and the entire ad should focus and elaborate on one message, such as "You will love this show; tickets going fast." "The entire thrust of the announcement must be geared to developing a response. A radio ad should be created as if you were addressing one person." Bob Schulberg suggests that a radio ad should be designed with the following tips in mind:

- Mention your organization's name several times.
- Include a deadline by which time the listener must respond ("Only two days left to purchase the few remaining tickets to the biggest event of the season!").
- The most important part of a radio ad is the ending. Everything should build up to it.
- Repeat the web address and/or phone number. End the ad with the last repetition, since the last sounds of an ad will linger in the listener's mind for a few seconds.[3]

Choosing the Media

Because message design makes some assumptions about which media will be used, presumably some thought will already have been given to the media that most effectively reaches the target audience. Media selection involves three steps: choosing the media categories, choosing specific media vehicles, and media scheduling.

Choosing Media Categories

The first step calls for allocating the advertising budget to the media categories, including but not limited to the Internet, email, newspapers, television, direct mail, radio, magazines, and outdoor media such as billboards and posters. Marketers choose among these major media categories by considering the following variables:

- *Target audience media habits*: Select the media used by target markets. Many concertgoers listen to classical music and public radio stations; they are likely to read the arts section of the local newspaper and certain weekly or monthly community publications.
- *Product or service:* Media categories have different potentials for demonstration, visualization, explanation, and reaching target audiences. Direct mail is still considered the most effective medium for describing and promoting a whole season of performances, followed by websites, email newsletters, and newspaper inserts. A color ad in a magazine may be the most effective medium to show off a newly renovated auditorium to specific audiences, whereas a YouTube video may be the most effective way to promote interest in a dance performance. Posters and billboards are often effective at building awareness, if well placed.
- *Cost*: Television is very expensive; newspaper advertising is moderately costly; email blasts are nearly cost free. What is most important, of course, is which medium will yield the highest return per dollar spent.

Media Exposures

Media categories must also be examined for their capacity to deliver reach, frequency, and impact. *Reach* is the number of different persons or households exposed to a particular media schedule at least once during a specified time period. *Frequency* is the number of times within the specified time period that an average person or household is exposed to the message. *Impact* is the qualitative value of an exposure through a given medium.

Media selection means finding the most cost-effective media to deliver the desired number of exposures to the target audience. Reach is more important when introducing new productions or services or seeking an undefined target market. One can extend reach by going on several radio stations at the same time (called *roadblocking*) or by advertising at different times of the day to reach the widest variety of listeners.

Frequency is more important where there are strong competitors, a complex story to tell, high consumer resistance, or a frequent purchase cycle. High frequency pays off for unplanned purchases, when there is low interest and loyalty, or when the product and/or the campaign is new. Advertising repetition deals with the problem of forgetting by putting the message back into memory. Many advertisers believe that a target audience needs a large number of exposures for the advertising to work. Too few repetitions can be a waste, since they will hardly be noticed.

Advertising expert Herbert Krugman favors three exposures to an advertisement: The response to the first exposure is a "What is it?" type of cognitive response that dominates the reaction. During the second exposure, most often people have an evaluative "What of it?" response. The third exposure constitutes a reminder, if a decision to buy based on the evaluations has not been acted on.[4] Here Krugman means three advertising exposures—meaning that the person actually sees the ad three times. This should not be confused with vehicle exposures, namely, the number of times the person has been exposed to the vehicle carrying the ad. If only half the readers look at the ads, or if readers look at the ads only every other issue, then the advertising exposure is only half of the vehicle exposures. A media strategist would have to buy more vehicle exposures than three in order to achieve Krugman's three "hits."[5]

Timing the Media Exposure

In timing media use, the advertiser faces both a macroscheduling and a microscheduling problem. The macro problem is that of cyclical or seasonal timing. Audience size and interest vary at different times of the year. Most marketers do not advertise when there is little interest, but spend the bulk of their advertising budgets just as natural interest in the offering begins to increase and when it peaks. At times when there is much habitual purchasing, such as for subscription renewals, lead time for advertising should be greater and advertising expenditures should be steadier, reaching their maximum before the sales peak. The timing of advertising for individual ticket sales is more complex. Interest must be stimulated in advance of the opening to fill seats before word of mouth and (hopefully) good reviews have a chance to help the advertising effort. Advance advertising is especially important when there is a low subscriber base and when there will be few performances of each production, such as with orchestral concerts and holiday shows. In recent years, marketers have found they can effectively start marketing their special shows farther in advance than ever before.

The microscheduling problem calls for allocating expenditures within a short period to obtain the maximum impact. For example, how should advertising be spaced during a one-week period? Consider three possible patterns. The first is called *burst* advertising and consists of concentrating all the exposures in a very short period of time, say all in one day. Choosing the day of the week depends on the performance day(s) the advertiser is trying to promote and on knowing when the target audience makes its purchase decisions. Are people more likely to decide on Monday or on Friday what they will do on Saturday night?

The second pattern is *continuous* advertising, in which the exposures appear evenly throughout the period. Continuity may provide the highest level of exposures and reminder value, but also carries the greatest costs.

The third pattern is *intermittent* advertising, in which intermittent, small bursts of advertising appear with no advertising in between (flighting) or continuous advertising at low weight levels is reinforced periodically by waves of heavier activity (pulsing). These intermittent strategies draw upon the reminder value strength of continuous advertising but at lower cost, and many advertisers feel they provide a stronger signal than a continuous approach.

These ideas apply to digital vehicles as well, since they are based on consumer behavior theories.

EVALUATING ADVERTISING EFFECTIVENESS

Said department store magnate John Wanamaker early in the twentieth century, "I know that half of my advertising is wasted but I don't know which half."[6] A century later, although we cannot clearly determine the effectiveness of all our ads, we have many tools and strategies at our disposal to help in this task.

The most important ad evaluation components are copy testing, media testing, and expenditure-level testing. Copy testing can be done both before an ad is put into actual media and after it has been printed or broadcast.

Copy Pretesting

The purpose of ad pretesting is to make improvements in the advertising copy to the fullest extent possible prior to its release. The copy should be evaluated to see how well it meets the following criteria:

- *Attention*: How well does the ad catch the reader's attention?
- *Comprehension strength*: How understandable are the words and sentences to the target audience?
- *Cognitive strength*: How clear is the central message or benefit?
- *Affective strength*: How effective is the particular appeal?
- *Behavioral strength*: How well does the ad suggest follow-through action?

An effective ad must score high on all these properties if it is ultimately to stimulate buying action.

To gather information about the strength of an ad, marketers may show a set of alternative ads to a panel of target consumers or advertising experts and ask them, "Which of these ads do you think would influence you most to buy the offering?" Or a more elaborate form consisting of rating scales may be used, asking respondents to evaluate the ad's attention strength, comprehensibility, cognitive strength, read-through strength, affective strength, and behavioral strength, assigning a number of points (up to a maximum) in each case. Alternatively, since advertisements are often viewed in a social setting, pretests with focus groups can often shed light on how a message is perceived as well as how it might be passed along. The focus-group technique also has the advantages that its synergism can generate more reactions than a one-on-one session. It is efficient because it gathers data from six to twelve people at once, and it can yield data relatively quickly.

Pretesting, however, requires subjective judgments and is less reliable than the harder evidence of an ad's actual impact on target consumers. Pretesting is more helpful for screening out poor ads than for identifying great ads.

Ad Post-Testing

The purpose of ad post-testing is to assess whether the desired impact is being achieved. One is tempted to use sales as the measure of success. But sales are influenced by many factors besides advertising, such as the nature of the offering itself, price, availability, and the competitive climate. The fewer or more controllable these other factors are, the easier it is to measure the effect of advertising on sales.

To post-test the effectiveness of alternative messages or media in influencing behavior, the advertiser may use techniques such as these:

- Place promotion codes in the advertisement. Each code should vary by message and medium.
- Ask target audience members to mention or bring in an advertisement in order to receive special treatment (such as a price discount or a free beverage).
- Ask individuals to call for further information (on which occasion they can be asked where they saw the ad, what they remember, and so on).
- Stagger the placement of ads so that this week's attendance or sales can be attributed to ad A while next week's can be attributed to ad B. This is also an effective method for assessing alternative expenditure levels.

To further test expenditure levels, staggering may be done by geographic area, so that the organization spends more in some neighborhoods and less in others. These tactics are called high-spending tests and low-spending tests. If the high-spending tests produce substantial sales increases, it appears that the organization has been underspending. If they fail to produce more sales and if low-spending tests do not lead to sales decreases, then the organization has been overspending. These tests, of course, must be accompanied by good experimental controls and must last sufficiently long to capture lagged effects of changes in expenditure levels.

Personal Selling

Personal selling is the most effective tool at the earlier stages of the consumer decision process, particularly in building up preference and conviction, although it is also highly effective at influencing action. Personal selling has three distinctive qualities not available through advertising.

- *Personal interaction*: Personal selling involves a living, immediate, and interactive relationship between two or more persons. Each party is able to observe the others' needs and characteristics at close hand and make immediate adjustments.

- *Cultivation*: Personal selling permits cultivation of relationships, ranging from matter-of-fact selling relationships to deep personal friendships. In most cases, the sales representative artfully woos the target audience, tactfully applying pressure to induce an action, while keeping the customer's long-run interests at heart.
- *Response*: Personal selling makes the target audience member feel under some obligation to respond, even if the response is only a polite "thank you."

Personal selling can be the organization's most expensive customer contact tool. Salespeople such as telemarketers must be well trained and highly motivated in order to function effectively for the organization and volunteers acting as advocates for the organization must be nurtured and cultivated over a long period of time. Yet, when personal selling efforts are well planned and well targeted, their costs are minimal compared to their benefits.

Personal selling has taken on an important new dimension with the power of social media, as consumers take on the role themselves of connecting with friends and acquaintances to spread the word about products and events of particular interest to them.

Piccolo Teatro di Milano has been hugely successful at harnessing its volunteeers' passion for the theater. The volunteers reach out to their contacts and arrange to bring groups ranging in size from ten to three hundred participants to see plays at the theater.

The personal selling strategy employed by Piccolo Teatro relates strongly to current consumer attitudes. Consumers believe one another more than they believe in companies. According to Nielsen Global Survey, fewer consumers rely on company-generated advertising. Around 90 percent of consumers surveyed trust recommendations from people they know and 70 percent of consumers believe in customer opinions posted online. Research shows that consumers trust strangers in their social network more than they trust experts.[7] Clearly, says Philip Kotler, "Marketing is not just something marketers do to consumers. Consumers are marketing to other consumers as well."[8]

SALES PROMOTION

Promotion offers encourage purchase of a product or service by incorporating some concession or incentive, such as coupons, premiums, and discounts that give value to the consumer. Promotions serve to gain attention and usually create a sense of urgency to "act now."

Organizations use sales promotion tools to create a stronger and quicker response. These tools can be used to dramatize product offers and to boost sagging sales. However, their effects are usually short run; they are generally not effective in building long-term involvement. Furthermore, people who take advantage of, for example, a $5 discount, may have been likely to pay full price with other compelling messages. The marketing challenge is to design an incentive that attracts the target audience, that the audience values, and that serves the organization's interests.

Planning Sales Promotions

In planning sales promotions, the marketer must analyze several factors.

Specify Objectives

The first step is to specify the objective(s) for which the incentive is undertaken. If the organization has excess capacity and wants to sell tickets for upcoming performances, it may announce ticket discounts for one week only. (Remember to take care in not using such discounts too often or broadly so as not to train people to wait until the last minute to buy tickets.) If the goal is to stimulate interest over the run of a production, marketers might stuff programs with discount coupons for patrons to share with their friends. If the objective is to promote trial among never-attenders, an orchestra may offer a free concert in the park, a special concert that samples the upcoming season, or a "singles" reception with performers after the program. If the objective is to encourage patrons to resubscribe early, special perks such as restaurant coupons may be offered until a specified date.

Specify the Recipients

The next step is to specify the recipient of the incentive. For example, when stimulating day of performance sales, some theaters "tweet" a special low price offer for students only. Since students have busy schedules and often plan last minute, this offer works well for them. Coupon books that offer discounts, such as, $100 worth of tickets for $80, are attractive to seniors who have discretionary time to attend multiple performances and are interested in attending more often than they could without some concession. Such offers need to be aligned with any senior discount the organization already offers.

Determine the Form of the Incentive

The marketer must also determine the form of the incentive—whether it will consist of free or discounted tickets, special events, or gifts. The more closely the incentive is tied to the organization's desired outcomes, the more effective it will be in the long run. Therefore, offering one free ticket with every five ticket stubs saved is likely to stimulate more frequent attendance than the offer of a mug or a T-shirt.

The marketer must determine the amount and duration of the incentive offer. Too small an incentive is ineffective and an overly large one is wasteful.

If 40 percent discounts are offered to new subscribers, the marketer must determine whether this offer will have to be repeated in subsequent years in order to keep those subscribers. In general, the marketer must carefully analyze whether the short-term value of the offer is worthwhile to the organization in the long run.

EXAMPLES OF SALES PROMOTIONS

Free Samples

Offering a free sample is commonly used in the business sector as an effective way to introduce a new product or to introduce new people to the offering. For arts organizations, a sample is usually inexpensive to provide, since filling an empty seat carries little or no cost. A theater may enclose two free guest tickets with each subscription so that patrons will invite their friends to attend. Organizations often have multiple opportunities each season to give away pairs of tickets to other nonprofit organizations as raffle prizes for their fund-raising events. This is an ideal way to reach new audiences who are buying or winning the tickets for a cause they believe in and support. Furthermore, this approach is totally devoid of any conflict of interest with people who have paid for their tickets through a regular channel.

Tie-in Promotions

Some promotions do not serve to stimulate sales, but may increase satisfaction of the patron's total experience. For example, tie-in promotions, such as discounts at nearby restaurants and parking facilities, are collaborative offerings between arts organizations and the business or public sector. Arts organizations may also collaborate with one another, such as through a centralized arts service agency. "Discover Jersey Arts," through its free membership program, offers discounts and promotions at local arts organizations.

Patronage Awards

Patronage awards are values given in proportion to one's patronage of the organization. Some companies have exclusive rooms where ticket buyers and contributors at certain levels may dine before a performance or have drinks during intermission. Other patronage awards may be the opportunity to have dinner with the artists after a performance or eligibility for the best seats in the house.

PUBLIC RELATIONS

Public relations has its roots in the times when the arrival of kings was heralded in advance by their messengers. Later, companies began to recognize the positive value of planned publicity in creating customer interest in the company and its products. Publicity entailed finding or creating events, preparing company- or product-slanted news stories, and trying to interest the press in using them. Recognizing that special skills are needed to develop publicity, companies began to add publicists and PR departments to their ranks. "The Internet has made public relations public again, after years of almost exclusive focus on media," says marketing expert David Meerman Scott.[9] "On the web, the lines between marketing and PR have blurred, and social networks allow people all over the world to share content and connect with the people and companies they do business with," says Meerman. In this sense, virtually anyone can do PR for an organization and its offerings.

An active, market-oriented public relations stance ensures that the organization has control over how others see it.

In the following sections, I focus primarily on working with the media for public relations, but many of these concepts apply to the organization's social media sites, blogs, and other direct communications with consumers. The challenges of PR departments have grown, says Chris Jones, chief theater critic of the *Chicago Tribune*, as "the media landscape is now so fractured."[10] Yet, public relations is a key component of any operation in this age of instant communications and inquisitive citizens.

Benefits of Marketing Public Relations

The appeal of public relations is based on its distinctive qualities.

High credibility: News stories and features seem more authentic and credible to readers than do advertisements. Public relations can reach many prospects who might avoid salespeople and advertisements. The message gets to the buyers as news rather than as sponsored communication. Some experts say that consumers are five times more likely to be influenced by editorial copy than by advertising.

Build awareness: Public relations can make a memorable impact on public awareness. It can bring attention to a product, service, person, organization, event, or idea. PR also serves to break through commercial clutter, to complement advertising by reinforcing messages and legitimizing claims, and to tell the product story in greater depth. It is effective in arousing attention, as it comes in the guise of a noteworthy and often dramatized event.

Low cost: PR requires a fraction of the cost of advertising, as the organization does not pay for the space or time obtained in the media. If the organization develops an interesting story, it could spread to other media as well.

The Tasks of Marketing PR

Public relations is the systematic promotion of organizational goals, products, images, and ideologies. Its activities fall into three distinct categories: image PR, routine PR, and crisis PR.

Image PR

Image PR tries to shape the total impression someone has of an organization. For example, image PR can:

- Revitalize, relaunch, and reposition the organization and its products.
- Build consumer confidence and trust: Rave reviews at home and on tour; stories about awards and sold-out performances; and about a company's directors, productions, and/or performers being in demand elsewhere may help to position the organization as a leader and expert.

Routine PR

Routine PR comprises most of the efforts by the public relations manager to promote individual productions, performers, and special events. Some of its goals are:

- *To introduce new products*: Advance media articles about upcoming performances are critical for building interest and excitement and for selling tickets before a show opens, reviews are printed, and word of mouth has a chance to spread.
- *To communicate new benefits*: A symphony that had trouble selling tickets for its Valentine's Day performance positioned its concert as the perfect prelude to love. It also offered a red rose to every woman who attended. In one day, the symphony sold two hundred additional tickets, all to young couples. Permanent benefits, such as a hall modified to accommodate handicap access, should be broadly publicized.
- *To involve people with products*: An organization may provide human interest stories about performers, playwrights, composers, and so on, to make people seem more a part of the action on stage.
- *To cultivate new markets—to reach preexisting target markets*: For example, a theater may choose to promote its lead actress in her nearby home neighborhood newspaper.
- *To tailor marketing programs to local audiences*: When the National Dance Theatre Company of Jamaica performed at the Brooklyn Performing Arts Center, the dance company's PR director worked closely with people in the Jamaican community. The Jamaican consulate mailed announcements in envelopes with its own return address and one of its representatives appeared on local radio stations to talk up the performances. Caribbean restaurants promoted the performances and offered to sell tickets, as did the Caribbean Chamber of Commerce.

Crisis PR

Crisis PR attempts to protect the organization and its managers, artistic personnel, and board members from problems that may shake the very foundations of its survival. Crisis PR helps the organization prepare to deal strategically with serious problems as they arise and to deal with the media and the organization's other stakeholders during and following a crisis situation.

PUBLICITY

Specific communications objectives should be set for different audiences for each season, production, and event. A clear, continuing program of releases and stories should be developed for each planning period. Not all stories that are interesting to the publicist or to the organization's administrators should be brought out if they do not help promote the organization's long-term interests. Timing publicity efforts well is crucial. More important than getting into the paper on a daily or weekly basis is to time "publicity breaks" that bounce off the other marketing components: the brochure, the radio spot, telemarketing, special events, email blasts, and so on, to create synergy among all aspects of the campaign.

A well-conducted publicity campaign begins months before the show opens; it is an ongoing effort, characterized by constant nurturing and planning for the future.

WORKING WITH THE MEDIA

Getting news items into the local press or on television or radio is itself a marketing task. As such, the publicist must start with the immediate audience—the media—and must understand what reporters are looking for in a news story. The PR director can assist reporters in this task by setting up interviews, by unfolding the human interest stories behind commonplace events, and by discovering the "news behind the news." The better the PR director is able to meet the reporters' needs, the more likely it is that the media will help meet the organization's PR goals.

A good publicist should be in regular contact with the key columnists, critics, editors, and producers in the community to build relationships and keep them informed about the organization's activities and motivations. Long-term cultivation of relations with media gatekeepers helps get news items reported and features covered.

Among the prime characteristics journalists will seek for any story are the interest of the subject to their audience; the possibility for dramatization through pictures, video, or live interviews; the clarity and exhaustiveness of any press release, including supporting materials; limited need for further "digging"; and the possibility for exclusive coverage—either for the entire story or for a specific angle.

Following are some tips for working with the media[11]:

- The key to an effective publicity campaign is advance planning. Space and time restrictions necessarily limit the amount of material that can be used by the media. The PR manager should prioritize the events and issues that best serve the organization's goals, and devote efforts to gaining media coverage accordingly.

- Know the media—read, watch, listen—and become familiar with each journalist's and media organization's style, orientation, strengths, and limitations. Know which regular columns and broadcast programs are appropriate for the events being promoted and which media will share information. Know what kind of audience each journalist addresses. Don't call everyone with each new idea; the story you pitch must be compatible with the type of program or publication you solicit, or else rejection is inevitable.
- Be selective. Not all events call for a full-scale publicity campaign, and if you overemphasize a minor story, you may have trouble getting coverage for a genuinely newsworthy program. Mike Martin, the former news director for KRBE-FM in Houston, Texas, puts it this way: "Remember, if you flood me with stuff that is not important, you're diluting my interest. Then I'll take anything you send me as probably not important. If you hit me with only important stuff, stuff I need, stuff I can use, that's important to the community, I'm going to pay more attention when I see your message [in my inbox.]"[12] "On the other hand," says former *Charlotte Observer* critic Tony Brown, "some journalists appreciate being kept up on a week-to-week basis. You never know when there is a slow week coming up and journalists might need a nugget of information to flesh out a column. Journalism is a capricious business."
- With specific stories, target one journalist at a time. Crafting a unique pitch to a particular journalist can work wonders.[13]
- Don't send email attachments unless asked to do so. Send plain text emails instead. If you are asked for other information, you can follow up with attachments, but be sure to clearly reference in the email what you are sending and why so the journalist will remember asking for it.
- Video clips are in great demand by journalists. Send them clips one to two minutes long of the artists speaking, scenes from the production, and prepack featurettes. *B-roll*, the supplemental or alternate footage made to share with the media, is highly popular with journalists, says Chris Jones of the *Chicago Tribune*. (*Note:* do not send big files!) Many journalists receive messages on their mobile devices, which have limited capacity. Jones says that if he receives a message with big files, he immediately deletes it.
- Create a "for media" page on your website, inform the journalist that new photos and/or video clips are available there, and insert the link to that page in your email message.
- Know each medium's own deadlines and preferred formats for submitting information for calendar listings and articles.
- Follow up promptly with media who request interviews or other information.

- When phoning a journalist to make an initial contact or to follow up on a mailed or emailed release, be brief and to the point. Figure out what to say in 90 seconds or less that will spark their interest. Media people get dozens of calls a day.
- When sending an exploratory email to a journalist, try to have three or four different story ideas so you have a better chance of providing something of interest. Journalists know the interests of their publics and may have personal preferences that affect their responsiveness to various ideas.
- If you have been turned down, even for an important story, do not argue with the journalist, as doing so will only alienate him or her. But, if you are able to discover a whole new angle to the story, test it out on some colleagues to make sure it is really different and meets the organization's needs, then contact the journalist again and make your pitch.
- Journalists are not responsible for selling tickets. Don't ask them to help you promote a show that is not selling well.
- Never tell a media person what she or he should write about. As you get to know the personalities of the various media people you deal with, develop an idea of what kinds of stories they like to cover. Send a written suggestion pointing out interesting aspects of a production, event, or human interest story.
- Consider that some journalists like to dig further to cover a story well; others do not. Journalists should be presented with not only the facts of the story but also a range of peripheral material that may fulfill their specific needs and interests. This material could include photos or photo opportunities, profiles of key figures in the story, lists of reference material, or links to new release materials.
- Don't criticize the critic or review the reviewer. The critic or reviewer has the task of evaluating your product on behalf of the public, not on behalf of the organization. The power inherent in the critic's position is one of the realities of the business. Yet, many critics and reviewers have extraordinary understanding of and sensitivity toward the pressures and issues affecting performing arts organizations. If you disagree with a reviewer, talk with him or her directly on a calm, professional basis. The dispute should not be taken to the editor, publisher, or station manager. This will make an enemy, rather than an ally, of the reviewer.
- Have one and only one person from your organization as your media contact, so that the media is not deluged with different requests and releases. However, make key people in the organization, such as the artistic and executive director, available to the journalist on request.
- Let the reporter know that a certain artist is available for an interview. The PR person may brief the artist in advance, but should not sit in on interviews, because journalists do not like to be monitored. Avoid rewarding

or punishing a journalist with access to a star. Says critic Tony Brown, "Punishment by withholding information is a dangerous game."

- Invite media people to rehearsals and send them a script.
- Schedule events to take advantage of slow news days such as holidays.
- Opening night invitations should go out to the media three to four weeks in advance. Arts organizations generally offer journalists tickets free of charge.
- In addition to sending press information about each show or event you are promoting, send a yearly press kit to reporters and editors listing your complete contact information. Include a short description of the organization, its offerings, and its upcoming annual program.

SHRINKING MEDIA COVERAGE FOR THE ARTS

Although the public relations function has grown in importance and impact in recent years, the media resources available to arts organizations have dramatically shrunk.

Creative public relations managers have found some new solutions to shrinking arts coverage, such as placing features in alternative newspaper sections. To prepare for an onstage kitchen scene at one theater, a "cooking lesson" was held for actors at a major restaurant chain and the event was publicized in the local paper's food section. For a corporate takeover drama, an interview with the star actor was published in the business section.

Capitalizing on holidays is a good way to promote the organization and its productions while helping journalists, who are always seeking new angles for covering oft-repeated and predictable events. For example, one Valentine's Day, a theater provided a feature on "marriages made backstage." On Mother's Day, one theater offered a free ticket to any mother accompanied by her offspring, utilizing both the holiday and price promotion strategies to gain publicity.

Arts publicists may find that in creatively seeking new outlets for publicity, they are actively building new audiences and giving current audiences a different and fresh perspective about the organization and its offerings. In addition to gaining visibility for specific arts events, such coverage has served other important goals: reaching new audiences and giving underwriting efforts, educational programs, and even capital campaigns an unexpected boost.

SOCIAL MEDIA VERSUS BROADCAST MEDIA

Through the use of social media such as Facebook and Twitter, arts organizations have the power to broadcast their own news and stories, to do so exactly when they want the news to go out, and spread the word in their own preferred way, without having it filtered by journalists.

However, says Chris Jones, arts organizations cannot announce their news on their own social media vehicles and then expect the media to report it. The media is interested first and foremost in *news*. If something is no longer news, journalists will lose interest, even though their reach is far broader than the organization's Facebook or Twitter feeds. Jones says that if the information is important, he will still report it, but he won't be happy about it. The organization has to choose, in each situation, whether to self-publish first or give the story to the media. This means that the people doing social media for the organization must coordinate with the PR manager so that their efforts are not at cross-purposes. It also means that the artistic department must be aware of and involved in this issue. For example, an artist might tweet on his own site, "I got a role at the Shakespeare Theater!" Artists need to be informed when they are hired that anything they announce on their own social media must first be approved by the organization's PR staff. In this way, the PR managers can control all the external flow of news.[14]

Other Media

There are many opportunities besides the widespread media for communicating with the public about the organization and its offerings. Consider asking for mention in souvenir programs for sporting events and other cultural events. Seek placement on bus shelters and benches; taxi panels; marquees at schools, banks, and other public and private buildings; community bulletin boards, and grocery bags. Send news releases to church and synagogue bulletins, chamber of commerce publications, and service club newsletters.

Companies and individuals that contribute to the arts organization can be instrumental in promoting the organization through their internal documents and on memo headers. Leaflets can be placed on college campuses and at shopping centers, doctors' offices, building lobbies, health clubs, libraries, and grocery checkout stands. Many arts organizations are more than willing to display brochures from other arts organizations in their own lobbies.

Public Relations Tools

Public relations managers have several tools at their disposal to utilize in their image-building and visibility-increasing efforts. The primary tools are the event, the press release, public service announcements, interviews, photographs, and video.

Events

Events are important tools for public relations managers. They can bring people together on sociable occasions and bond people to the organization. They can also stimulate word of mouth about the organization and its offerings. Public relations managers help reporters identify natural events and stories that occur in the life of an organization. PR managers also make news happen, in effect, by creating events. Several advantages can be gained by creating events. An event can be created to obtain news coverage, such as having a music director participates in the formal groundbreaking of a new performance hall. Its timing can be arranged for the convenience of the reporting media and the event's success is measured by how widely it is reported. An event can be created to celebrate the organization's history or triumphs, or to dramatize a particular program or personality. A theater's gala thirtieth-anniversary party broadcasts that it has a long and successful history; an arts organization may sponsor a book or record signing by a featured playwright or singer.

One dance company promoted its New Year's eve party, following a performance of the *Nutcracker*, by keeping the Nutcracker, the Mouse King, and Maria dancers constantly on the road with personal appearances in malls and businesses. Collaborative PR was done with the cooperation of a major department store, which turned its whole store into a *Nutcracker* setting, using costumes and sets made especially by the ballet for promotional purposes. The pièce de résistance of the publicity effort was having a wooden, life-size replica of the Nutcracker climb to the top of an office building near the theater, followed by fireworks at the stroke of midnight. People called all the following week to buy tickets for the remaining *Nutcracker* performances.[15]

Events make images more vivid, more attractive, more impressive, and more persuasive than reality itself. When used effectively and sparingly, events can go a long way to help imprint the organization's personality on the target public's mind.

A celebrity constitutes an event. Operatic arias do not attract crowds; Placido Domingo and Renee Fleming do. Star appearances are events because they spawn other events. When a star dines at a local restaurant or signs copies of a recent book or compact disc, it becomes an event for the media, who, in describing the event, mention and thereby promote the upcoming performance, elevating the organization's visibility and prestige.

Managers should also focus their efforts on building new interest in the artistic experience itself, which can carry over from composer to composer and from playwright to playwright, rather than only from star performer to star performer.

The Press Release

The press release is the basic tool for communicating with the media. A press release's job is not to sell or entertain, but to inform; it should be concise, to the point, and free of background "filler." In writing a press release, the PR manager should avoid hyperbole, but should take advantage of favorable reviews or notices from leading critics. An endorsement from a reputable third party always lends credibility to the publicist's claims.

Send the press release via email. Chris Jones says that mobile is his vehicle of choice and when in his newspaper's office, he rarely even checks his "snail mail" box anymore.

Each press release is competing for attention with dozens of others each day. Unless the message is immediately discernible, the release will be passed over in favor of one more readily understood. A press release should contain the following elements:

- Strong subject line: Include the organization's name and the message's key point in just a few words.
- Contact name and phone number: Include your own name or the name and contact information of someone who will be available to answer follow-up calls.
- Exclusivity notice: Situations that call for exclusivity are rare. If you do have a story that lends itself to an exclusive cover pitch, send the prospective journalist a standard press release, then contact him or her in person or over the phone to offer exclusive information. An alternative is to type "EXCLUSIVE TO:" (in all caps and underlined) followed by the name of the person and the name of the media organization immediately below the contact name on the unique press release. Never mark anything "exclusive" if you intend to send the same information—even in an altered form—to anyone else. If you do, you will damage your own credibility and that of your organization.
- Release date: If you want publication of the release to coincide with a specific event, or if there is a tie-in with a particular date, this should be noted. Either "For Release on or After (date)" or "For Immediate Release" should be typed in. If the release is about a special announcement—say, the naming of a new artistic director—do not provide the journalist with any information that you would not want publicized in advance. The journalists' job is to report anything newsworthy as soon as they hear it, and of course their preference is to be the first to do so.
- Headline: The headline should include as much vital information as possible (without assuming paragraph proportions) and should clearly indicate the tenor of the story to follow. It should be written in the present tense, even if the report is of a past or future event, and should use the active rather than the passive voice. For example: Royal George Theater Announces Extension of Record-Breaking Show: "The Pianist of Willisden Lane." In general, copy that is clever or funny should be avoided.
- Body copy: Follow the journalist's rule of thumb, the inverted pyramid. Include all the critical information (who, what, when, where, and why) in the lead or opening paragraph, then move through the rest in descending levels of importance, concluding with the least essential points at the end. This structure is helpful to copy editors, who, when pressed for space, can

simply cut copy from the bottom and be assured that the crucial information will remain intact.

To get feedback on press releases, ask journalists via email, phone, or at the next performance to let you know if they found the release useful, whether they want more like it, and if they want you to call them to supply further information. The answers can help you evaluate your publicity material.[16]

Public Service Announcements

Although airtime on radio and TV is expensive to purchase, the broadcast media are usually quite willing—even eager—to give airtime for public service announcements (PSAs).

As an incentive, a performing arts organization may offer to trade tickets for announcements. The station may then use the tickets as prizes in contests or as gifts to advertisers. In addition to filling otherwise empty seats, it is an opportunity for the organization to attract new audiences.

Major symphony orchestras raise huge sums of money annually through symphony "radiothons," during which an FM classical music radio station devotes an entire weekend to the symphony for the playing of live music by orchestra members, the playing of recordings on request (with contributions accompanying each request), feature interviews with music personalities, and offers of special perks for subscribing and for various levels of donations. Such radiothons are highly effective in building visibility for the orchestra and its spokespeople.

Photographs and Video Clips

Photographs and video clips are more than just visual supplements to features and reviews; often they tell a complete story themselves with only the aid of a caption. When pitching a story to the press, the publicist should be sure to alert the photo desk to any photographic opportunities that may exist.

The publicist should identify a compelling visual angle for every event or program connected with the organization. Whether the media sends a photographer or not, the organization should have its own photographer on hand to record such events as an opening-night party, a rehearsal with a visiting celebrity, a special announcement by the artistic director or president of the board, or an enthusiastic standing-room-only audience. The organization should keep an archive of such materials for future promotions.

THE PUBLIC RELATIONS MANAGER

Many arts organizations handle all of their public relations in-house; others choose to hire an outside professional PR agency. One of the biggest mistakes an organization can make is to underestimate the amount of time and planning that goes into successful public relations. Therefore, an organization that chooses, for economic reasons, to do publicity in-house, using already overextended personnel, may be working against its own best interests. Either adequate time must be allotted for an in-house staff person to handle myriad details on a timely basis, or a professional PR person or agency should be hired. In any case, the ultimate responsibility for the planning and direction of all publicity efforts belongs in-house.

When hiring an outside publicist, the organization should look for someone with a proven track record with the performing arts—someone who clearly understands the special needs and opportunities of this field. In order to avoid any gaps or overlaps in service, a written agreement should spell out all responsibilities and fees. One member of the in-house staff should be designated to act as the PR liaison to avoid the confusion that inevitably results when more people are directly involved. Managers should meet with the publicist on a regular basis; they should assemble a calendar indicating deadlines for press releases, photographs, PSAs, opening-night invitations, and so on, and help the publicist to develop story ideas and identify new audiences that may be targeted for specific productions or other offerings. While current productions are being publicized, planning of PR strategies for upcoming productions (and seasons) should be under way.[17]

There are several precautions every publicist should take. If other people in the organization have to approve the publicity material, the publicist should ask them to initial it. When the organization quotes someone in a news release, the person should be sent a copy ahead of time, along with a note asking her to initial the release and send it back. Third, the publicist must keep copies of everything written, including memos and interview notes, particularly if others are being quoted.[18]

Crisis Management

Crises are an integral part of organizational life. No organization, regardless of its size or the nature of its operations, is immune to crises. As stated by one observer, "If you are not now in a crisis, you are instead in a pre-crisis situation and should make immediate preparations for the crisis that looms on the horizon."[19] Crises can be caused by a range of internal and external factors including mismanagement; inappropriate policies, strategies, and practices; scarcity of financial resources; swift economic, legal, or political changes; credit squeezes; significant changes in the nature of market competition; sharp drops in attendance and in sales levels; labor strikes; and loss of credibility with consumers. Because crises tend to be highly publicized, sharply affecting the organization's public image, crisis management has grown as a specialty area within the public relations function.[20]

CHARACTERISTICS OF A CRISIS

Not all stressful situations are crises. A crisis creates a new situation that usually requires drastic measures to correct. It may call for replacing top managers, drastic budget cuts, major programming changes, or other major steps.

Typically, crises brew for some time before they surface and eventually explode. It is crucial to correctly diagnose and understand the cause of a crisis. Many organizational and environmental problems that are not financial in nature tend to surface as financial crises. Under most circumstances a financial squeeze is a consequence rather than the cause of a crisis.

Successful crisis management can sensitize managers and board members to the strengths and weaknesses of the existing planning system, highlight the need for better controls, and boost the morale of managers, who gain a sense of accomplishment by coping with the crisis.[21]

MANAGING A CRISIS

Effective crisis management is a difficult, urgent, involved, complex, and time-consuming task. It demands objectivity, flexibility, creativity, persistence, commitment, courage, teamwork, and willingness to change and adopt unconventional and unpopular options. It calls for tough and drastic decisions and important sacrifices.[22]

When considering matters that affect the general public or the community, such as closing down a symphony or theater and leaving patrons stranded with unusable tickets, managers are ethically obligated to disseminate complete and accurate information as quickly as possible. Board members must be fully informed about the crisis and about policies in place regarding who the spokespeople will be and what will be said publicly.

Dealing with the Press

When a crisis occurs, the relationships an organization has developed with the media may result in fairer (or more favorable) reporting of crisis events and in more objective interpretations of the organization's decisions and actions.[23]

Planned communication with the press that is honest, open, cooperative, and responsive to the reporters' deadlines is essential for bringing some degree of control back to the organization and for helping the organization through its difficulties, rather than further complicating them.

Stratford Sherman recommends the following guidelines for dealing with the press in times of crisis.

- The chief executives should be responsible for press relations, speaking often for the organization, and should delegate enough authority to make the PR spokesperson a credible source.

- If you err, admit it candidly. Avoid hedging or excuses. Apologize and explain how you're going to make things right.
- Consider the public interest in every operating decision. Your reputation depends far more on what you do than on what you say.
- If you want your views represented, you have to talk. Reporters are paid to get stories, whether you help or not. When you don't respond, they must depend on other sources. Release the bad news yourself, before a reporter digs it up.
- Provide journalists with names of follow-up sources—both inside and outside the organization.
- Respond quickly. You won't appear to be credible if you seem to be stalling. But tell the truth—or nothing. Nobody likes a liar.
- Know whom you're dealing with. Do your homework on journalists before you talk to them.[24]

Preparing for a Crisis

Not every crisis can be anticipated. But an organization prepared to deal with crises is in a better position to deal with emergencies and handle media inquiries when they do occur.

Anthony Katz recommends the following steps to crisis planning.

- Conduct issues management programs. Discuss how the organization can best approach its publics when an issue arises. Situations such as a musicians' strike, an outraged community reaction to an avant-garde play, or the retirement or firing of a well-liked conductor can adversely affect an organization's image, ticket sales, fund-raising capacity, and reputation.
- Develop a communications plan. The keys to crisis communication are speed, accuracy, thoroughness, consistency, and credibility. Include updated media contacts and other stakeholder lists in the plan.
- The organization's messages and actions in a crisis must fully account for the attitudes and opinions of its key audiences.
- Conduct mock crisis drills. Practice company actions and responses to various scenarios utilizing role-playing that represents sensitive community groups.
- Plan for direct communications with target audiences to get your messages across without the filters of the media and with the certainty of reaching those audiences.
- Consider postcrisis communication. Crises pass, but their effects often linger. Crisis preparation, therefore, should include steps for identifying those audiences that will require follow-up communication as events wind down, and for ensuring that such communication is actually carried out. For example, plans should be developed for addressing disgruntled donors,

newly lapsed subscribers, and employees and performers whose morale is low.[25]

An added benefit of crisis planning is that through the process, the organization is likely to become more sensitive to issues stewing under the surface, and thus may identify and focus on its problems and perhaps avert escalation to crisis proportions.

CHAPTER 14

HARNESSING AND LEVERAGING THE POWER OF DIGITAL MARKETING METHODS

THE INTERNET, EMAIL, AND SOCIAL MEDIA—KNOWN AS NEW WAVE TECHNOLOGY—have irrevocably changed the daily lives of consumers. They have also irrevocably changed the work of marketers. The emergence of new wave technology marks the era that Scott McNealy, chairman of Sun Microsystems, declared to be the age of participation. In this age, people create news, ideas, and entertainment, as well as consume them. New wave technology has enabled people to turn from being *consumers* into *prosumers*. It enables connectivity and interactivity of individuals and groups.

New wave technology consists of three major forces: cheap computers and mobile phones; low-cost Internet; and open source, which allow individuals to express themselves and collaborate with others.[1] The advances and broad adoption in recent years of smartphones and tablets have made mobile the key growth area for marketers. Says Chuck Martin on the Harvard Business Review Blog Network, "Consumers no longer *go* shopping, they always *are* shopping."[2] Marketers are learning that to get out real-time messages that respond to what consumers are doing "now" is a high priority.

In a 2012 survey of 3,800 performing arts attenders, Patron Technology found that 66 percent of respondents reported buying their tickets online. Among the respondents, 42 percent said that email is the most effective media type for keeping them updated about arts organizations; 13 percent said that websites are most effective. The traditional media of newspapers and direct mail received highest ratings from only 23 percent and 16 percent of respondents, respectively; 55 percent reported that they visited social networks weekly, an increase of 308 percent in five years. It is important to note that 80 percent of respondents were 45–75 or older, so the growth in use of new technology is not reserved for the younger generations.[3]

New Marketing Paradigms

The new information age has changed perspectives on how and when people make their choices and their purchases. Before the high-tech information age, all contacts between a business and a customer were initiated and controlled by the marketer. With traditional methods, marketing is driven by the marketer; the marketer chooses what information to disseminate and when, where, how, and to whom it is distributed. Historically, customers knew only what marketers chose to tell them.

In the new paradigm, customers choose to enter the engagement process and define the rules of the engagement. Because of the response mechanisms built into new technology with email, blogs, and social media, marketing has become a two-way exchange. Along with these advances come certain service expectations that customers never had before. Because online patrons can define the features and options they want for many products and readily access a world of information on virtually any topic, they are naturally impatient with organizations and marketers who do not offer a wide range of information and who are not flexible or responsive to their needs and preferences. Customers also expect marketers to make the search and purchase processes as convenient, hassle-free, and quick as possible.

Websites

An organization's website is the simplest form of Internet-based marketing, and is very effective if done correctly. A well-designed website can communicate to the public the brand image of the organization and the art form itself. Using video and music the public can view the dancing and hear the music without setting foot in the venue, albeit in a distanced way. The website can be an important entry point and ongoing direct connection for building a relationship between the consumer and the cultural organization.

Website Benefits

Websites offer myriad benefits to both the organization and the customer.

Benefits to the Organization

Although the website places much responsibility on the shoulders of arts managers, the organization receives many benefits that more than compensate for the effort required. First of all, the web presents everyone's site equally. On the web the smallest arts organizations can do many things as well as multimillion-dollar companies. Highly detailed information can be presented without concern for printing costs and mailing weight. The web is colorful, easily updateable, and ideal for spreading news. It makes frequent general and personalized communication with patrons easy. An online presence is important for raising the company's profile locally, nationally, and internationally. It is cost-effective, both in terms of reducing costs and increasing revenues. Effective websites can be highly valuable for developing new audiences and for improving interaction with current customers.

Web marketing delivers useful content at just the precise moment a consumer wants and needs it, but also tells people things they didn't think to ask. It encourages browsing, not just searching. Marketers can tell their organization's story in ways that are compelling and useful with text, pictures, and video.

Benefits to the Customer

The Internet offers arts patrons an opportunity to browse in a relaxed setting, quickly obtain virtually any information desired, decide at leisure when and what to attend, and act instantly—at any time of day or night—to order tickets. The Internet offers the conveniences not available through any other media. The ease of online ticket purchasing 24 hours a day, seven days a week, has made traditional box office hours obsolete for most patrons. In fact many arts organizations are finding that the most popular time frame for ticket purchasing and surfing for information on their sites is from 9:00 p.m. to 2:00 a.m. Patrons can also "try before they buy" with video and audio clips or photos, features that help break down barriers inherent in traditional media. The website can help patrons plan their visit with transportation, parking, and restaurant information. Patrons can email their questions and requests to the organization and, one hopes, receive a timely, personal response. These features are of course major benefits to the organization as well.

The London Symphony Orchestra (LSO) offers visitors to their website a vast array of LSO-recorded ringtones that people can download for their cell phones. This idea resulted from the orchestra's use of text messaging to alert students to short-notice availability of concert seats. Socializing at pubs has stimulated huge numbers of ringtone downloads; when people hear their friends' phones ring, they want classy ringtones too.[4]

SITE DESIGN, CONTENT, AND EVALUATION

The marketer's goal is to encourage customers to go beyond visiting a site just to check out the organization's hours or to purchase a ticket. This means that marketers should review the effectiveness of promotional campaigns used to bring people to the site and update the site frequently, ascertaining that they budget for updating the site frequently, not just for the initial design and technology.

The site should have a marketing focus with technical know-how, meaning that the marketing department should have ownership of website management. When designing a website, marketers should think in terms of what online patrons want and need. They should ensure that all sections can be used by all browsers, and that the links work across the site. The site must download quickly; if it takes ten seconds to download, a marketer may lose visitors before they get to the home page. Once there, can they quickly find what they are looking for? With one click from any page, visitors should be able to access schedules, the ticket purchase page, and any special promotions; everything else on the site should be no more than two or three clicks away.

An effective performing arts website generates interest and excitement. The site should be attractive and eye-catching and should visually represent the organization well. However, designing a website is not just about having the best-looking site possible; it is primarily about getting information across in a clear, concise, engaging manner.

Write especially for online readers, adding more content than is practical in brochures and advertisements. Include detailed information about the organization, the productions, and the artists—anything that will add value to the visitor's experience. Arts patrons are willing—sometimes even eager—to read more than the standard 75- to 100-word descriptions about productions used in other media. Companies can engage patrons by providing video clips and photos and by posting the program content online. Quality information fosters investment by patrons in the organization and its offerings.

A major challenge in online marketing is to attract people to the website. Whenever possible, the organization's name should be used as its web address (URL) so people can easily guess where to find it. Organizations should consider having multiple URL addresses, all of which drive the patron to the same site, especially if an organization's name presents users with a spelling choice, such as "theatre" versus "theater." Relevant organizations can be asked to exchange links to one another's sites. Possible link partners are arts councils, tourist information agencies, and corporate sponsors. The organization's website address should be put on everything it produces: print and radio ads, direct mail, programs, booklets, posters, Facebook page, and so on. Regular email bulletins should be sent that encourage readers to click-through to the website with hyperlinks.

Once people have come to the site, they should want to stay, browse, and return for future visits. Organizations should take advantage of the opportunities Internet marketing provides to update frequently—even daily—so that the site is continually fresh and encourages frequent visits. The web should offer the visitor an experience that, in comparison to static media, such as brochures and ads, is more thorough, more personal, more involving, faster, and easier.

As soon as performance reviews come out, the best quotes can be excerpted and posted at the top of the home page. The organization can also include entire reviews on the website or provide a link to review sources. The website should offer current ticket availability and special offers for particular performances or for special market segments, such as students or seniors. Keep people engaged with such items as a brief interview with an artist, event programs, site-related quizzes, and interactive activities.

Of course the website should also feature a full complement of basic, rarely changing information: the organization's mission, history, and people (artistic and administrative personnel); full descriptions of each of the season's offerings, along with schedules, venue information, maps, driving directions, parking, and local restaurant information; donor and volunteer information; a learning center; and an FAQ (frequently asked questions) section. The website should also offer links to useful resources such as in-depth information about the works the organization presents or produces, links to YouTube videos from the organization or its artists, and sites for purchasing specific books, CDs, or DVDs that relate to the organization's offerings.

The website is an important resource for encouraging people to sign up to join the organization's email list. There should be a button for this purpose clearly visible on the home page and on many other pages of the website. A brief message on the site should reassure visitors that the organization will keep them up to date with information of value to them and will not share their email address with anyone else. When asking people to sign up to join the email list, the request should be simple and should not include options to join the organization's Facebook site, Twitter, or other media at the same time; those invitations can be extended later on. Email is still the most important medium and organizations must not confuse people with multiple options or scare them away by asking for too big a commitment.

The organization's website should provide contact information for each relevant department in the organization. Someone in each department should respond to these contacts with a personal message in a timely manner and all customer comments and complaints should be carefully considered.

WEBSITE EVALUATION

The first step in evaluating the organization's site is to have key managers agree on the purpose and mission of the site. Is it a channel for selling tickets, acquiring members, soliciting donations, providing education? Is it an information resource for donors? Patrons? Artists? Is it part of an integrated multichannel user experience strategy or single-channel strategy?

Arts marketers should conduct research to learn what is important to their website visitors. What part of their experience most influenced their level of satisfaction and their resulting behaviors? Arts marketers should test the usability of their site by inviting 10–15 people of different ages and comfort levels with using the Internet to purchase tickets and seek other information online. Watch these subjects as they go through the process to identify gaps in navigation, misunderstanding of buttons and functionality, and problematic descriptions or jargon. When these people are asked to provide feedback, marketers may find, for example, that the 10- to 15-minute time limit allowed for completing a ticket purchase transaction online is inadequate for many people who have interruptions in the process and are frustrated by having to start over, which means more of a time commitment and possibly losing selected seats.

Managers must keep a constant watch on their organization's reputation. A great monitoring service that will provide information about what is being written about an organization is Google Alerts, a free service that scours the web and delivers information matching keywords the organization sets up.

Understand what leads to conversions, especially ticket sales. Analyze the clickstream, which is the route prospects take to get to the goal, such as clicking through from an email message to a landing page on the organization's site. Understand where visitors drop out of the process and what is and is not working. Then revise it.

Organizations should also analyze their metrics. Google Analytics has become popular for the cost-free, extensive intelligence it uncovers about an organization's digital audience. There are many other tools for this purpose, but some are very costly. An organization may apply a metrics tool to its site, then use the information to fine-tune its marketing efforts, gauge the impact on the metrics, fine-tune again, and repeat.[5]

An organization should establish a baseline reading of its performance. These metrics will help it assess its current performance levels to be measured against when changes are made. Determine the transaction volume: how many ticket sales, donations, email sign-ups, and referrals come through the website? Also determine the activity level: what is the number of unique website visits? How long do visitors stay on the site? How many pages do they visit?

By tracking how many pages users view on its website per visit, a marketer can test the site's *stickiness.* Julie Aldridge, executive director of the Arts Marketing Association of Great Britain, and consultant Roger Tomlinson suggest that if many people visit fewer than three pages, the marketer should determine if the home page is confusing and if it provides the right information and links. If people are viewing six to ten pages, marketers should check the conversion rate to see if people are doing what the marketers want them to do and are not just spending a lot of time because they are lost. Also, check the site's effectiveness at bringing people back for more visits, and actively employ methods to increase that percentage.[6]

Online Ticket Sales

It is not the art itself that is at issue for many nonattenders or infrequent attenders, but how and when information and tickets are made available. There are low-cost, simple ways for even the smallest organizations to offer ticket sales online.

THIRD-PARTY TICKETING SERVICES

Many arts organizations have a ticketing contract with a third-party ticketing service. Some services, like Ticketmaster, require that the visitor navigate through a wide variety of offerings from multiple organizations on their site. They also charge fees to both the patron and the organization. Sometimes these fees are so high that they either deter people from buying tickets or, at the least, decrease the patron's satisfaction level with the experience. Furthermore, services like Ticketmaster use their own logo on their website, are not integrated with the arts organization's database, and do not provide the organization with comprehensive patron information, so the organization cannot readily access all-important patron data. Often these third-party ticketing services are not set up to take any payments other than for single tickets, so if a patron would like to subscribe or make a contribution, that must be done as a separate transaction directly with the arts organization.

INTEGRATED TICKETING SERVICES

Some organizations are restricted in their choice of ticket sales vendors by the venue in which they perform. But arts organizations are far better off in the long run if they invest in ticketing services that serve their marketing and fundraising strategies. These systems typically appear to the customer to be part of the arts organization's own website. The ticket provider uses the organization's home page design template so that the transition is seamless and patrons are unaware that they are going to another site when they click on "Buy tickets." In addition to the upfront costs of installing these systems, which vary depending on the sophistication of the organization's hardware and software, the organization is usually billed a reasonable per ticket fee, which is typically passed on to the patron as a handling fee. The patron does not pay any fees directly to the ticketing service, and the arts organization has complete control over all charges to the patron. Ideally, the ticketing service is integrated with the organization's database so that the organization captures full patron information on every transaction.

Features of Online Ticketing

Every arts organization's website should have one-click access to ticket purchasing information highly visible on the home page and easily accessible from every other page. The ticket information should also be one click away from information about each production, schedules, layout of the venue, reviews, and other performance-related information.

With most ticketing systems, patrons can view a diagram of the hall and sometimes view the stage from various seating sections. Purchasers select a preferred section or price category and are then offered the best available seats or a selection of available seats. They are also given the option to reselect if acceptable seats are not available on the chosen date. In the United States, Great Britain, and some other countries, the customer provides credit card information (sites are guaranteed to be secure), and tickets are mailed to them if the performance date is at least ten days away. If the performance is closer to the purchase date, tickets are held at the box office to be picked up as the customer arrives at the venue for the performance.

Other features that are becoming more commonly available on arts organizations' websites that greatly enhance the customer experience are ticket exchanges online, discounted tickets "authorized" by a key code or access number, group sales with an automatic discount or perks for volume purchases, and subscription ordering and renewal. Organizations can offer a special code for concierges to receive a commission for tickets purchased on behalf of a guest.

A discussion of other aspects of ticketing systems can be found in chapter 9.

BUILDING E-LOYALTY

Websites are far more than just publishing tools; they are communication tools, offering opportunities for two-way dialogues with customers and automated personal relationships.

Superbly designed sites like Amazon.com and bbc.co.uk have given customers high expectations for their online experiences. People expect easy navigation, up-to-date information, easy purchasing options, fast service, and prompt responses. However, many websites frequently disappoint, frustrate, and anger customers as a result of being too slow, difficult to navigate, or unresponsive. Says Laurie Windham, in her book *The Soul of the New Consumer*, "The web has created an impatient customer with a short attention span and a low tolerance for mistakes."[7] Another factor complicating online marketing is that with the inroads the Internet has made into consumer-buying patterns, companies must now target consumers online based on their activity and shopping habits, which can also change rapidly. Constant research and investigation are keys to tapping into consumers' needs and wants.

Julie Aldridge and Roger Tomlinson suggest that arts marketers design their websites to appeal to people at each level on the *loyalty ladder*—whether they be new visitors to the site, infrequent patrons, frequent attenders, or advocates who are excellent candidates for spreading the word about the organization. In designing and updating the site, the current interests and needs of each of these groups should be addressed to help encourage them to continue their relationship with the organization.[8]

ONLINE EDUCATION

Websites offer arts organizations myriad opportunities for educating their publics, whether novice or knowledgeable, young or old. The organization can offer in-depth information about each production, biographies of performers, program notes, and even a basic course in the particular art form, such as describing the instruments in an orchestra or opera styles through recent centuries.

> DSOkids.com at the Dallas Symphony Orchestra
>
> The Dallas Symphony Orchestra (DSO) features an extraordinarily creative and playful yet highly educational learning center for children at www.DSOkids.com. There are enough colorful, graphically exciting, and fun games and material to keep even an adult involved for hours at a time. In "Beethoven's Baseball" the site visitor can make a "hit" by answering a question and identifying the correct composer. This game builds familiarity with composers and shares interesting, humanizing anecdotes about their lives. Clues help the player both succeed and learn. In the "Time Machine" game the player scores by placing composers in the correct time period. Composers' names pop out of the machine at the click of a mouse, and the player is entitled to three hints for each composer. Another option on the site provides information about musical instruments. One can click on the name of any instrument in the orchestra and hear it play. There is also an extensive, cleverly designed section for teachers, filled with information that is also of interest to parents and any adult who wants to learn more about orchestras, orchestral music, and the DSO, and to access links to other informative websites.

BLOGS

A blog is a discussion or informational site published on the world wide web, consisting of discrete entries ("posts") typically displayed in reverse chronological order (the most recent post appears first). Most blogs are interactive, allowing visitors to leave comments and even message each other, and it is this interactivity that distinguishes them from static websites. Blogging can be seen as a form of social networking. Bloggers do not only produce content to post on their blogs, but also build social relations with their readers and other bloggers. Many blogs provide commentary on a particular subject; some function as more personal online diaries; others function more as online brand advertising of a particular individual or company. A typical blog combines text, images, and links to other blogs, web pages, and other media related to its topic. The ability of readers to leave comments in an interactive format is an important contribution to the popularity of many blogs.

Companies that blog have 55 percent more website visitors per month than those who do not, according to online marketing specialist HubSpot. Of the people surveyed, 71 percent say blogs affect their purchasing decisions either somewhat or very much. These statistics make it clear that blogging is a critical piece of a company's inbound marketing strategy.

Arts managers can stimulate people's interest in the organization and greatly enhance their experience with the performances and their understanding of the art form by writing regular blogs and posting them on their website. Organizations should offer a daily or weekly blog of commentary by the artistic director, artists, or others, such as performers, musicians, dancers, singers, or set or costume designers who can bring alive the creation and analysis of the production for site visitors. The artistic director's blog can discuss controversial issues or present an interesting perspective that is often available only to people working in the field. A blog can be used to showcase the writer's expertise on a given subject and is an asset that positions him or her as a thought leader.

Key people in the organization should be asked to write down the questions people ask them as this provides a rich supply of blog topics. Guest bloggers should be invited —people who will want to gain exposure on the organization's blog. This is a great way to build a strong relationship with industry influencers and to offer readers broadened, interesting resources.

Businesses that blog 16–20 times per month get more than twice the traffic than those that blog fewer than 4 times per month. It is okay to blog once a week or twice a month if the posts reveal great content that engages an active audience. Make some posts short and easily shareable and some longer and comprehensive.

Optimize Blog Posts

Blogging is a great tool for driving search engine traffic. To optimize blog posts, make sure their titles incorporate industry keywords that people enter in search engines. The more blog posts an organization publishes, the more indexed pages it will create for search engines to display in their results. This helps to improve the organization's rank in search engines so the organization is found when people search for specific information. An organization should offer separate blogs when targeting niche audiences by tagging content with appropriate keywords.

Place "calls to action" in the content of the blog or in a sidebar by hyperlinking keywords to appropriate offers. Extend the reach of blogs by including social media sharing buttons for Facebook, Twitter, and the like.

As with all marketing efforts, managers should measure the organization's results—the number of visits to the blogs; the number of click-throughs to pages on the website; and the number of tickets purchased, donations made, and friends connected.[9]

EMAIL MARKETING

Email is a highly effective and efficient communications tool for arts organizations. It is used for motivating ticket sales among a wide variety of market segments, for enriching the experience of people who already have tickets, for following up with patrons after their attendance, and for sharing all kinds of information in a timely and nearly cost-free manner.

Email marketing cannot completely replace other marketing media, but it does eliminate the need for certain costly marketing efforts and greatly enhances the marketing department's total communication plan.

PERMISSION MARKETING

People want email from the companies that interest them. When patrons sign up to be part of an organization's email list, they are engaging in a transaction, even though no money is changing hands. They are trading something they value—their privacy and personal information—in return for the marketer's promises to protect this information by not sharing it with others, and to send them valuable, relevant, and timely messages. As a result, when consumers receive opt-in email, messages they have given the marketer permission to send, their reaction to these messages differs dramatically from their response to junk mail. People tend to sense a personal relationship with individual arts organizations, and typically read opt-in arts emails carefully. Says Seth Godin, in his book *Permission Marketing*, "The most important part of the permission troika—anticipated, personal, and relevant—is anticipated."[10]

Godin, a renowned marketing expert, dismisses traditional advertising practices as interruption marketing. He concedes that permission marketers also rely on interruptions to introduce themselves to a broad base of customers. But the introductory ads can be quite simple because they do not need to sell the product. All they need to do is ask permission to say more. From that point on, all participation is voluntary.

EMAIL BENEFITS

Email provides several important benefits not found in other forms of direct marketing.

Email is nearly cost free. If the email message is created in-house, it costs only staff time for writing copy, uploading images, and formatting. If an organization contracts with a professional email service, such as Patron Mail or Constant Contact, it costs only a few cents per patron. Printed and mailed marketing materials, in contrast, easily cost many thousands of dollars, averaging $1 per patron or more, depending on the quality of the paper and whether the material is sent via first or third class.

Email is an important vehicle for stimulating visits to the organization's website—just one click away. It allows the organization to create mass customization—to communicate with many people at the same time in a personalized way. By generating regular, direct, two-way communication with key customers, the organization can develop an understanding of customer needs—and respond to them. Such response develops a sense of involvement and trust in the organization on the part of the customers.[11]

Arts patrons often forward emails of interest from arts organizations to their friends. And the forwarding does not necessarily stop there: friends often forward to other friends. This is what is known as *viral* marketing—what is commonly a word with negative connotations becomes a happy consequence for arts marketers who know that their interested and loyal patrons are their best resource for spreading news.

Email is an immediate vehicle. As with all media, the messages need to be strategized and crafted, but email delivery is instantaneous, whereas marketing pieces sent by regular mail require time for printing, stuffing, labeling, stamping, sorting, and mail delivery.

Email tends to generate a much higher and quicker response rate than postal mail marketing. An average response rate for mailed pieces is 1.5 percent. (However, this response rate does vary widely with the quality of the list. One can expect a significantly higher response from a well-targeted mailing list.) The Florida Grand Opera tracked response rates to an email request for updated patron information. Less than two days after the message was sent, 27 percent of recipients opened the message, and 37 percent of those people responded with the requested information.

EMAIL STRATEGY

Email, like all other marketing tools, requires strategic and creative planning. Sending out an occasional email message or blasting patrons with frequent email promotions will not sustain interest and loyalty for very long.

The marketer should always keep in mind the recipient's perspective. Why would people want to receive this information? How will they benefit? One of the strengths of email is the ready ability to target messages. To personalize email campaigns, be sure to segment email lists properly to ensure that the right message is making its way to the right inbox. Make sure the call to action reinforces the value in taking the action. Create compelling and appealing subject lines and content that matches the target audience's preferences and behaviors.

Importantly, test email campaigns. Determine which headlines, subjects, content, and links work for each customer segment. Make sure the benefits provided dovetail with other marketing offers, such as those contained in subscription brochures and advertisements.

"At the heart of all email marketing," says Eugene Carr, "is relevance and value."[12] The success of email campaigns is dependent on the nature and quality of the offer itself. Email communications can offer value by providing more detailed information than is available to the general public, early notice of events or offers, timely reminders of special programs, and private offers or discounts. Email recipients want to feel special. It is up to the marketer to deliver something of special value to the customer in each email message.

Building Email Lists

A successful email program depends on building and maintaining a high-quality list. Quality, of course, means having an extensive list of current addresses of those who opt-in to receive messages from the organization. The organization needs to regularly update the email list as new names are added, people's addresses change, or people choose to opt out.

The marketer must promote its email newsletters and announcements and communicate benefits at every possible opportunity the organization has—and creates—for this purpose with current and potential patrons. At the Mark Taper Forum and its sister theater, the Ahmanson, roughly 85 percent of the email addresses on the theaters' list are collected through online transactions or by box office and phone staff at the time of sale. The remainder come from people as they sign up on the website or enter the theaters' online "Win Tickets" promotions. Email addresses can also be collected through all ticket order forms and by box office personnel. If the box office personnel are very busy, such as during the hour before a performance, they can ask patrons to fill out a simple form and give it to an usher when showing their ticket for entry. A program booklet insert can be headlined with a request that patrons provide their name and email addresses and drop the form in a box specially provided for this purpose in the lobby. Forms may also be distributed in the lobby for those who do not have program books or who have left them at their seats. To encourage people to fill in the forms, the organization may offer a drawing for a ticket giveaway, a compact disc recording, dinner for two at a local restaurant, or other gift of value to the patrons. At one organization, I taped small golf pencils to brightly colored index cards and taped a card to the back of every seat in the hall. I arranged for a charismatic manager to give a curtain speech before the performance, encouraging patrons to fill in the requested contact information on the card so we could email them valued information and then to give the card to an usher who was walking up and down the aisles collecting them.

All mailings and advertisements should encourage people to visit the organization's website, where a version of "Join our email list" should be boldly visible on the home page and accessible with one click from other website pages. An effective approach for collecting email addresses is to have a sign-up pop-up screen appear the first time a patron visits the site.

Many organizations (and some countries) have a policy that prohibits the sale or trade of email lists. It is a good practice to follow. Arts organizations can collaborate with one another by sending a special email to their own audiences about the other organization's events. Organizations should ensure that they are providing value for their patrons by making an offer they couldn't get elsewhere, and making sure the offering is congruent with their patron profile. If people respond to the partner organization about the offer, then the partner organization can request permission to add their email addresses to their own list.

Shortly after people sign up for an organization's email list, they should be sent a message welcoming them. If they have signed up via the website, the welcome message should go out automatically, within seconds if possible. On the website sign-up page, patrons can be asked brief questions about what interests them. The more data the marketer has about the patrons and their interests, the more the organization can segment mailings.

The Messages

The email marketing plan involves the selection and timing of various types of messages. The content, style, and frequency of the messages should be driven by what the marketer hopes to achieve. The two primary types of messages in use by many arts organizations are information-oriented e-newsletters and action-oriented e-postcards.

E-newsletters

In his book *Wired for Culture*, Eugene Carr says that "regular communication from an institution to its members is at the heart of a long term loyalty building strategy."[13] Depending on the length of the organization's season and the number of productions it performs, an e-newsletter can be sent weekly, monthly, or every two months. But whatever the frequency, these newsletters should be regularly scheduled and consistent in look and type of content. An organization might produce several versions of each newsletter, sending different versions to different target groups.

The newsletters are typically filled with information designed to educate the recipients; to reinforce their relationship with the organization; and to promote the organization, its activities, and its mission. Each newsletter should consist of a series of short articles of 50–100 words, with photos (when appropriate) and links that direct readers to the organization's website for more information. The newsletter, or at least one article, may be signed by the artistic or executive director, giving the recipients the good feeling of being contacted by an important person in the organization.

E-postcards

E-postcards are used to supplement e-newsletters, targeting segments of the organization's list with specific offers. The goal of the e-postcard should be to motivate an action—to purchase tickets in response to an announcement of a renewal deadline approaching or a performance that is nearly sold out. The postcard may also be used to announce an upcoming educational program, a free lunchtime concert at the library, or special parking arrangements while an area near the concert hall is under construction. It may provide a link to a recent review or an interview with a key performer. It may link the patron to information that will facilitate a greater understanding of the art form in general or of a particular production, composer, playwright, or choreographer. It may provide an opportunity for patrons to give the organization feedback on their experience at a recent performance or it may be used for market research so the organization can learn more about its e-patrons and their attitudes and preferences. The postcard should be highly focused, with a limited amount of copy, and should have a simple and direct visual approach. If the marketer tries to accomplish multiple goals in one e-postcard, most likely the main goal will not be achieved.

FastNotes at the Los Angeles Philharmonic

The LA Phil devised a solution to the problem experienced by many audience members of trying to scan program notes in the few moments between the members' arrival and the start of the concert. FastNotes is a brief set of program notes emailed to interested parties a week or so before a concert. The notes include links to iTunes or YouTube where FastNotes subscribers can hear a brief passage from the music to be played. Deborah Borda, the Philharmonic's president, hopes that in addition to providing information about the composer and the work being performed, the notes will give audience members a sense of how concert programs take shape and why she and music director Gustavo Dudamel decide to juxtapose certain works and composers during the same evening. Just one day after FastNotes was announced, the service had already enrolled about two thousand subscribers.[14]

Subject Lines and Content

Eugene Carr points out that there is a real art to subject lines, as the marketer has much to accomplish in only eight to ten words.[15] First, establish an identity to ensure that the organization's name or "brand" is unmistakable. Make the purpose of the message clear. For example, "Seattle Opera Single Tickets on Sale Today," or "Symphony Holiday Tickets: Special Web Offer." The format of the subject lines and body content should be consistent so that readers will instantly recognize an organization's messages. Digital marketing expert Larry Freed says as much time must be spent writing subject lines as the rest of the email copy.[16] The subject line should be clear and informative; not clever and glib. Avoid exclamation points and all caps.

As audience attention is very limited, the message should be considered as an extended headline. Place important information before the scroll feature is needed.

Personalize messages. For example, don't send messages repeatedly to people who have already bought a ticket asking them to buy a ticket.

Every email communication should be short, sharp, and relevant. The content should be compelling and its meaning must be immediately clear to the reader. Avoid using attachments; people often will not open them because of concern about viruses or because of the inconvenience. Instead, provide links to key pages on the organization's website where viewers can investigate in-depth information.

Test everything—the day and times for sending, which subject lines get better responses, which offers work—and then adjust these efforts accordingly. Studies show that a majority of subscribers responding to email do so within 24 hours after a message has been sent. This fast reaction cycle enables marketers to send out campaigns in stages, measuring the response on the first wave of emails to quickly adjust the campaign for the next waves, as needed. Test the timing, subject lines, and special offers by sending half of the emails to different groups at different times and evaluate open rates and follow-through rates. A 20 percent open rate is standard in the United States, 30 percent is common in the United Kingdom where typically fewer messages are sent. Focus on all the key data: deliverability, open rates, click-through rates, and ultimately, the most important, conversion rates.

Email Etiquette and Policies

Arts marketers should be aware of the policies that constitute good email etiquette. When sending an email to a number of people, make sure the email program has a "blind carbon copy" (bcc) feature so the list of email addresses is hidden from the recipient's view. If this feature is not used, patrons could receive an email with all the other recipients' email addresses visible. However, the bcc feature should be used cautiously as an organization's server could malfunction if a message is sent to too many people at once. When sending emails to large numbers of people, organizations should use an email program capable of merging the database of recipients with individual email messages.

The organization's privacy policy should be published at the bottom of each email message and on the organization's website. Each email message should include an option to "unsubscribe," with instructions for simple and quick execution. Once a person opts out of receiving further emails from an organization, the only email that person should receive from it is the unsubscribe confirmation.

MOBILE MARKETING

Mobile is the screen of choice for users today. Nearly 90 percent of American adults who are live arts consumers report owning a smartphone and 70 percent of them report that they use their phones to look up arts events. According to Google, 70 percent of mobile searches end in an action (visit, purchase, download, etc.) and 50 percent end in a purchase within one hour.

In a special report on mobile marketing published by *Musical America,* Dina Gerdeman said, "We are connecting all the time. With mobile technology, consumers are empowered to take the most efficient journey. If your email isn't readable, people are not going to click through to your website."[17] Limit the word count of email messages for the mobile audience. Be brief and to the point. Use website analytics to get a clear picture of what devices people are using when they visit the sites, so offers can be catered to them. Responsive web design allows organizations to set up one design that works on all screens, no matter what the size. The framework of a responsive website will shrink and expand, and even rearrange itself, depending on the size of the screen that's displaying it—from iPhones up to and including big screen HD TVs. Device detection is built into the code. This means that the content is managed once; there is no work duplication.

"But," says digital expert Luke Wroblewski, "starting with a desktop may be an increasingly backwards way of thinking about mobile marketing. [There are] three key reasons to consider mobile first: mobile is seeing explosive growth; mobile forces you to focus; and mobile extends your capabilities."[18]

Consumers are trending to expect to make any online purchases with their mobile devices, including tickets for arts performances. This feature is becoming more and more important for both savvy marketing and good customer service.

Marketers must keep mobile simple. Says James Orsini, president of Single Touch Systems, "We should focus on convenience."[19] The key approach for marketers to consider is to be less interruptive and more helpful. "The best mobile success stories are those where strategy, technology and creative marketing minds come together to offer the simple, compelling, mobile service solutions for which consumers are yearning."[20]

APPS (APPLICATIONS)

Apps are highly effective at making mobile a direct marketing channel. Apps enable one-to-one communication with patrons via push notifications that alert users to new content. When done well, apps become powerful tools to contact the most-engaged patrons. A mobile website is a passive presence, while an app is an interactive communication channel.

Although tech experts have predicted that apps would become so popular that they would replace websites as the go-to source, this has not proven to be the case. Customers prune apps over time to essentials such as commuter train info and movie show times. Furthermore, separate apps need to be developed for each device, so this is not cost- or labor-effective for most arts organizations.

Only a handful of large performing arts organizations will have a big enough following to keep a sustained audience for their app. They may be able to count on their most loyal patrons to add their app to their mobile devices and use it regularly, but it is not likely that infrequent users will add the app. Is using apps to develop narrow but very important loyalists worth the time and money it requires? Every organization must address this issue for itself.

Social Media

According to the "2011 Digital Buzz Blog Facebook Statistics," Facebook is viewed 700 million minutes every day; 72 million links are shared, 144 million friendship requests are accepted, 216 million messages are sent, and 30 billion pieces of content are shared. Fans are 28 percent more likely to continue being loyal users. The average fan is 41 percent more likely to recommend the product or service to friends. As of June 2012, there were 500 million Twitter users averaging 190 million tweets per day.[21] The important takeaway for marketers is that if they're not communicating with their customers via social channels as well as through email, they're not reaching their audience where it lives, works, and plays.[22]

Marketing expert Mohan Sawhney defines social media as real people having real conversations about real things, people, and ideas. Given the rampant and continually growing popularity of social media, it should be front and center of marketing strategy.[23] It is not enough to market to a consumer; we must market to his or her social groups. People connect online to explore mutual interests; share ideas, wisdom, and expertise; build relationships; have fun; provide and solicit answers to questions; and conduct business.

Social media has four basic functions: to monitor social channels for trends and insights, to respond to consumers' comments, to lead changes in sentiment or consumer behavior, and to amplify word-of-mouth effects, the last of which it does with great potency. This means that brands must function as both listeners and hosts.

Types of Social Media

Social media can be classified into two broad categories. One is the *expressive* social media, which includes blogs, Twitter, YouTube, Facebook, and photo-sharing sites, among others. The other category is the *collaborative* media, which includes sites such as Wikipedia (a free encyclopedia built collaboratively by volunteers from all over the world), Rotten Tomatoes (movie reviews), and Craigslist (classified listings for jobs, housing, and personal items).

With expressive social media, consumers are increasingly influencing other consumers with their opinions and experiences. As this occurs, the influence that business advertising has on shaping buying behavior diminishes accordingly. Organizations must take a holistic approach to answer the tough questions that executives face: How does an investment in social media help achieve my business goals? What are the benefits and how do I manage the risks? How do I write a company-wide social media policy, enforce it, and update it? How do I get everybody on the same page?[24]

PUTTING SOCIAL MEDIA TO WORK

To ensure that social media complements broader marketing strategies, companies must coordinate data, tools, technology, and talent across multiple functions. In many cases, senior leaders must recognize the importance of supporting and even undertaking initiatives that may traditionally have been left to the chief marketing officer. Said one chief executive officer, "We're all marketers now."[25]

Social media can be used to solicit patron input after the performance experience. This ability to gain insights from customers in a relatively inexpensive way is emerging as one of social media's most significant advantages. Comments can be aggregated and prominently displayed on a dedicated website, the organization's Facebook page, or can be tweeted. It is useful to add links to the organization's Tweets and Facebook messages. Photos are effective for building interest.

Says Scott Monty, the global digital and multimedia communications manager for Ford Motor Co., "If you're deciding to interact in the social space, realize it's a conversation, not a forum. It's about give and take, people getting to know people. It's not a forum to unleash platitudes and marketing spin," he says. "The ability to humanize your brand is absolutely critical to achieve success in this space."[26]

Social Media at Diablo Ballet

For Diablo Ballet of Walnut Creek near San Francisco, social media has come into the forefront as a low-cost and effective vehicle to get audiences involved in the arts in a new and interactive way. "Doing social media (Facebook, Twitter. YouTube, Pinterest) takes a huge amount of time," says Diablo's marketing director, Dan Meagher. "It's worth it, though."[27]

Meagher found a huge community of dancers and dance fans on Twitter. In just seven months, Diablo had more than 1,500 followers and added about 50 new fans each week. Diablo offers dance news stories, dance quotes, "did you know" facts, and things happening with the ballet. Diablo posts dance quotes on Twitter and finds that five minutes later, they have 25 retweets (people posting the quote on their own Twitter page).

For one performance, Diablo offered a Twitter Night, as management and artistic staff were eager to see how the average person would describe dance in 140 characters. Said Meagher about this experiment: "The *text-perts* saw things in the dance pieces that none of us ever considered. All the text-perts were in the back of the theater in a secluded area nowhere near other patrons. Most of the audience didn't even know the text-perts were there."[28]

Diablo Ballet also uses Facebook, but because Diablo posts consistently throughout the day on Twitter, more people find them there.

The company also videotapes rehearsals to put on YouTube. They use a basic Flip Cam and don't do any editing, to retain the sense of it being live and real. With these videos, Diablo is able to show great choreographers working with the dancers in the studios, which allows viewers to get up-close and personal.

Meagher is convinced that his intensive social media strategies are a major factor in helping to sell tickets. "Social media works because it capitalizes on word of mouth—the best salesman and the only kind of advertising everyone trusts. You can't buy this. . . . It's a cumulative process. All these marketing efforts feed into each other." Concludes Meagher, "Social media has opened up a new world to the arts, and we need to harness this power now."[29]

Pick Seatmates via Meet and Seat

The Dutch carrier KLM offers its customers a program it calls "Meet and Seat," allowing ticket holders to upload details from their Facebook or LinkedIn profiles and use the data to choose seatmates. KLM is betting that ticket holders would be willing to share their profiles in exchange for a chance to meet someone with a common interest or who might be going to the same event. After selecting the amount of personal information they wish to share, passengers are presented with seat maps that show where others who have also shared their profiles are seated. They can then reserve the seat next to anyone who seems interesting—provided it is available—and that person will receive a message with their profile details. While it is not possible to "reject" a person who has chosen to sit with you, you can select another seat until two days before the flight. Those feeling awkward about moving can delete their data and select new seats using the standard anonymous online platform.

For the arts, a "meet and seat" offer can appeal to singles (young and old) looking for companionship, art aficionados who seek others to talk with about the performance, and people looking to share transportation or a meal after the show.[30]

Social Media Is for Engaging, Not for Selling

Social media is good for awareness; not for driving sales. It can assist in revenue, but is not tied to it. It is cheap, but not free. Some organizations simply look at the number of their online adherents to gauge their success in social media. But, says Douglas McLennan, a social media consultant and editor of *ArtsJournal*, impressive numbers alone do not tell the full story. "I could have a million followers of my social media, but if I say something to them and it doesn't make a ripple, nobody decides to do anything with it, nobody retweets it or comments on it, then it doesn't really mean much." One of the biggest mistakes arts groups make, he said, is treating social media as just another way to blast information on their activities—essentially as a vehicle for free advertising. Instead, he said, they need to think of social media as a "connector of people," a catalyst for building an online community with similar interests and engaging in an ongoing, back-and-forth conversation with them. In other words, social media is not purely a marketing tactic, a transactional vehicle that only kicks in when the organization wants to sell tickets. Says Stacey Recht, associate director of marketing for Hubbard Street Dance Chicago, "It's actually part of our mission. We're engaging people in their lives and the way that they interact with each other. We're part of the conversation, and we're always promoting, encouraging and sharing dance."[31]

Twitter is an excellent medium for giving a brand a personal touch. A dancer may tweet from backstage, telling "insider" stories about what is out of view to the public. At one European theater, to build interest in a forthcoming production of *Romeo and Juliet*, an actor playing the role of Juliet tweeted about her relationship with Romeo as if it were unfolding hour by hour as she reported it.

Some organizations are allowing, even encouraging patrons (usually in specified seats in the rear of the hall) to tweet during performances. In efforts to attract young audiences to their performances, some organizations are responding to the near perpetual convenience, customization, and connection that technology has fostered. At one focus group of potential audience members, a young man said, "Sitting in the dark unable to talk to my friends either in person or virtually is not my idea of a good time."[32]

Is it audiences who need to change their behavior or should performing arts institutions change their outdated rules and rituals? "Back in the nineteenth century, pretty much anything was considered acceptable, people would hoot and holler in the theatre, talk in their boxes at the opera," says Paul DiMaggio, a Princeton University sociology professor who studies nonprofit institutions. "It was not until the late nineteenth century that the conductors, with help from the patrons who paid for the opera or the orchestra, took it upon themselves to demand certain behavior from the audience." These became places where decorous behavior and fancy dress were expected, and unwritten rules, such as not clapping between movements of classical music pieces, were enforced with shaming.[33]

Some arts managers believe they are making the performance-going experience more enjoyable for certain segments when they allow texting during the show. As an added benefit, these patrons buzz about the show to their contacts and hopefully stimulate interest in attending. But a frequent criticism of this practice is that when people are texting or tweeting, they are not mentally present in the performance, which requires complete attention to be fully appreciated.

The Mann Center for the Performing Arts in Philadelphia, an outdoor summer concert venue, is allowing photography on a limited basis during shows, except for orchestra concerts. Patrons are encouraged to take pictures on camera phones and submit them to the organization's website and Facebook page. Some of those photos could end up on the big screen during the concert. This approach allows audience members to be interactive and have extra engagement.

Social media best practices are just starting to emerge. In fact, new capabilities and uses for social media abound and refresh continually, so that marketers must be vigilant to stay current as new media and new uses surface and as "old" ones lose favor and popularity. In fact, during the time it has taken me to research and write this book, I have eliminated some of the ideas that were "hot" just a year or two ago and have done my best to keep up with this rapidly evolving field. So, if any reader finds that some of my content in this chapter is obsolete, please consider the examples I give in terms of their strategic intent, which endures, and not in terms of their tactical approach, which changes rapidly.

Tracking and Evaluation

Among the outstanding benefits of Internet, email, and social media marketing is the ability to readily track a wide range of information about patron behavior. Many web hosts provide the hosted organization with a site where managers can check at will such information as how many people visited the website by day or by the hour, the patrons' host server addresses, how many pages on the website each patron visited, and how long the patron spent there. Managers can also view the page at which each patron entered the organization's website and at what page he or she exited, which is especially helpful in evaluating the effectiveness of special marketing tactics. Managers can track the number of people who opened the message, the number of click-throughs to the website, the numbers of new email subscribers and of people who are opting out of their email subscriptions, and even the number of destinations to which an email is forwarded.

Marketing managers can easily measure the success of email messages by determining increased sales or inquiries around the time a particular email was sent; by counting the number of sales at a unique price or the sales or inquiries going through a unique telephone number or email address, available only to email subscribers; or by asking callers how they heard about an offering or event.

The organization may decide to promote an offer in two different ways to small groups of recipients on its email list, then test to see which promotion received the best response rate. The best performing offer can then be sent to the rest of the email list.

Organizations may also email short surveys to their patrons to monitor whether emails are read and valued. The survey might ask patrons how often they read the messages, assess their level of satisfaction with the content, and request or test ideas to improve the email communications.

Arts marketers shouldn't be reluctant to experiment. Not everything will work every time and people may be gravitating to new digital media. Experiment with new approaches and make sure to track the impact instantly; an approach can be changed quickly if it's not working. The cost, and therefore the risk, is low. And results can be virtually instantaneous, a boost to other flagging efforts. At the time of this writing, Facebook, YouTube, and Twitter are the most popular social media being used by performing arts organizations. Foursquare, a location-based site, was popular until other more widely used sites adopted location-based features. QR codes have distinct advantages—they allow a person to use the QR reader on their smartphone to click-through to a specific website, which is quick and convenient, but some high-tech experts say that this methodology has become cumbersome in the era of voice-activated searches and will become obsolete before long.

INTEGRATING ONLINE AND OFFLINE MARKETING

As powerful as digital marketing is, it cannot replace traditional marketing methods. It is essential to integrate online marketing with all other marketing methods: direct mail, telemarketing, public relations, personal selling, and advertisements. Special web and email pricing offers should be strategically integrated with the overall pricing plan. Connected consumers require a seamless transitional experience from device to device, and marketing strategies specific to a single device are fundamentally flawed.

The consumer experience, not the technology itself, must be the primary consideration. According to marketing expert Mohan Sawhney, the way "to do good digital marketing [is to] do good marketing digitally."[34] Says Scott Forshay, "It is important to remember that experience is not a product of technology; it is a product of emotion. From positive emotions come connections, and from connections come relationships."[35] It is in communicating with flexibility and responsiveness that the battle for consumer hearts and minds will be won.

Online marketing is a gift for the arts. How arts marketers use it will determine their success.

Chapter 15

Building Audience Frequency and Loyalty

For more than three decades after public relations expert Danny Newman introduced his Dynamic Subscription Promotion campaign in 1961, the widespread application of subscription drives created a substantial and loyal audience base for hundreds of performing arts organizations.

Yet, in recent years the public's interest in full-season subscriptions has waned significantly. Not only are many arts marketers less successful at attracting new subscribers, but each year, at many performing arts organizations, fewer current subscribers renew. As a result, marketers have had to work harder and more creatively than ever before to attract sizable audiences. For the most part, arts marketers have successfully filled the seats of lapsed subscribers with single ticket buyers. However, it is far more challenging and costly to market performances one by one. Subscribers are the lifeblood of most performing arts organizations. Even though their numbers have declined, they are the most loyal attenders and most generous donors.

Subscriptions

The full-season subscriber is the ideal ticket buyer, guaranteeing an audience and an expected revenue source. Historically, claims Newman, hundreds of stage companies closed quickly because their economies were based on the hope of selling most of their capacity to the general public through single ticket sales. Consistently strong single ticket sales could happen only if all the shows produced were major hits—something that is extremely unlikely for a nonprofit organization with a mission of artistic exploration.[1]

History of Subscriptions

In the 1960s and 1970s, backed by the Ford Foundation and its creation, the Theatre Communications Group, Mr. Newman helped more than four hundred performing arts organizations thrive by rooting them in the subscription concept—selling tickets not to one show but to a full season of performances—and by teaching managers how to attract subscribers in significant numbers. The momentum gained by early successes with subscription campaigns encouraged the inception of numerous new professional theaters, dance companies, symphonies, and opera companies, and before long, subscription drives became the backbone of most every performing arts audience development campaign. In the United States many organizations large and small rely on subscribers to guarantee an ongoing audience for their programs, and focus a great deal of their marketing efforts and resources on building and retaining subscriptions.

In the results of the John S. and James L. Knight Foundation's Classical Music Segmentation Study in 2001, the largest discipline-specific arts consumer study ever undertaken in the United States with nearly twenty-five thousand completed surveys and interviews, researchers became acutely aware of changing attitudes toward subscribing, especially among younger audiences. Among ticket buyers in the 18–34 age cohort, 15 percent are highly inclined to subscribe, compared to 56 percent of those aged 75 and older. About half of the subscribers are 65 or older. Among single ticket buyers, 36 percent are former subscribers who have opted out of subscription packages but who remain in the audience. In the 1980s, 80 percent of the audiences at the South Coast Repertory Theater in Costa Mesa, California, were subscribers; today 40 percent are subscribers. In response, the theater has reduced the number of performances of some of their productions and when a show is popular, sometimes the show is extended for an additional performance or two.

Furthermore, the traditionally accepted concept that people who enjoy attending an organization can gradually be encouraged to subscribe has not generally held true. One orchestra found that 68 percent of its single ticket buyers have been attending for two to eight years or more and have not been responsive to subscription offers.

Clearly, says consultant Alan Brown, "subscription marketing is becoming an increasingly dysfunctional marketing paradigm."[2]

The Rationale for Subscriptions

In his book *Subscribe Now!* Danny Newman presented several compelling explanations for the value of a strong subscriber base.[3]

Single Ticket Buyer Attitudes and Behavior

Single ticket buyers typically attend only the biggest hits of the season. For the esoteric shows with limited appeal, for productions without big-name performers, and during bad weather, reliance on single ticket buyers often means playing to halls with too many empty seats. Not only does this hurt the organization financially and morally, but it also deprives it of the opportunity to inspire and educate. Through repeated exposure to a variety of offerings, people develop a rising threshold of repertoire acceptance.

Artistic Benefits

A strong subscriber base gives artistic directors more latitude to experiment than they have when dependent on single ticket buyers. With a subscription package, people are buying tickets for some programs they would not have attended otherwise, guaranteeing an audience for unfamiliar or unpopular repertoire. Even if there are one or two programs in a season that the subscriber does not like, typically he or she takes it in good spirit and renews the following season.

Influence of Critics

For organizations without a strong subscriber base, especially smaller, grassroots companies without big name productions or star performers, critical acclaim can be a matter of life and death. Large subscription audiences greatly reduce the power of the critics—both professional critics and social contacts—to close a play with bad reviews. The real power belongs to the subscribers, who spread the word about the shows they like and who cast their vote at renewal time each year on the basis of their reaction to the entire season.

Economic Benefits

Subscribers provide the organization with guaranteed revenue, which is often paid many months in advance of the season. This early revenue stream helps the organization maintain a cash flow for ongoing expenses, even during the off season.

Subscribers require much lower marketing expenditures than single ticket buyers do. TCG, in its annual survey of its member theaters, reported in 2012 that over the previous several years, single ticket marketing expense as a percentage of single ticket income averaged 21 percent, whereas subscription marketing expense to subscription income averaged 11 percent.[4] (It is important to note that single ticket marketing expenses are much lower today than they were during Danny Newman's era, thanks to the Internet, email, and social marketing.)

These numbers do not tell the whole story. The marketing costs for attracting a new subscriber may be very high—in some cases they total as much as 50–100 percent of the first year's subscription revenue. However, one must consider the *lifetime value* of the subscriber. The cost of renewing subscribers is minimal, so over time those patrons who continue to renew provide the organization with significant earned revenue garnered at relatively low cost. Also, many subscribers bring in other new subscribers and single ticket buyers from among family and friends. The 80/20 rule applies to subscribers at many organizations; 20 percent of ticket buyers purchase 80 percent of the tickets.

Furthermore, subscribers, motivated by their sense of commitment to the organization, are prime donors to the organization. In fact, most contributions from individuals come from subscribers.

LIFETIME VALUE OF SUBSCRIBERS

An analysis of the lifetime value of a subscriber clearly indicates that over time, the costs of attracting and retaining subscribers are low compared to the benefits. Consider the following example.

An orchestra runs a subscription campaign and generates 100 new subscribers at a rate of $200 per subscriber (for five shows). In the first example (see table 15.1), the campaign cost is $5,000 for printing, mailing, telephoning, and advertising, or $50 per new subscriber. There is generally a large attrition rate after the first year, but attrition drops significantly in following years. Assume the orchestra follows a typical pattern of retaining 50 percent of new subscribers after the first year, 80 percent after the second year, and 90 percent each successive year. Subscribers who stay with the organization over time continue to generate revenue each year at a minimal cost to the organization. Assume that the cost of renewing subscribers averages $10 per year per patron and that the cost of a subscription remains at $200 for the next five years.

After five years, only 32 of the original 100 subscribers are still subscribing. But, from this one subscription campaign alone, the organization has generated revenue of $51,600 at a cost of only $8,580. The organization still comes out far ahead financially even if the campaign cost exceeds the first year's campaign revenue.

Now imagine that the 100 subscribers and $20,000 revenue are obtained at a cost of $20,000 instead of $5,000, as in the first example. (see table 15.2).

Although at first the campaign costs exceed revenue earned, over time the revenue grows to exceed the costs by a continually greater amount. One must also consider that many subscribers will remain with the organization beyond five years, that many will bring in new subscribers, and that many will become

Table 15.1 Lifetime value of subscribers: Example A

Year	Percent renewing	No. of subscribers	Revenue ($)	Cost ($)
1		100	20,000	7,000
2	50	50	10,000	500
3	80	40	8,000	400
4	90	36	7,200	360
5	90	32	6,400	320
		Total	**51,600**	**8,580**

Table 15.2 Lifetime value of subscribers: Example B

Year	Percent renewing	No. of subscribers	Revenue ($)	Cost ($)
1		100	20,000	$24,000
2	50	50	10,000	500
3	80	40	8,000	400
4	90	36	7,200	360
5	90	32	6,400	320
		Total	**51,600**	**25,580**

regular contributors. In other words, the lifetime value of this one subscription campaign is much higher than the chart indicates. Furthermore, in each of the five years, the organization will conduct its annual subscription campaign, bringing progressively more subscribers into the organization.

Says marketing expert Philip Kotler, "A business is worth no more than the lifetime value of its customers."[5] And, said the character Moss in David Mamet's play *Glengarry Glen Ross,* "What did I learn as a kid on Western? Don't sell a guy one car. Sell him five cars over fifteen years."[6]

LIMITATIONS OF SUBSCRIPTIONS FROM THE ORGANIZATION'S PERSPECTIVE

Despite their many benefits to the organizations, subscriptions present some significant limitations. The thought that "I might not enjoy this" undermines the perceived value of a subscription. As a result, a subscription program tends to work best with "safe" repertoire—repertoire that is within the range of people's expectations. Subscribers expect the works to fall within a certain stylistic range; they don't want their avant-garde theater to perform traditional productions of the classics and vice versa.

Orchestras in particular, with their high fixed costs, are constrained by a financial model that is largely dependent on subscription sales. Therefore they have little room to experiment—they have no R&D capacity like other industries—and even less room to fail. "Until this equation fundamentally changes," says consultant Alan Brown, "subscription marketing will continue to be the sweet honey that sustains orchestras and a slow-acting poison that impedes their long-term sustainability."[7]

In the performing arts, it is extremely helpful to be able to set different capacities for different productions according to demand. It doesn't make sense to offer the same number of performances of every concert or play when the marketing director knows that some productions will be a hard sell and others could sell out more performances. Also, both the audience and the performers have a far better experience when the hall is full than when it is half empty. The structure of a subscription-based season, however, makes variations in the number of performances of each production difficult to arrange. Some organizations have "dark" weeks between productions during which a show can be extended if it realizes enough single ticket demand. And occasionally, an organization—typically a theater because of its flexibility compared to symphonies, operas, and dance companies—will move the successful play to another venue, if the performers, venue, and budget allow for it. But when a show is not well received, nonprofits will continue to offer all the scheduled performances, unlike Broadway (for profit) theaters that are fully responsive to audience demand.

At a session titled "Are Subscriptions the Past, Present or Future of Opera?" directors attending the 2004 Opera Europa conference summarized the varying attitudes of European performing arts managers. Some companies still rely heavily on subscriptions; some, like Théâtre Royal de la Monnaie in Brussels, believe that having more than 50 percent of the house subscribed is "dangerous," whereas others have experimented with changing or eliminating subscriptions in order to avoid bringing in the same audiences time and again.

Hanover Opera lost five thousand people—half its annual audience—when it boldly dropped subscriptions. "But," said artistic director Oliver Kretschmer, "we continued to do work we believed in and began to make more contact with the audiences, and after two years our audiences started to change. The old types never came back but new ones started coming in." Hanover's dramatic policy change was driven by its managers' opinion that, according to Kretschmer, "the subscription system is not the system of our time. We are always sold out now, but younger people do not necessarily want to subscribe." Andreas Homoki of Berlin's Komische Oper said that the advantage of having no subscription system is that the audiences are less conservative and dull. However, he added, the risk is that "you can have a perfectly good production that ends up with only 300 people in the audience." Yet, some companies prioritize the high level of income they can consistently depend on from subscriptions. Says Ulrike Hessler, director of public relations and development at the Bavarian State Opera, "Subscribers are not our favorite audiences; they like 'easy art.' But they bring in money."[8]

Subscriber Benefits

Some people are willing to pay well in advance for an entire season of performances, to risk disappointment over some productions while committing to artistic exploration and supporting an organization they believe in, and to eliminate the role of the critic and other opinion leaders from their ticket purchase decision. But these factors, which constitute Newman's rationale for subscriptions, are actually *costs* from the customer perspective—costs people are willing to pay as long as they receive benefits of value to them.

The benefits of greatest value to subscribers are seating priority, ticket exchange privileges, and sometimes, discounts. Of course, most subscribers want to be sure to see all the shows in the season as they have developed loyalty to the organization. Many organizations offer other subscriber benefits, but most often these are additional perks, not reasons to subscribe.

Seating Priority

A popular subscriber benefit is seating priority. In heavily subscribed organizations, people wait years to work their way up to preferred seats. Some subscribers who consider dropping their subscriptions for a season or two do not do so because they are afraid of losing "their" seats. The best seats are still a draw for many people, and long-term subscribers and generous contributors should always be granted first seating priority.

Ticket Exchange Privileges

Subscribers generally expect ticket exchange privileges as a benefit for committing far in advance to attend on specific dates. Heavily subscribed theaters have typically required that tickets be exchanged for the same production only. However, performing arts organizations can best meet their subscribers' needs and preferences by allowing them to exchange tickets for other productions as well, according to availability. If the subscribers already have tickets for the other available production, they may bring their friends to use their "extra" seats, thereby bringing new patrons into the organization. Also, people who travel frequently or spend winters away may be more likely to subscribe if they can count on such flexibility. This benefit has become so important to subscribers that many US orchestras, which have long held out against this policy, are now offering their subscribers the privilege to exchange their tickets for any season performance. Box office personnel complain of being burdened with multiple exchange requests, but they must learn to consider this a minor inconvenience compared to the great advantages of having a strong subscribership.

Many organizations charge a ticket exchange fee, but in recent years, some organizations have eliminated the fee to make ticket exchanges more palatable for their patrons. It makes the most sense to offer subscribers free (or very low cost) exchanges and charge single ticket buyers for their exchanges. Each organization must analyze its own patrons' characteristics, expectations, and behavior to determine which policy and what exchange fee would work best for them.

Discounts

The discount is one of the most common promotional instruments for selling subscriptions because so many people are drawn in by a bargain, even if they can afford a higher price. Many performing arts organizations offer subscribers five plays for the price of four, one play free, a discount off the single-ticket price that may range from 10 percent to as high as 40 percent, or some similar offer.

In recent years, as many subscribers have been lapsing their subscriptions and it is increasingly more difficult to attract new subscribers, arts organizations are luring potential subscribers with deep discounts. This strategy may have the short-term effect of increasing the subscriber base, but unless the artistic programs continue to be highly attractive, people who subscribe because prices are low are highly likely to lapse when renewal time comes around.

Frequent performing arts attenders tend to be in higher income categories than the general public, so their attendance is generally not dependent on getting significant discounts—except for students and other low-income groups, for whom special price offers should be made. Each organization should carefully research what level of discount will provide the desired incentive so that it will not be overly generous. An organization can often greatly increase its ticket income by reducing the discount that it offers to subscribers.

When designing discount offers, the organization should estimate the implications of each offering. For example, if an organization were to offer new subscribers a 50 percent discount for the first year, either the offer would have to be repeated year after year, at great financial cost to the organization and at great risk of annoying other subscribers who were not offered the same benefit, or the organization would have to expect to lose many of these new patrons when the price goes up. On the other hand, if the organization offers all new and resubscribing subscribers a 10–20 percent discount each season, the offer may serve as a viable incentive to subscribe that helps build a loyal audience base at low cost to the organization. Some organizations are experimenting with offering a deep discount the first year and reducing the amount of the discount in subsequent years.

Guaranteed Seats

Subscriptions are easiest to sell when there are many attractive shows being performed over the course of the season, or when one show is so attractive, the only way to guarantee tickets is to subscribe. At the Lyric Opera of Chicago during the 2012–2013 season, performances of *A Streetcar Named Desire,* starring the renowned Renée Fleming, were available to subscribers only. This strategy anticipated that demand would be high for tickets to the limited performances of this production and the opera wanted to accommodate its most loyal patrons. Also, since the subscriber base had been eroding over recent years, this strategy served to retain some subscribers who may have lapsed their subscriptions without this incentive. In organizations with a low subscribership or with an intimate hall that has uniformly good sight lines, it is harder to make a compelling case for a campaign that says: "Subscribe and guarantee your seats!"

LIMITATIONS OF SUBSCRIPTIONS FOR AUDIENCES

People who subscribe report that they *like* to plan in advance. But many people are attending less frequently and are buying smaller packages. For some, this is because they have become more spontaneous in their lifestyles or have difficulty scheduling in advance. Research shows that the majority of former subscribers say that their primary reason for no longer subscribing is that they prefer to select specific programs to attend.

The preference for selecting which shows to attend has come to take precedence over priority seating and other subscriber benefits. And as more people lapse their subscriptions and more seats become available, guaranteeing one's seats in advance becomes less urgent. This is an advantage for ticket buyers, even though it is a worry for marketing managers. Increasingly, patrons are putting off ticket buying until the week of the performance.

The change in buying patterns is primarily a generational shift, part of a seismic shift in how people spend their leisure time and dollars. Many in the arts express concern that younger people are not interested in their offerings and will not attend, while in fact, the younger generation eagerly attends when marketers make offers available that match their lifestyles.

ATTRACTING SUBSCRIBERS

The marketing director must carefully consider which benefits prospective subscribers value and prioritize aspects of the offer around these benefits. Tactical benefits like discounts and complimentary ticket exchange privileges typically are not *reasons* for people to subscribe. These benefits make the offer more attractive once people have the *desire* to see a series of performances. As I cannot emphasize enough, a season of highly desirable plays is the most important factor in attracting and retaining subscribers.

Design Your Own Series

Many organizations have discovered that they achieve the greatest success in retaining and attracting subscribers by being flexible—by listening and responding to their customers. According to Jim Royce, marketing director of the Center Theater Group in Los Angeles, the purpose of his Design Your Own series is to meet customers' expectations, not the theater's. When create-your-own subscriptions were first offered at the South Coast Repertory Theater in 2010, they attracted one hundred subscribers. Two years later, there were seven hundred create-your-own subscribers.[9] For many years, while subscriptions were in high demand, the Lyric Opera of Chicago had the luxury of dictating and enforcing rigid subscription plan parameters. The only flexibility allowed patrons in those days was exchanging to another performance of the same opera within the same series. During its heyday, the Lyric Opera sold more than 100 percent of capacity, since tickets that were turned back to the box office by subscribers who could not use them were resold to eager single ticket buyers. In more recent years, as the subscriber base had eroded and people have become increasingly selective, the Lyric has been offering a wider variety of packages with different options as to the number and choice of operas. The organization also added an option to switch from one opera in the package to another opera not part of that package, and the option to design your own package.

Money-Back Guarantee

In response to some people's concern about avoiding the risk of not liking the performances, the Mark Taper Forum in Los Angeles offers new subscribers a money-back guarantee. "It's simple. Just attend your first performance. If you're not satisfied with your subscription, we'll refund your money for the balance of the season." Says director of marketing and communications Jim Royce, "It makes a lot of people feel much more comfortable about buying subscriptions. And it works. We have had fewer than 500 out of 67,000 cancel." Royce attributes his organization's success in selling and retaining subscriptions to a focus on building long-term relationships with single ticket buyers, reengaging former subscribers, upgrading donors, gaining referrals from current subscribers, and providing current patrons with superior service.[10]

Designing the Offer

The target market for each offer should be clearly identified so that the right messages and language are used to attract the target audience. Designing a subscription offering is a relatively straightforward task for an organization that produces only a few productions a year. Organizations such as large orchestras, which may offer dozens of different concerts in a single season, need to divide their season into different packages of balanced offerings, creating combinations of concerts that will appeal to the various subscriber segments. This is a complex task that is best accomplished with much experience and detailed records of past sales and ticket exchanges.

Extended Payment Plans

The actual cost to consumers of a subscription series is not only the price of the tickets but also the timing of payment. Traditionally, subscribers are expected to make their payment in one lump sum when they place their order, typically in the spring before the fall performance season begins. However, people prefer paying on a monthly basis for other large expenditures such as their health club membership, mortgage, and insurance payments.

In the San Francisco audience survey, we tested the premise that some single ticket buyers would purchase a subscription series if they had the option to pay for their subscription over a period of several months. The encouraging result was that 23 percent of single ticket buyers at the symphony, ballet, and theater reported that they were very or extremely likely to purchase a subscription if an extended payment plan was made available. These people represent a broad range of demographic characteristics in terms of age, income, length of attendance, and marital status. Any plan that would convert even just a small percentage of single ticket buyers to subscribers is definitely worth considering and implementing.

To administer this option, the organization should charge a modest service fee—$10 should be more than adequate—and a credit card number for each participant should be kept on file with the understanding that this account will be charged monthly over a period of time, which can range from three to six months or even longer, depending on how expensive the organization's ticket prices are and when the person makes the first payment. It is best to allow the patron to decide the time frame for payments.

The arts managers I have spoken with who offer extended payment plans report they have never had anyone fail to pay, even when payments are due well after the season has begun. It is highly worthwhile to offer extended payments even if an occasional patron defaults.

The Subscription Campaign

A subscription campaign requires involvement at every level of the organization, including staff, members of the board of directors, and volunteers. The roles and responsibilities of each will vary according to the size and structure of the organization. In larger organizations, the marketing staff will do most of the work; in small organizations, board members and other volunteers may play a more central role, with one or two employees managing administrative details. Qualities to seek in persons managing the campaign are initiative, leadership, and the ability to inspire, motivate, supervise, provide incentives, and otherwise involve the staff and volunteers. The campaign leadership should have direct contact with each volunteer at regular intervals throughout the selling period. The knowledge that there is an ongoing reporting procedure will have a positive effect on sales.

Setting Goals and Structuring the Campaign

For each subscription campaign, the marketing director should set specific goals for the number of new subscribers, for renewals, and for bringing back lapsed subscribers. Objectives should be set realistically so that each campaign's goals are attainable. The target increase may be a small increment, or there may be a major drive to substantially increase the number of subscribers, based on the appeal of an upcoming award-winning play, a new venue, a star performer, or other attractions that may serve as incentives for new subscribers. When an organization enjoys a large subscriber increase in one year, it rarely can sustain that level of growth in succeeding years. In this case, the organization should focus on renewing as many of the first year subscribers as possible. If the organization had lost a significant percentage of subscribers recently, it should investigate the reasons and focus on efforts to bring them back.

The campaign should be formulated in detail, considering all opportunities for audience growth, with strategies for targeting each group, including personal selling, mass media, and social media. Marketing managers should study the effectiveness of various tactics used in the past when deciding how much effort and budget to commit to each. For example, when an arts organization trades lists with other organizations, the manager should code the mailings with the source of each contact so that the effectiveness of each source can be measured.

Timing the Campaign

Renewal of past subscribers is the first step in the annual subscription campaign, but efforts to gain new subscribers must begin long before the renewal effort is over. A successful organization generally renews between 60 and 90 percent of its subscribers, meaning that it must compensate for attrition before it even begins to increase the subscriber base.

The subscription drive is usually initiated in late winter or early spring for a season that starts in the fall to allow the marketer enough time to follow up with patrons who have not responded before the season begins. The campaign runs through the season's first production so that single ticket buyers can be encouraged to apply the cost of their ticket to the price of the season subscription. There is evidence that starting the campaign earlier actually generates a better response rate overall. And organizations can and should extend the campaign period by offering partial subscriptions at various times as the season is running, such as during the holiday season for the winter and spring.

RENEWALS AND RETENTION

Past attenders are far more likely to respond to marketing offers than people who have never attended; it is much easier to sell a fifth ticket to people who have already purchased four than to sell someone a first ticket to an organization. The cost of getting a subscriber to renew is a fraction of the cost of recruiting a new one. TRG's extensive database of ticket-buying statistics at many organizations demonstrates that the cost of sale to new subscribers is 25 percent the value of the sale, while the cost of renewing subscribers is just 3 percent. Furthermore, once people have renewed, their renewal rate over subsequent years is much larger, creating much higher value for the organization. Whereas 69 percent of all subscribers renew, 88 percent of subscribers who are also donors renew. In contrast, the cost of sale to single ticket buyers averages 20 percent; merely 23 percent of single ticket buyers purchase tickets for another performance.[11]

When people subscribe because of a star performer or other similar enticement, they are less likely to renew the following year if the season is less exciting. Also, the more gimmicks that are used to lure new subscribers, such as deep discounts, the less likely they will be to renew as they may have been coming for the "wrong" reasons. For every person who expresses annoyance at being contacted several times by mail or phone, there are likely to be many others who express appreciation for being reminded. Assume an organization begins its renewal campaign on March 1. A second notice should be mailed to those who have not responded on April 1, followed by a third on May 1. By May 15, a telephone campaign should begin to follow up with the recalcitrants.

A deadline can be set and enforced after which the subscriber's seats cannot be guaranteed. If the "official" campaign deadline was May 31, the organization may want to extend a grace period to those who have technically allowed their subscriptions to lapse. They may be told that they are considered members of the subscriber family, that it is difficult to accept their withdrawal, and that the door is still open for them to retain (or improve) their seat location and other benefits for a certain period of time. At this point, it is appropriate to have someone special in the organization (a dancer, singer, or actor?), rather than regular telemarketing personnel, make the contact. Even when a person does not respond to such efforts, he or she should not be considered irretrievably lost and should be approached again in early fall amid the excitement of the new season getting under way. Of course, the person's former seat locations will have already been given to others, but the best seats available at that late date can be offered.

The marketing director should coordinate timing with the artistic director to avoid scheduling the more "difficult" works either for the first production, when new subscribers may be getting their first impression of the organization, or during the renewal period. This will avoid giving subscribers too convenient an opportunity to express their dissatisfaction and abandon the organization. The marketing director should also coordinate schedules with the fund-raising director so that subscribers, especially those new to the organization, have a chance to attend two or three performances before being solicited for contributions.

When a new subscriber signs up or a subscriber renews, an acknowledgment with thanks and a special welcoming letter should be sent within three to five days with notification as to when the tickets can be expected. When the tickets are mailed, a special "helpful hints" flyer should be inserted, highlighting important points on ticket exchange information, parking, special educational programs, dining information, and so on.

Renewing First-Year Subscribers

Increasing first-year renewals by even 10 percent can amount to significant subscriber growth over time. From the time people first subscribe, the organization should welcome them into the subscriber "family," provide them with in-depth information about the organization, its policies, its artists, and its programs. Newsletters and other special information should be sent on a regular basis. New subscribers should also receive at least one customer service phone call during the season asking them to evaluate various aspects of their experience and to express any concerns or suggestions they may have. This requires that callers are fully informed about the organization and capable of carrying on a meaningful conversation with the subscriber, and are not just reading a script. The organization should follow up on all customer requests. If a patron has missed one or more productions, the organization may offer extra tickets to an upcoming production.

At renewal time, all first-year subscribers should receive a personalized letter inviting them to renew. Sometimes the renewal effort requires several contacts by mail or phone, and some people actually express appreciation for being reminded. For the others who clearly express annoyance at repeated contacts, telemarketers should ask people their reasons for lapsing, record this information for the use of the marketing department, try to find other ways to satisfy that person with alternative offers, and put that household on the "do not call for subscription" list, at least for that season.

Renewing Long-Term Subscribers

If fewer than 50 or 60 percent of long-term subscribers are renewing, the organization must carefully determine the reasons for customer dissatisfaction and make dramatic efforts to improve its products and/or services.

Even though most subscribers renew "automatically," their commitment and loyalty should not be taken for granted. The organization can plan "anniversary" events for subscribers after five years, ten years, and twenty years or more of subscribing. Once a year, subscribers can be invited to enjoy a complimentary dessert with the artists after a performance. On occasion, they can be invited for backstage tours and preseason previews and offered an occasional souvenir. The point is for the organization to show subscribers how special and important they are. The benefits need to be strategized so they have value for each customer.

Reaching Out to Lapsed Subscribers

Some people who are unfamiliar with the organization and its offerings are willing to try it out for a season and then find it is not for them. Some people are lured by various benefits and premiums, by a special event or performer, factors that apply for one season only. These are common explanations for the high attrition rate among first-year subscribers. Among long-standing subscribers who do not renew, most report their reason for lapsing is their dissatisfaction with the programming. Others have schedule conflicts and can no longer conveniently plan in advance, or want to select what to attend on a show-by-show basis.

Some of these factors are in the control of the arts marketer; some are not. The organization should contact all lapsed subscribers to determine the reasons for dropping their subscription. Some people can be encouraged to resubscribe if the organization can remedy their concerns; many others will purchase tickets on a show-by-show basis. It is important for the marketing department to treat lapsed subscribers as valued patrons, even if they purchase occasional single tickets, a pattern that is becoming more and more the norm.

Increasing Subscriptions at the Arena Stage

From its peak number of subscribers in 2002 until 2007, the Arena Stage in Washington, DC, lost 40 percent of its subscriber base. Chad Bauman, then director of marketing and communications at Arena Stage, made significant efforts to turn around this serious erosion. First, he conducted focus groups with specific target audiences, including current subscribers, lapsed subscribers, multishow buyers, and single ticket buyers. Using the information he gleaned from these groups, he created and implemented several strategies to rebuild the subscriber base. The result was that from 2008 (during the economic crisis) through 2012, subscriptions increased each year and subscription income increased 115 percent over those four years.

Bauman articulates the formula for his success as:

Great artistic product + best seats + best price + outstanding customer service = more subscribers.

Artistic product: If your customers are not satisfied with the artistic product of your organization, you will not see an increase in your subscription base. This is by far the most important aspect of the offer.

Best seats at the best price: Being able to get the best seats in the house at the best possible price is a powerful value proposition for subscribers. If you have a robust subscription base, oftentimes the only way to get the best seats in the house is by subscribing. Make sure to message that in your sales materials. Also, be very careful of undercutting your subscriber average ticket price, particularly at the last minute. A substantial last-minute discount may provide a lift to an underperforming production, but the long-term side effects could be much worse.

Outstanding customer service: Customer service gives marketers the perfect opportunity to shine. Be proactive in finding ways to provide exceptional service. For example, if inclement weather is on the way, instead of waiting for subscribers to call you to exchange their tickets, why not send them an email alerting them to the approaching inclement weather and offering to make the exchanges on their behalf? And if you don't already, find ways to thank your subscribers throughout the year. For example, there is a theater on the west coast that partners with a winery each year to give subscribers a free bottle of wine when they renew their subscriptions as a way of thanking them for their support.

Arena Stage made additional strategic changes that significantly added to its success in increasing subscriptions:

Lengthen the subscription campaign: Prior to 2009, Arena Stage announced its season in March and continued to sell subscriptions until October, providing for an eight-month subscription campaign. Now, it begins the subscription campaigns in January and sells through March of the following year, thereby lengthening the campaigns to fifteen months. Bauman suggests that marketers avoid delaying the start of their subscription campaign at all costs. Each week is valuable and lost weeks cannot be replaced.

Don't forget about upgrades: Our goal, says Bauman, is no longer just to renew our subscribers; we want to upgrade them as well year after year. Primarily, "we focus on getting subscribers to increase the number of plays on their subscription, but you can also have them upgrade into better seats, add parking to their orders, or increase their annual fund donation. We are even experimenting with add-ons for café meals to great success. In fiscal year 2013, almost ten percent of our subscription base upgraded into larger packages, amounting to nearly $175,000 in additional revenue. Furthermore, full season subscribers have a renewal rate 25 percent points higher and give donations that are four times larger than partial season subscribers."[12]

Speak to subscribers like you know who they are—because you do: Gone are the days when you can create one beautiful season brochure that speaks to all of your patrons, and then mail it over and over again until you beat people into submission. Subscription renewals and solicitations should be highly targeted. You know what types of productions each patron likes and on what nights they like to attend. If you sell café meals and parking through your box office, you even know if they park a car and what they like to eat. You know if they are a full price or discount buyer, how many shows they attend a year on average, and how many people are usually in their party. So why are we still wedded to one-size-fits-all solicitations? Our job is to get the right offer in front of the right prospect at the right time. And we have all the data we need to accomplish that.

Develop a sales pipeline: Until a few years ago, we mailed subscription solicitations to traded lists. Then we started to look closely at our response and tracking reports and we found that list trades were not working, not even close. It would have been just as effective to drop season brochures out of a helicopter over the city. And list trading was considered a "best practice" that every major arts organization in the city bought into. However, we were not measuring efficacy. The failure of these campaigns is easy to understand. Most of these targets had never seen a show at Arena Stage. Why would they invest hundreds of dollars when they had never stepped through our front doors? We changed tactics and concentrated our efforts on trading lists for single tickets, primarily to our most popular productions. This in turn would create an influx of new single ticket buyers. Once they had their first experience at Arena Stage, we would send them an offer to return to a second show. Once a patron had seen two or more shows, the likelihood that they would then respond to a subscription solicitation quadrupled. Don't waste time and money mailing to poor prospects. Instead concentrate your resources on developing more multishow-buying patrons, as those will be your best leads in your next subscription campaign.

Testing and failing: The only way to succeed is to fail. The key is to succeed on a grand scale, and fail on a small one. Aggressively measure the success of every campaign, no matter how small. And test something new at least every week. Tactics will change from year to year, and you'll need to adjust in order to maximize return on investment. As we doubled our subscription revenue over the past four years, we actually started to *spend less* as we grew more efficient. For example, I like to test new offers in our telesales room. Over the period of a week, we may have three or four offers in the telesales room. By the end of the week, after a thousand or so calls, we usually have a clear winner among the offers tested. That offer is then rolled out in an email solicitation, and if it responds well, then we'll include the offer in a large direct mail campaign and then test it against the current control package to see if we achieve a better ROI (return on investment).

Concludes Bauman, "Sometimes it isn't the [subscription] model that is dying, it is how we apply the model that is responsible for our underwhelming results."[13]

LOYALTY

People are as loyal as *they* think they are. People who have attended one performance at an organization each year for the past several years may see themselves as loyal attenders. The fact that they do not subscribe should not be seen as a failure; their annual ticket purchase should be viewed as a success. Of course, the organization rightfully values different levels of loyalty and involvement differently. But subscription is just one type of customer relationship, and it is clear that organizations need to build value around other kinds of marketing relationships as well, particularly those that reward loyalty without a large commitment.

Also consider that loyalty is a two-way street; patrons deserve loyalty from the organization as well. A couple may be loyal subscribers for ten or fifteen years, drop their subscriptions for one or two years, then resubscribe. When they renew after a hiatus, sometimes the organization will treat them as first-year subscribers with the lowest subscriber seating priority. Rather, they should be welcomed back with as close to their former seat location as possible.

Some organizations believe that they win customer loyalty by offering a loyalty award program. Says Philip Kotler, "Some programs are disloyalty programs, as when an airline says the points will be lost unless the customer flies within two months."[14]

Organizations can garner repeat business by "bribing" people with special offers. Loyalty must be earned—and retained—on a continual basis.

When creating loyalty benefits, consider how to attract those who may require several seasons to accrue the status of loyal attender, not just those who accrue the benefits in one season.

A serious concern about shrinking subscriptions is that many fewer single ticket buyers than subscribers make donations to the arts organizations they attend. As it becomes more important to garner donations from occasional ticket buyers, marketing and development must work together to develop and implement strategies that will maximize both return visits and contributed income.

Building Primary and Selective Demand

In efforts to build their audiences, there are two types of demand that may be stimulated by marketers: *primary* and *selective*. When marketers stimulate primary demand, they attempt to enlarge the whole "pie" of their category's consumer base by increasing the total number of people who attend the symphony, opera, dance, or theater—or, in a broader sense, all of the performing arts. When marketers stimulate selective demand, they seek a larger piece of the current "pie" for their specific organization.

Arts marketers can attempt to expand demand in several ways:

1. Current attenders can be encouraged to attend the same type of event more frequently. This strategy is the basis for subscription series and is the driving force behind "second stage" or other additional performances that current patrons are encouraged to attend.
2. Enlist current attenders to bring friends, to give gift certificates on various occasions, and to spread word of mouth about the performances in person and through social media.
3. Other events and art forms can be suggested to current attenders. This strategy aims to introduce current arts attenders to other art forms or other organizations. Thus, a patron of classical theater may be encouraged to attend a modern dance or experimental theater performance. This strategy can be carried out by individual organizations, collaborating organizations, or umbrella organizations such as theater guilds that promote all their members' offerings. Arts organizations are not in competition with

each other to the same extent, for example, as Coke and Pepsi, since cultural attenders report the greatest satisfaction the more variety of shows they attend.

4. Nonattenders can be converted to attenders. The goal is to increase the overall number of persons who attend performing arts events. This may be accomplished, for example, by exposing new people to classical music by way of free concerts in the park, "flash mob" performances in outdoor spaces or train stations, or by providing special events for target groups.

Of these strategies, the first is the easiest to accomplish and the second is highly effective as people respond best to word of mouth from peers and aspiration groups. The third is more difficult and requires not only a change in consumer behavior but also cooperation between organizations. The fourth strategy is the most difficult, as it necessitates changing basic attitudes and influencing tastes among the nonattending public.

Primary demand stimulation is of great interest to arts marketers but is difficult for organizations to undertake on their own. The most effective means for stimulating primary demand is through organizations with a broader scope such as foundations, community-wide arts associations, and schools, and through collaborative efforts between arts organizations.

Even under situations of the toughest competition, arts marketers can make worthwhile efforts. For example, during the World Cup soccer playoffs and finals, in countries where soccer is hugely popular, other forms of entertainment—movie houses, theaters, and restaurants—sit nearly empty. Imagine the performing arts banding together with an appeal to arts aficionados. The advertisement could show actors, dancers, and musicians kicking a soccer ball forlornly on stage and looking out at a sea of empty seats. This may serve as a humorous reminder that the rest of the entertainment world doesn't stop when Barcelona's famed Lionel Messi is on the field.

ALTERNATIVES TO SUBSCRIPTIONS

A highly significant and growing portion of the arts-going public is unlikely ever to become full-season subscribers. Yet arts organizations can offer many alternatives to full-season subscriptions that build some level of frequency and commitment—offers that are effective when they are designed to meet the specific needs and interests of their target markets. Organizations may choose to offer one or several of these alternatives, according to opportunities identified in the marketplace.

But arts marketers must be careful *not* to build a portfolio of multiple offers just to see what may be attractive to their customers. Too many offers can be confusing to people who do not take the time to understand each and compare the differences. Most importantly, marketing offers should be based in customer research and highly targeted to the segments for which they are intended. This means listening to customers to learn what might be appealing to them, rather than creating multiple new offers to which marketers think customers will be drawn.

Some of the options available to marketers are miniseries, flex plans, memberships, and group sales. Although more and more organizations adopt some of these alternatives, most often they are offered—and offered for renewal—only after the full subscription campaign has run its course, to try to up-sell small package buyers to larger subscriptions, and so as not to tempt full subscribers to switch to a reduced commitment. The best way to make sure current full-season subscribers renew without being tempted by smaller packages is to send a renewal letter out well before the brochure is mailed—or even printed. Most subscribers will gladly renew based on a letter that includes information about the upcoming season. Some may require a second or third reminder by letter, email, or phone, but a great majority of subscribers can be renewed by most organizations before the season brochure goes out. It benefits both the organization and the patrons to fully value the preferences of all ticket buyers and to invite smaller package buyers to renew at the same level, to not expect that they will move up to a larger package (although the opportunity and accompanying benefits should certainly be offered).

THE SEASON BROCHURE

Traditionally, a brochure that offers only the opportunity to subscribe is sent to thousands of people at a high cost to the organization. Except for the awareness and interest these brochures build, they are wasted on the people who clearly do not wish to subscribe but want to attend certain events. So instead of a *subscription* brochure, the marketing manager should offer a *season* brochure, offering miniseries, flex plans, design-your-own series, single tickets, and other alternatives to subscriptions in addition to the subscription package. Of course, benefits need to be delineated so people are clear, for example, that subscribers get first seating priority. This approach is far more sensitive to the patrons than policies that say, in effect, "If you don't subscribe, we don't want to hear from you until we are so close to the performance dates that we do not expect to sell any more subscriptions, and then we will eagerly take your money."

By *allowing* and *encouraging* people to purchase exactly what they want upfront, the organization is simultaneously validating their personal preferences, thereby increasing customer satisfaction and involvement, reducing costs later in the season for single ticket marketing, and most likely selling more tickets than they would have otherwise sold. The exception to this recommendation is any highly targeted offer that should not go out to a general audience.

MINISERIES

Miniseries are packages of a small number of performances—often consisting of three to five programs. At many organizations, they can be designed around specific programming, such as full-length story ballets, modern repertoire, a piano series, or music of the Romantic period. This serves to attract potential patrons with specific interests and eases the decision process for people who may be attempting to select from a long list of programs. Additionally, miniseries can be packaged around the lifestyle characteristics of certain consumer segments. The Chicago Symphony's Afterwork Masterworks series consists of one-and-a-half hour programs without intermission, which begin and end early to appeal to people coming directly from work. Some theaters in cold climates offer a "snowbird" series of fall and spring programs to accommodate people who head to warm climates for the winter, or a nonspecified "choose-your-own" series that includes the benefits offered other mini-subscribers.

FLEX PLANS

Flex plans are designed to appeal to people who want to select exactly which programs to attend and plan their attendance at their own convenience. A patron who purchases four flex ticket vouchers may use them for four different shows or all at one performance. Flex plan buyers order their tickets whenever convenient with the understanding that they will get the best available seats at the time of their order. Some organizations allocate some seats for flex plan buyers for each performance, once all subscribers have been seated. With a flex plan, patrons retain certain subscriber benefits and the organization enjoys a far higher level of frequency and loyalty than with single ticket buyers. The organization also has a viable opportunity with this segment to sell additional tickets and gift certificates.

Some arts marketers report that a number of their flex plan buyers are unhappy with this program as they reach the end of the season with ticket vouchers remaining. The marketing manager should do everything possible to satisfy these customers. The vouchers' expiration can be extended to the following season, patrons can be sent an email or postcard reminder notice mid-season of remaining flex tickets, and they should be offered alternatives to flex plans for the upcoming season that better meet their lifestyle.

MEMBERSHIP PLANS

A membership plan is an even more flexible option than the flex plan. Rather than committing to a certain number of seats for the season, members pay an annual fee that makes them eligible for a meaningful discount on tickets to individual performances of their choice and for the opportunity to purchase tickets in advance of the general public.

The membership concept gives people a sense of belonging and provides them with a range of benefits, without requiring commitment to specific programming or frequency of attendance.

The Lincoln Center Theater (LCT) abandoned subscriptions altogether in favor of a membership program. The intent of this major change was twofold: to give managers more flexibility in scheduling plays and extending runs when there is high demand, and to attract to each play people who really want to be there, not just people who happen to have subscription tickets. Because members select which shows to see, their response at individual performances is noticeably more positive than that of many subscribers, who may not have had a particular interest in a production. LCT undertakes various strategies to attract nontraditional theatergoers to the plays, especially production-specific audience outreach through highly targeted media and in specific neighborhoods. There is a modest annual fee for membership; members have the right to purchase their tickets before the general public and for a much lower ticket price. Quite a few members say that the low ticket prices encourage them to take risks and to see some plays that they might not otherwise choose. Whether or not they like every play they see, the fact that it was their *choice* and not determined for them as part of a subscription package leaves them with a better feeling about the experience.

Some organizations offer both subscriptions and membership plans and often reach out to specific target audiences with their membership plans. The typical annual membership fee of $30–$40 is more than paid back by the time members attend their second show each season. Importantly, the membership offer does not erode the subscriber base at all. Subscribers like to plan in advance; they like to guarantee "their" seats, and do not value the lower ticket prices enough to give up their subscriber benefits.

Group Sales

Group sales provide arts organizations with the opportunity to target specific audience segments in significant numbers for individual productions. Groups may attend for social, fund-raising, or educational purposes; many organizations, corporations, and schools center activities on a performing arts event. Some special interest groups have a natural affinity for the subject of a play, with a key performer, with the nationality of a composer, and so on, and are ideal target markets for group promotions. Many people who are unaccustomed to attending the performing arts find the experience comfortable and enjoyable when they are among friends. Therefore, group sales offers are a wonderful way to attract people who ordinarily would not attend on their own, while they efficiently sell multiple tickets at one time.

Group sales require a great deal of advance research, planning, and frequent follow-up. Groups typically plan their programs many months or even a year in advance and often require several meetings to make decisions. A person who is knowledgeable about the organization and its offerings, is highly organized, and has good follow-through capability should be put in charge of this task.

Group Sales at Piccolo Teatro di Milano

At Piccolo Teatro di Milano, nearly half of the tickets are sold to "planned" audiences, consisting of groups of adults and students of 10–300 participants each. Students include anyone from primary school to postgraduates. Adult groups receive a 35 percent discount from single ticket prices; students receive a 50 percent reduction. For all the group sales public, and in particular for students, a wide range of educational activities is arranged every season, including various training programs for students and their teachers. During the workshops, the shows being staged at Piccolo are used to illustrate the different theater disciplines, including the text, staging, and direction, to bring the language and theater experience alive.

At Piccolo, the groups are coordinated by "animators," volunteers who select appropriate shows for their groups and purchase tickets at reduced prices. Animators are season ticket holders or teachers who devote themselves to organizing groups of colleagues, students, or friends, for no fee. Animators are motivated by their love of the theater and by a strong sense of belonging to Piccolo Teatro. Events are organized for them every year, such as a preview of the new theater season, a Christmas-time evening, various presentations, and fun evening events. The animators and their groups are looked after by a team of six marketing staff members who take bookings and handle seating arrangements; they also organize meetings, lessons, and proper training courses for the teachers, students, and adult group members.

Leveraging the Database to Increase Audience Frequency

The modern database is as powerful to the marketer as a Stradivarius is to a violinist. The database contains information of huge value to the organization, and needs to be put to full use. Through the database, organizations know what types of shows people like, where they like to sit in the venue, their preferred day of week for attendance, how often they attend, how many tickets they typically buy, and much more. Marketers can and should utilize this information to reach out to their customers on a *personal* basis, not just promote the shows with mass media and semicustomized mass emails. For example, box office personnel and volunteers can be put to work telephoning and sending individual emails to patrons who are likely to be interested in an upcoming show based on their past behavior.

Recommendations based on a person's past likes and dislikes give people the license and comfort to do something new, without it being or feeling completely new. I think of this approach as *beyond the familiar.* It takes people just a step beyond their comfort level, which is usually a small and doable reach. People are creatures of habit. Even though organizations often promote the arts as *special* (they are!), they should do their best to make arts attendance a routinized activity for people, like going to the movies, attending sporting events, and going out to restaurants.

Valuing the Single Ticket Buyer

A large percentage of performing arts attenders *like* being single ticket buyers and their ranks are growing each season. Yet many arts marketers and managers resist focusing on this segment as a viable, rich source of income and growth.

Kim and Mauborgne, authors of *Blue Ocean Strategy*, say that the business universe consists of two distinct kinds of space, which they describe as red and blue oceans. Red oceans represent the known market space where boundaries are defined and accepted and the competitive rules of the game are well understood. As the space gets more and more crowded, increasing competition turns the water bloody. Blue oceans denote the unknown market space, untainted by competition. In blue oceans, demand is created rather than fought over. In some cases, a blue ocean is created when a company gives rise to a completely new industry, as eBay did with the online auction industry. But in most cases, a blue ocean is created from within a red ocean when a company alters the boundaries of an existing industry.

The luxury brand Tiffany & Company created a blue ocean and significantly enlarged its market by shifting from demographic to behavioral targeting; repositioning from "baubles for the world's elite" to "timeless gifts for special occasions and special people." By changing its marketing and branding strategies, Tiffany developed new and viable market segments for its products.

Say Kim and Mauborgne, "There is no consistently excellent company; the same company can be brilliant at one time and wrongheaded at another. Likewise, there is no perpetually excellent industry; relative attractiveness is driven largely by the creation of blue oceans from within them."[15] The authors' research shows that blue ocean creators offer a leap in value for both buyers and the company itself.

An obvious blue ocean opportunity for arts marketers is to focus on building audiences among single ticket buyers, a major source of audience growth in recent years and in the foreseeable future. TCG reports that from 2003 to 2012, subscriber attendance declined steadily; the number of subscribers declined 22 percent. But during this same period, average single ticket attendance increased 6 percent. During the five year period 2008–2012, single ticket income growth exceeded inflation by 9.2 percent.[16]

It is increasingly crucial for arts marketers to develop messages and offers that meet the needs, wants, interests, and concerns of occasional ticket buyers. This means treating single ticket buyers as *valued* patrons and also means that marketers must redefine what they consider to be a *loyal* attender—even when more and more people want to select exactly which shows to attend and make their ticket purchase decisions close to the performance date.

SINGLE TICKET MARKETING COSTS

Thankfully, single ticket marketing via the Internet and email is inexpensive. Arts organizations have not yet realized the low cost of high-technology marketing on their bottom line, as this medium has been added to other marketing media; it has not replaced them. Before long, it is likely that the marketplace will be ready for arts marketers to reduce expensive direct mail and newspaper advertising as more customer segments adopt the Internet, email, and social media as their primary and preferred sources of information, promotions, and ticket purchase. In the meantime, frequent contacts with customers via digital media are effectively added to other ongoing communications without significant additional cost, except the personnel required to manage these media.

And, if single ticket buyers have the opportunity to purchase tickets for what they want to see, when they are motivated to make the purchase—including through the season brochure—an organization's marketing messages serve a much broader audience more efficiently.

ATTRACTING SINGLE TICKET BUYERS

Arts organizations can employ many different strategies for attracting single ticket buyers and building their frequency of attendance. In addition to customized and personalized digital marketing methods, which were discussed at length in chapter 13, and appealing to initiators and responders, discussed in chapter 3, here are some options designed to meet this segment's needs and lifestyles.

Ticket Exchange Option for Single Ticket Purchase

Arts organizations have been so effective at communicating the scarcity of their tickets (even when this is not the case!) that many potential patrons do not even consider trying to buy tickets within a short period of time before a performance. A common marketing refrain is "Subscribe now! Guarantee your seats! Guarantee good seats!" As a result, many performing arts organizations face empty seats that would have sold with a different marketing approach.

Communicating scarcity is a mixed blessing; marketers hope that this approach will create a sense of urgency for people to purchase tickets in advance, but it also deters people from even trying to see if seats are available. Sometimes people don't buy early because they don't know in advance what their schedule will be. Then, when it comes close to performance time, they may think they won't be able to get tickets at all, much less good seats.

When respondents to the San Francisco study were asked their reasons for attending fewer performances than in the past, about 40 percent overall reported that they have difficulty scheduling in advance. Many people are eager to attend a performance and would like to purchase tickets early enough to get good seats, but face the risks of missing the show and of losing the monetary value of their tickets if other circumstances take priority.

The option to exchange tickets will stimulate advance ticket purchase among those who are interested in attending but uncertain as to whether they will be able to attend and do not want to risk availability at the last minute. Ticket exchange privileges mean that no sale is final and therefore no patron needs to be "stuck" with an unusable ticket.

In the San Francisco survey, we asked single ticket buyers if they would be more likely to purchase tickets if they could exchange their tickets for a different performance, for a handling fee. On average, 25 percent of respondents indicated that they would be very likely to take advantage of this ticket exchange option. Marketers consider any strategy that is anticipated to increase sales by 5–10 percent as highly attractive; a 25 percent response is phenomenal. Surprisingly, we found a strong response to this offer among people with lower income levels. Obviously, the value of the ticket is worth a great deal to them.

Marketing managers have resisted offering ticket exchange privileges to single ticket buyers for two reasons: (1) ticket exchange is a subscriber benefit, and (2) it incurs box office expense in terms of personnel time. However, with a fee attached to a ticket exchange offer, there are no negative implications to the free subscriber benefit and the fee will defray any costs incurred.

The appropriate ticket exchange fee is dependent on each organization's average ticket price and the exchange fee charged to subscribers, if any. A lower fee will be likely to attract more single ticket buyers and increase their satisfaction. A higher price makes the free ticket exchange privilege for subscribers seem more valuable. Organizations may offer a per-order fee rather than a per-ticket fee to minimize the cost to people who have purchased multiple tickets.

The lead time required for the exchange can vary, but usually is in the range of 24–48 hours before the performance.

Some managers say they are willing to make a ticket exchange when a patron phones with a problem, but they are not willing to advertise the offer. Getting the word out loud and clear is key to the success of this venture, meaning that the ticket exchange offer should appear in all ads, brochures, and the website. People *expect* ticket sales to be final; they will not even consider ordering a ticket in some cases unless they know of the new policy. Typically, the last words a patron hears or reads when purchasing tickets are "All sales are final. There are no refunds or exchanges." Imagine the customer satisfaction and potential for increased future sales if instead people hear: "No sale is final. If you are unable to attend, please phone us at least twenty-four hours before the performance and, for a small fee, we will be happy to exchange your tickets for another performance this season. We hope you enjoy the show."

Positioning on Occasions

Arts marketers traditionally capitalize on occasions that feature a composer's birthday, a music director's anniversary with the organization, or the opening of a new performance hall. Such events are worth noting, but they focus on occasions relevant to the organization, not to most ticket buyers.

Arts marketers should consider ways to attract patrons for their *own* special occasions. For many people, attending a performing arts event is special and infrequent. Price may be an obstacle when it comes to a usual evening out; but to celebrate certain occasions, this expenditure is not a problem.

Test Drive the Arts

Test Drive, founded by Morris Hargreaves McIntyre Consulting of Great Britain, allows people to try new things for free initially and then gradually build up to paying full price. Test Drive is not really about price; it's about breaking the inertia of the nonattendance habit. Conventional marketing communications have failed to reach, persuade, and trigger bookings from many people. Test Drive finds new ways to make a connection. Once people have experienced the art, repeat attendance becomes far more likely.

Some of the biggest Test Drive programs use indirect promotion via media channels. Organizations in Northern Ireland and Victoria, Australia, have media partners that actively promote Test Drive opportunities via the "What's On" sections in their printed publications and websites. This ensures a steady stream of online registrations for the program. Other schemes, like at Malmö Opera in Sweden and Circa Theatre in Wellington, New Zealand, run free prize draw competitions with local newspapers. Readers enter to win a first prize of a VIP night at a performance, including taxi, tickets, meal, and champagne for four. Runner-up prizes, a pair of tickets, are offered to every entrant, so that, effectively, everyone wins. The winners really value the tickets as they have been won rather than given. The newspapers and the arts organizations both benefit.

Other organizations have worked with sponsors to offer Test Drive opportunities for the sponsors' staff, or have worked with a major employer, such as a hospital or local company.

Some organizations have targeted the lapsed audiences found on their own databases with Test Drive offers. Those who haven't booked in even five years have had great response rates.

When implementing this program, marketers should choose accessible entry-level productions, even if few seats remain, and give free ticket recipients the best available seats in the house.

The key to the success of this program is in the follow-up. Follow up with a new offer—an offer that demands a commitment, such as "two shows in two months for 25 dollars." Make a second offer to those who did not respond the first time, but then cut your losses and focus on other people with more potential. Obtain and use feedback and box office data to track, measure, and evaluate the impact, then refine the program accordingly.[17]

Developing a Relationship with Single Ticket Buyers

Eugene Carr, founder and president of Patron Technology, recommends ways for marketers to engage new audiences, a segment that is difficult to bring back for another performance. Says Carr,

> Identify people who purchase tickets to one of your events for the first time, and send them a targeted message by e-mail or phone. Consider placing a welcome note inside their will-call envelope or attach one to their seat. I recently attended a concert where new audience members found personalized invitations taped to their assigned seat inviting them to an intermission reception. For just that small amount of effort, you can make a great first impression and begin cultivating a relationship with those new patrons.[18]

Frequent Ticket Buyer Programs

People who do not want to risk paying to attend something unfamiliar may be tempted with a version of the airlines' frequent flyer program.

The policy of rewarding frequent users with free tickets has been adopted successfully by many arts organizations. This program encourages people to experiment and attend unfamiliar programs. The experienced marketing director will usually know before the season starts which programs are likely to have excess capacity and are good candidates for free ticket offers. Patrons taking advantage of this offer may discover that they enjoy something they have not as yet experienced, such as Handel's operas or modern dance.

Typically, people are offered one complimentary ticket to be chosen from among a list of specific performances, after they have purchased tickets for four to eight performances, depending on the organization and its offerings. The organization should address the issue of expiration when designing the program. If single ticket buyers attend only one performance a year, it is unlikely that the frequent ticket buyer program would motivate them to go four times in one season just to get a free ticket. However, these patrons may choose to go to two performances per year for two years to take advantage of the program and should be offered extended time to earn this reward.

Gift Tickets with Subscription Orders

Performing arts organizations can attract new audience members and offer a benefit to subscribers at the same time. When people subscribe to a series of four or five shows (or more, depending on the organization's production schedule and goals), the organization can give subscribers a gift of one complimentary ticket to be used by a guest at a performance of their choice, according to availability. A subscribing couple will receive a pair of guest tickets. In this way, the subscribers can host friends for a performance, at no cost to them. In turn, the guests are likely to offer to host a dinner before or after the show. This becomes a wonderful, social way to thank subscribers and to bring in audience members, many of whom would not have attended without this invitation.

In order to make this program a win-win for both the subscribers and the organization, the marketing department must be sure to collect the guests' contact information before their tickets are distributed so the guests can be added to the mailing list and be followed-up with appropriately.

Activating Infrequent Attenders

TelePrompt, developed by Andrew McIntyre in Great Britain, is an initiative to activate infrequent attenders. Typically, half the single ticket buyers on a database only attend once a year. Research shows that they generally get the brochure and mailings but tend only to purchase tickets for major shows or ones they already know they'll like. With TelePrompt, patrons are telephoned periodically and given a menu of shows about which they can order more information—information that is more in-depth and persuasive than the general direct mail pieces. TelePrompt uses a deliberately soft-sell approach that is designed to build up a trusting relationship. Typically, 70 percent of patrons called join the plan. For every dollar invested by the organization, patrons have spent $3 at the box office. The program could be further developed to promote value-added offers such as backstage tours, preperformance talks, incentives for patrons to introduce new people, or to convert seasoned TelePrompt patrons to cost-effective email.[19]

Targeting Single Ticket Buyers for Subscriptions

Jim Royce, marketing director at the Center Theater Group in Los Angeles, claims that the best subscriber prospects are recent single ticket buyers. Royce suggests phoning first-time attenders within a few days after they attend their first performance. During these calls it is important for the callers to *listen* to what patrons have to say, rather than just give their usual "spiel." The patrons' feedback should then be used to help design a customized series that meets their interests and needs. This strategy will require an advanced selection process of people to call and extra training for qualified phone personnel, but will be well worth the effort. To this purpose, some organizations enlist their own performers and volunteers to contact audience members.

In most cases, says Royce, the patron will not subscribe, but that should not stop the marketer's efforts to keep the organization top-of-mind. Marketers should send postcards, emails, and other promotions as appropriate and consider following up with another phone call in a few months. At the least, people are likely to purchase tickets to one or two shows.

Cultural Tourism

Whether traveling for business or for pleasure, a huge percentage of tourists attend arts and cultural events. Large institutions get the word out to travelers through the mention they receive in guidebooks, hotel room magazines, and advertisements in local papers. Of course, many travelers search the web for activities in their destination city and often order event tickets before they leave home. In this way, the web has opened up a world of possibilities for all organizations, no matter what their size.

A highly effective approach for exposing people to all that is available in a region is through the sites of local arts councils and tourist bureaus. The "Artsopolis" website (www.artsopolis.com) is a comprehensive source for all cultural events, classes, workshops, auditions, organizations, venues, and individual artists in Silicon Valley. It is a project of Arts Council Silicon Valley in partnership with the San Jose Convention and Visitors Bureau and the Norman Y. Mineta San José International Airport. The site makes it easy to search the calendar or to search by type of event from 13 different categories.

Each organization should keep tourists in mind when designing its own site and anticipate the questions that newcomers to both the organization and the city may have. Arts marketers should also consider strategies for attracting tourists, business travelers, and conference-goers by working with local hotels, conference planners, corporate officers, and tourism agencies and implement them as part of their marketing efforts. Tourism organizations often have periods of time when they need more events to encourage visitors to come. Arts marketers should reach out to them and offer to help them meet their goals.

CHAPTER 16

FOCUSING ON THE CUSTOMER EXPERIENCE AND DELIVERING GREAT CUSTOMER SERVICE

GREAT CUSTOMER SERVICE IS DEPENDENT ON THE MARKETER MASTERING the art and science of patron knowledge and understanding. The science is in the database and all the market research undertaken to better understand customers' attitudes, preferences, interests, and behavior. The art is in how this information is used in designing and improving the customer experience.[1]

All the strategies, tactics, and principles for marketing the arts can be encapsulated in one phrase: *focus on the customer experience.* Too many marketing departments focus primarily on their products and services, without realizing that the total customer experience is what matters most in attracting, retaining, and delighting customers. Focusing on the customer experience requires marketers to think holistically about every single customer touch point and every stage in the customer life cycle. It is the marketing director's responsibility to ensure that every employee in the organization understands how he or she impacts the customer experience. And it is the responsibility of marketing to orchestrate the customer experience across every aspect of the organization's functions and through all stages in the customer's buying cycle.[2]

It also means that relationship marketing is key. The marketer must take advantage of every opportunity to get close to the customers; to seek regular, direct contact with them; to anticipate their needs; and to develop a reputation for responsiveness—in other words, to build strong relationships.

Consumer Power

Being a marketer in the new information age means rethinking the role of the customer in the exchange process from passive receivers of the marketer's offers to initiators of the contact and even active cocreators of the offer.

Say marketing experts Prahalad and Ramaswamy,

> Business competition used to be a lot like traditional theater: On stage, the actors had clearly defined roles, and the customers paid for their tickets, sat back, and watched passively. . . . Now the scene has changed, and business competition seems more like the experimental theater of the 1960s and 1970s; everyone and anyone can be part of the action. What's more, that dialogue is no longer being controlled by corporations. . . . Consumers can now initiate the dialogue; they have moved out of the audience and onto the stage.[3]

It is ironic that the performing arts, the very industry from which this metaphor was drawn, may be among the slowest to understand and respond to this dramatic change in how customers expect to do business.

For example, commonly, subscription brochures and telemarketing calls are not set up to accept single ticket orders. If a patron wants to buy tickets for two of five shows in a season, he or she has to wait for several months until single tickets go on sale. This situation is not uncommon in the performing arts, but to me it is as ludicrous as if a department store salesperson told me that if I want black socks only, I could purchase them in three months if any remain; if I want them today, I must also purchase blue and brown ones. Instead, the orders should be taken and patrons should be informed that their orders will be filled after subscribers are seated, rather than asking them to wait a few months, when they may have lost interest or when less desirable seats are available.

Say Smith, Clurman, and Wood in their book *Coming to Concurrence*, "With the emergence of self-invention, marketers must do more than simply customize; marketers must facilitate self-customization. Finding profitable ways of giving power to consumers is now the only way to succeed at giving consumers what they want."[4]

Customers increasingly want to shape their experiences themselves and with their companions. Performing arts managers and marketers must understand that by involving their customers as cocreators of the marketing experience, they have the best chance of broadening their audience, building loyalty and satisfaction, and increasing frequency of attendance.

Customer Relationship Management

CRM has been seen as the panacea of marketing since the sophistication of databases has allowed marketers to capture detailed information about individual customers. CRM involves examining a customer's past purchases and certain demographic information so that the organization can improve customer acquisition, cross-selling, and up-selling with highly targeted specific offers. Yet, Frederick Newell, in his book *Why CRM Doesn't Work*, accuses CRM of falling far short of serving customers well. Newell claims that CRM projects are more concerned about which customer segments are going to deliver the most value to the organization and about internal efficiency in handling customers, rather than being concerned about the real needs of the customer. As a result, some customers may be treated even worse than before.[5]

Similarly, the authors of an article titled "Preventing the Premature Death of Relationship Marketing" say, "Ironically, the very things that marketers are doing to build relationships with customers are often the things that are destroying those relationships."[6] Companies ask their customers for friendship, loyalty, and respect, but too often, they don't give those customers friendship, loyalty, and respect in return. As a result, many marketing initiatives seem trivial and useless instead of unique and valuable.[7]

Newell calls for a change, suggesting that companies *empower* customers, not *target* them. Newell advocates replacing *customer relationship marketing* with *customer management of relationships* (CMR) with a goal of humanizing relationships and delivering better solutions to customers.

Newell summarizes the journey from CRM to CMR like this: from the company being in control to putting the customer in control, from making business better for the company to making business better for the customer, from tracking customers by transaction to understanding their unique needs, from treating customers as segments to treating them as individuals, from forcing customers to do what the company believes they'll want to letting them tell the company what they care about, from making customers feel stalked to empowering them, from organizing around products and services to organizing around customers.[8]

A key factor in facilitating this process is to focus on consumer lifestyles, attitudes, and insights, rather than just on transactions, as is common in CRM. In today's marketing environment, companies will be better off if they stop viewing customer engagement as a series of discrete interactions and instead think about it as customers do: a set of related interactions that, added together, make up the customer experience. Organizations must listen constantly to consumers across all touch points, analyze and deduce patterns from their behavior, and respond quickly to signs of changing needs. The fact that nowadays lifestyle is central to people's choices of products, services, and experiences is good news for arts marketers. Arts events provide innumerable opportunities and associations for meaningful lifestyle fulfillment.

GREAT CUSTOMER SERVICE

A patron's experience with a theater, symphony, dance, or opera performance does not begin when the curtain rises, nor does it end with the last applause.

Rather, the total experience begins when a potential patron first becomes aware of an organization's offering. The marketer should consider this the beginning of a long-standing relationship between the customer and the organization. Meeting the preferences of current and potential patrons, whether they are sophisticated, educated audience members, new attenders, or people who want specialized and individualized services, means more creative, out-of-the-box thinking and sometimes more staff to implement these services.

Every contact the patron has with the organization's personnel, including box office personnel and ushers, and with the organization's communications, whether mass advertising, the website, Internet, email, social media, or direct mail, affects the person's satisfaction and involvement level. Also, every organization must have policies in place for handling any problems that may arise with customers.

The following story is one that, in my experience, is unsurpassed about how a theater director turned an unfortunate incident into an opportunity to build a loyal customer.

From Dismay to Delight in North Carolina

A long-standing subscriber to the North Carolina Blumenthal Performing Arts Center purchased eight tickets to take her family to a performance as part of her special birthday celebration. When the family arrived at the theater, other people were sitting in several of "their" seats. Unfortunately, the seats had accidentally been double sold. Management decided to allow the people who arrived first to retain their seats, and they placed the woman's family in good seats, but scattered throughout the hall in pairs.

Two days later, Judith Allen, then president of the center, received a letter from the woman saying that her birthday celebration was ruined when her family had to be split up and seated around the hall. As a result of her great disappointment, the woman was canceling her subscription and was writing a letter of complaint to the editor of the local newspaper.

Ms. Allen sent the woman a bouquet of flowers with a note of apology. After receiving no response, Ms. Allen sent the woman another note acknowledging that although she couldn't recreate the woman's birthday celebration, she would like to offer her eight complimentary seats together for another production. This time, the woman replied, saying that it was clear to her that Ms. Allen truly cared about her feelings and about reaching out to her personally. Along with the woman's appreciative acceptance of the ticket offer, she enclosed her subscription renewal for the following season. When the family members arrived for the performance, a staff member greeted them at the door and escorted them to their seats. Shortly thereafter the woman sent the Center a contribution of $500, and the next season she increased her contribution to $1,000.[9]

Bill Gates once said famously, "Your most unhappy customers are your greatest source of learning." A commonly accepted principle of marketing holds that people who have had a bad experience remedied are more satisfied than those who never had a bad experience. (This does not mean that bad experiences should be orchestrated.) Marketing researchers have long claimed that people who are happy with an experience tell three others; unhappy people tell eleven others. Now, in the age of the Internet and social media, bad news travels far and wide and quickly, and marketers must take special care to listen to their customers and to search online for any comments with potential negative impact to their organization. Clearly, turning around a customer's negative experience should be a high priority for marketers. Furthermore, acquiring new customers can cost five to ten times more than the costs involved in satisfying and retaining current customers. In this era when it is increasingly difficult to attract new audiences and develop among them some level of loyalty, the most crucial task of managers and marketers is to nurture and build relationships with current attenders one by one. Deepening the bond with current patrons, increasing their frequency of attendance, and going well beyond expectations in serving them are the best ways for the organization to guarantee that it will grow and maintain a healthy base of patrons and contributors.

Marketing Director Jim Royce of the Center Theatre Group in Los Angeles says that great customer service is crucial. He would like the words "no" or "sorry we can't do that" eliminated from his employees' sales vocabulary. In every instance, says Royce, an alternative must be offered, even if they physically cannot do a requested action. For example, his box office personnel may say, "We're sold out, however I have seats for this other event or performance." Royce says that if anyone sends him a complaint letter—about anything—he replies personally with an apology and thanks-for-letting-us-know letter, along with a complimentary certificate for two tickets, usually for a preview performance of an upcoming show, with the idea of getting these people back in the theater again. If people are angry, Royce gives them a refund.[10]

Royce's policy is in keeping with the "finish strong" principle. Behavioral science research shows that the ending of an encounter is more important than the beginning because it is what remains in the customer's memory. This means that customer service personnel should get unpleasant news out of the way early in the conversation and enhance the customer's experiences during the process, ending with a highly satisfactory result, so that people will have positive recollections of the process after it is completed.[10]

Businesses, including the performing arts, have the opportunity to encourage customers to communicate directly with them, before any negative comments end up on a review site like Yelp, before they cancel subscriptions, stop donating, or before they spread the word about their dissatisfaction to their friends (on Facebook or otherwise). A service called "TalktotheManager" aims to foster a closer connection between businesses and their clientele. As a substitute for comment cards or other less reliable means for collecting feedback, TalktotheManager lets patrons anonymously text their comments directly to the manager via their cell phones. This service is most commonly used by restaurants, which begin by signing up with Seattle-based TalktotheManager and receiving a dedicated phone number for the service. They can then download preprinted stickers and signs that encourage their patrons to text feedback to that number. Managers can be notified each time a customer sends a comment, and they can use the service to send a response. Arts organizations can adopt a similar approach, and may request that the patrons identify themselves so managers can intervene directly with them.

Measuring Customer Satisfaction

A customer-centered organization always asks: How satisfied are our customers with our offering? In what ways can we make them more satisfied? In what ways can we create and market satisfaction for other potential audiences as well?

For arts organizations, creating and sustaining satisfaction is not adequate; attending arts events is optional, unlike more "necessary" purchases like telecommunications plans or airline tickets. I put the word "necessary" in quotation marks because many of us who love the arts and thrive on them find attending performances to be necessary for our well-being. Even so, we typically have many options for satisfying that need. Explaining the philosophy of his company, a perceptive executive once said that the aim of his company went beyond satisfying the customer—the aim of his company was to *delight* the customer. This higher standard may well be the secret of the great marketers. They go beyond meeting mere expectations and inspire the customer to rave.

Many organizations institute customer satisfaction surveys to gather an understanding of the general needs of their customer base. However, say Gilmore and Pine in their book *The Experience Economy*, such techniques do not go far enough, as they measure the *market*, not individual customer satisfaction. The surveys are typically designed to ease tabulation, not to gain true insight into customer-specific wants and needs, so people who fill out these surveys gain little or no direct benefit. Furthermore, customer satisfaction surveys rarely ask for information about the particular needs and wants of the respondents. Rather, they ask customers to rate how well the organization and its personnel are performing on a series of predefined categories. As a result, managers gain too little insight into what people truly want and need.[11]

Dave Power III of J.D. Power & Associates says, "When we measure satisfaction, what we're really measuring is the difference between what a customer *expects* and what the customer *perceives* he gets."[12] What organizations must seek to analyze is *customer sacrifice*, the gap between what a customer settles for and what he or she actually wants. At a certain point, the customer believes that the sacrifice is too great—whether it be the ticket price, the time commitment, lack of interest in the offerings, disappointing customer service, and myriad other factors—and decides not to purchase tickets or renew a subscription. Some factors, of course, are out of the marketing manager's control, but it is the responsibility of the organization to ascertain that each customer receives as close to what he or she expects, needs, and wants as possible.

Mark Graham Brown, an expert in performance measurement, advises people to be wary of the most common measurements, such as customer surveys (they usually fail to ask the most important questions), customer complaints (only about one in ten unhappy customers complain), and customer loyalty (they are based on faulty assumptions since loyalty is driven by many factors other than satisfaction).

Instead, Brown suggests employing a "customer aggravation index." He advises the use of focus groups to identify the things customers hate, rank them according to severity, and keep track of how the organization performs on them. It is a meaningful metric that predicts loyalty and is being used by some companies in the business sector.[13]

Pine and Gilmore also suggest going beyond what people expect by *staging the unexpected.* This doesn't mean offering an improvement that customers may then expect to become institutionalized. It means instigating customer surprise by staging memorable experiences. Some examples may include moving patrons to a better seat location when there are empty seats available just before the curtain rises; offering patrons two-for-the-price-of-one tickets to a preview performance of the upcoming show, available during intermission; or offering complimentary beverages (compliments of a sponsor, of course). At Chicago Shakespeare Theater, actors in costume mingled with patrons before the performances of one production. Similarly, members of a dance company warmed up for the performance on stage with the curtain open, so that patrons felt like "insiders" to what is typically out of view. With such surprises, audience anticipation and excitement is enhanced.

Pine and Gilmore say, "Companies must realize that they make *memories*, not goods."[14] Performing arts organizations, by their nature, make memories with their core product: the works of art on their stages. Arts organizations need to realize the importance of engaging their audiences and creating satisfying, memorable experiences with the business, off-stage aspects of their work as well.

Capturing and Using Service Quality Information

Where does one begin once a commitment has been made to offer great customer service? The answer, of course, is to *listen* to the customers, to truly understand their experiences with the organization and its competitors. Researchers—staff members, outside consultants, or both—should also adopt the role of the customer to have first-hand experience with each encounter with the organization.

Baldridge.com, which features expertise in high performance institutions, recommends nine ways to get closer to customers.

1. *Interview them.* Sit down with current, former, and potential customers and ask about their interests, requirements, and expectations.
2. *Survey them.* Ask customers what is most important to them in addition to how satisfied they are.
3. *Conduct focus groups.* Pull together representatives of a specific customer segment to discuss their needs.
4. *Get feedback on recent transactions.* Sometimes the transaction process is enough to ruin customer satisfaction.
5. *Collect information from your sales and service people.* Be specific about seeking information that will help identify and affirm customer requirements and expectations.
6. *Ask for complaints, comments, and suggestions.* Communicate your desire for information and make it easy for customers to speak their minds.
7. *Form customer advisory groups.* Choose eight to ten key customers to meet semiannually to discuss your strategies, potential products or services, industry trends, and their preferences.
8. *Involve customers in your processes.* Invite key customers to speak to senior leaders and department heads.
9. *Observe customers using your products, services, and programs.* This seems to be the best way to discover their unarticulated needs.

The final step is to close the loop on these information-gathering approaches by designing and refining processes to collect, analyze, validate, and use the information gathered from these listening posts.[15]

Create a Service Strategy

To determine a service strategy, ask: What attributes of service are—and will continue to be—most important to our customers? On what important service attributes is the competition strongest/weakest? What are the existing and potential service capabilities of our organization?

Managers need a profound understanding of customers' requirements to increase satisfaction and secure loyalty. According to prizewinning customer service strategist Noriaki Kano, the relative importance of these requirements differs for each individual and is unequal among consumers. Kano characterizes "relative importance" as three factors: *basic*, *performance*, and *excitement*.

Basic services or features do little to improve satisfaction unless they fail, in which case they can cause serious dissatisfaction. We expect our online ticket purchase transaction to function effectively and efficiently. When it does, we don't feel more satisfied with the organization because that is what we expected. When it doesn't, we feel frustrated and dissatisfied.

Performance services or features are those that produce customer satisfaction. At a theater, this may take the form of, for example, spacious legroom between rows of seats, an adequate number of rest room facilities to prevent long lines during intermission, convenient parking facilities, or ticket exchange options.

Excitement services or features are the unexpected "wows" that customers experience. It may be when performers interact with patrons before the show begins or an offer for people in balcony seating to move down to empty seats on the main floor just before curtain time.

The Mark Taper Forum in Los Angeles offers a concierge service for upper-level patrons and donors. The concierge has a private number and the ability to book hotel rooms, restaurant reservations, theater tickets in London and New York, and provide other services. Over time, the concierge service will be available to wider groups of patrons. A concierge service is applicable for only the large arts organizations that have an infrastructure and sizable audience base to make this offer worthwhile.

Before you can offer any of these services or features, says Kano, you must be absolutely certain that you understand who your customers are and what each customer group requires.[16] Every organization, large and small, should start planning its service strategy by investigating how it can improve procedures that already exist.

It is crucial to measure customers' service expectations, as expectations are the basis for their satisfaction level. Service quality may stay the same or even improve; expectations may continue to rise even more. Measure the relative importance of service quality attributes to customers. When is good, good enough? The organization should prioritize its customer service efforts according to what matters most to the people it serves.

Steps in Service Recovery

Systems should be put in place for handling customer complaints and they should be clearly explained to all employees. Details should be developed for responding to a variety of situations, but in general, always be sure to offer a sincere apology, offer a fair fix for the problem, and treat the customer in a way that shows the organization cares about the problem and helps the customer to solve it. If possible, offer recompense equivalent to or higher than the burden the customer has endured. The manager should ascertain that the organization gives the promised level of service recovery rather than one that falls short. The story earlier in this chapter of how Judith Allen recovered the loyalty of a very unhappy patron is exemplary in all these regards.

Managers should try their best to anticipate and plan for problems. With a controversial work on stage or a new marketing offer, analyze what negative reactions may result and develop a plan to deal with them. Whenever possible, of course, deflect the potential problem before it occurs.

INTERNAL MARKETING

Implementing a customer service strategy requires internal marketing—the marketing conducted with staff at every level within in the organization. To establish an overall service strategy, leaders should employ the following methods.

CREATE A VISION

A broad vision is worth believing in, is challenging, provides emotional energy, generates commitment, and helps create a sense of teamwork and belonging that is sustaining on difficult days. Short-term pressures encourage service mediocrity; keeping a vision in mind helps employees focus on the big picture.

STRESS PERSONAL INVOLVEMENT

Management guru Peter Drucker defined a leader as someone who has followers. A leader with a strong, clear vision, who listens to his or her employees, is the one who gains trust and succeeds. To be effective, people need to be able to think on the job, be creative, venture outside their routine, and be encouraged to take initiative. This is true even for the lowest-paid employees such as the box office personnel and ushers, those who have the most direct contact with the patrons. These employees play a crucial role in providing excellent customer service, in developing goodwill, and in bringing issues of concern back to their managers. Granting employees some autonomy and allowing them to show what they can do require trust. Employees may make some mistakes, but overall this strategy will pay off well.

CONDUCT EMPLOYEE RESEARCH

Employee field reporting involves asking employees to report what customers are saying and doing. Are people who have phoned the box office hanging up without buying tickets? Are they unhappy with available seat locations? Are they asking questions that employees cannot answer? Ask employees to rate service quality as they see it. The employees who are in direct communication with customers can help reveal why problems occur and what to do to solve them. Ask such questions as: What is the biggest problem you face day in and day out trying to deliver a high quality of service to your customers? If you were the organization's executive director and could make changes to improve service quality, what changes would you make? What other problems do you encounter and how would you suggest managing them? Provide a suggestion box so employees can comment anonymously.

SHARE VALUES

Schedule sessions on a monthly or bimonthly basis to share ideas and experiences and to evaluate them together. Bring in all levels of employees, including upper-level management. Reaffirm every employee's role in quality improvement. Skits with role-playing may help bring alive the situations effectively.

BOOK DISCUSSION GROUPS

Management can select a variety of service quality books and ask employees to discuss one chapter per week in group meetings. It is important for discussion facilitators to be able to conceptualize how to make the principles apply to their organization.

QUALITY IMPROVEMENT TASK FORCES

The best way to help people learn quality is to engage them in solutions to real organizational problems. Employees and managers can be invited to conduct mystery shopping at their own organization or that of competitors. It is often a revelation for these people to see how they respond to an experience themselves.

EMPLOYEE SUGGESTIONS

Encourage employees to make suggestions within and without their departments to improve customer service. Also ask: If you could make one change to improve employee motivation, what would that be? Give prizes such as recordings, books, or gift certificates for the best suggestions of the month. Employee performance reviews should take into consideration the degree of involvement in customer service improvement.

SETTING AND MEETING EXPECTATIONS

In this era when it is increasingly difficult to attract new audiences and develop among them some level of loyalty, the most crucial task of managers and marketers is to build and nurture one-on-one relationships and deepen a bond with current attenders. Some marketing experts suggest that organizations have to exceed expectations to keep customers coming back for more. But rather than setting such a high and possibly unattainable goal, work to set the appropriate expectation and then meet it consistently.

CHAPTER 17

AUDIENCES FOR NOW; AUDIENCES FOR THE FUTURE

ARTS MANAGERS FACE TWO MAJOR CHALLENGES TODAY. FIRST, THEY MUST REACH outward to their communities, with a goal of creating relevance, understanding, and accessibility and making art an integral part of people's everyday lives. Second, managers must look inward to professionalize their management and marketing, to approach their tasks strategically in light of a continually changing environment, and to learn how best to be responsive to the needs and interests of their publics. Says marketing expert Mohanbir Sawhney, "It is not the biggest, the smartest, nor the richest organizations that survive; it's the most adaptable."[1]

Also consider the wise perspective of Ivor Royston, M. D. Said Royston, "There are those who make things happen, there are those who watch things happen, and those who ask: What happened?"[2]

ENGAGING AUDIENCES FOR NOW AND FOR THE FUTURE

The concept of *engaging audiences* has been widely discussed in the arts industry in recent years. However, says marketer Doug Borwick, the lack of understanding of the term is so pervasive that the arts are in danger of losing the power that engagement represents in the fog of meanings that surround it.

Audience development, claims Borwick, consists of strategies and tactics designed for immediate results (sales, donations, etc.). In contrast, the word *engagement* implies relationship. *Audience engagement* is about deepening relationships with current stakeholders and expanding reach over time. It may result in new modes or venues of presentation and means of explaining the arts to the public. *Community engagement* strategies are designed to create and maintain relationships with individuals and communities (many of whom may not be currently affiliated with the organization). Community engagement is dependent on the establishment of trusting, mutually beneficial relationships over time; the arts and the community become equal partners. The desired end results are deepened relationships; expanded reach for the arts organization; and healthier, more vibrant communities. Robert Lynch, president and CEO of Americans for the Arts, says, "The challenge is not whether to build communities or audiences but how to build communities and audiences together."[3]

How do arts organizations work to better serve their communities? Rick Lester, the late founder of TRG Arts, said,

> Improved audience engagement encompasses every decision an organization makes—especially those influencing what we put on stage. It takes bold leadership from the top, not a well-meaning marketing manager filled with energy and optimism—and often no political clout—to change the course of an institution. It seems to me that if organizations want a different culture (in the broadest sense of the word) to notice, embrace, and become engaged with your work—the faces on your stage must reflect the faces you hope to attract into the venue. The faces in your offices and boardroom must also reflect this community. Your theater, gallery, or concert hall must become a center for gathering and activity—for all of your community.[4]

Arts organizations must become active, involved members of their communities. They should have representatives on the local chamber of commerce participate in the Fourth of July parade and perform for patients in local hospitals. When a community event is being staged, such as the grand opening of a new store, or when a young public official is speaking before a crowd, performing artists and presenters can lend their expertise by helping to stage events and by coaching public speaking. In return, the artists and managers will benefit from gaining awareness, familiarity, and gratitude; which will translate into new audiences, new funders, and new sources of gifts-in-kind. Such exposure will weave the arts into the very fabric of the community. Such activities will create new opportunities for showing people how art is integral to our everyday lives, our experiences, and our feelings, which in turn will make the artistic experience compelling to a larger number of people.

Says consulting group Morris Hargreaves McIntyre,

> For several years now, arts organizations have been exhorted to face up to the realities of a rapidly changing environment, with changing public expectations and changing public behavior; and to embrace new ways of engaging with their existing and potential audiences that are both interactive and personalized. If we want to move with the times and be relevant to the new generation of cultural consumers we must embrace personalization and we must invite a greater level of interactivity.

Morris Hargreaves McIntyre defines interactivity as:

> a two-way dialogue with customers and stakeholders that result in deeper engagement, the ultimate manifestation of which is co-creation. Interactivity is the consumer's response to the organization's artistic vision. And personalization is equipping the customer with the ability to tailor and personalize the service experience to their needs. Both are possible routes to a bigger goal: deeper engagement.[5]

Most marketers think that the best way to hold on to customers and build their involvement is by interacting as much as possible with them. But not every patron will want to have a relationship with an arts organization, or such an involved one. Many patrons are happy to attend an organization's performances once or twice a year, when the mood and programming strike them. According to Karen Freeman, Patrick Spenner, and Anna Bird, only 23 percent of consumers in their study of more than seven thousand consumers say they have a relationship with a brand, so the marketer must learn to identify the people who actually want to engage and market differently to them. Also, contrary to common belief, only 13 percent of respondents to their study cited frequent interactions as a reason for having a relationship. Of the consumers who say they have a relationship with an organization, 64 percent cited *shared values* as the reason. Instead of relentlessly demanding more consumer attention, arts managers should treat the attention they win as precious. Engagement is about *quality* of relationships, not *quantity* of interactions.[6]

To build engagement, *relevance* and *value* are key, overriding concepts for marketers to focus on with all their plans, strategies, and tactics. There are three main components to the concept of value: the three Cs. Arts managers need to *comprehend* what value people are seeking, *create* value to meet those needs, and then *communicate* it effectively.[7] They must understand what people value and create value that breaks down barriers and builds enjoyment. This value must be communicated effectively; otherwise there is no value.

If organizations can communicate value more effectively, they can increase perceived value.

The English National Opera (ENO) has launched an initiative to attract younger audiences. "Come in shorts, armor, jeans, pumps, anything!" said artistic director John Berry. For this scheme, entitled "Undress for the Opera," one hundred £25 tickets were made available for performances of four operas in the coming season: *Don Giovanni*, *La Traviata*, the world premiere of the 3-D film opera *Sunken Garden*, and Philip Glass's *The Perfect American*. The tickets were for the best seats, and included a preperformance introduction to the evening's opera, a downloadable synopsis, and the opportunity to meet the cast and creative teams after the performance. Chief executive Loretta Tomasi also promised club-style bars, themed cocktails, and a relaxed atmosphere. "Lots of people are put off by the way opera is presented—they think it is too stuffy, too posh, too expensive. We want to change that perception," said Berry. Thirty percent of ENO's current audience is under 44; Berry's aim is to increase that figure to at least 40 percent. The special tickets will be on sale one month before each performance.[8]

Reflecting its town, which is both artsy and outdoorsy, for the 2013–2014 season, the Boulder Philharmonic's theme is "Nature & Music: The Spirit of Boulder." The goal of the season is to connect a wider segment of the community to orchestral music inspired by nature through a series of unique partnerships, including the City of Boulder Open Space & Mountain Park; the Geological Society of America (based in Boulder); the University of Colorado; local filmmakers, composers, and dance ensembles; and others. Says marketing consultant Holly Hickman,

> This Nature & Music initiative is part of the orchestra's mission and vision, which is focused on developing community-centric programming that reflects the unique qualities of the Boulder area. Subscriptions are up 70 percent since 2008; the number of single ticket buyers increased 10 percent in the 2012–2013 season; and the organization enjoyed three consecutive years of increasing donations (2010–2012). The Boulder Philharmonic is programming specifically to appeal to its community's interests, and the community is responding to the organization's increased relevance.[9]

Marketing in a Dynamic Environment

Marketing is central to an organization's functioning and well-being. Says Philip Kotler,

> Marketing is about the marketing of meaning embedded in the organization's mission, vision, and values. By defining marketing in this way, the state of marketing is elevated into being a major player in designing the organization's strategic future. Marketing should no longer be considered as only selling and using tools to generate demand. Marketing should now be considered as the major hope of a company to restore consumer trust.[10]

Every organization needs objectives and a clear and compelling strategy. Managers must ask themselves: Do we have a good plan? A feasible plan? Are the tactics aligned with our strategy? Did we spend enough money? Too much money? Do we have good controls in place?

SUSTAINABILITY

As the performing arts industry is straining under significant deficits, managers, consultants, researchers, and analysts have devoted much effort to determining the causes of the decline, hoping to identify solutions. Some experts predict a systems collapse, especially in the orchestra world. Extrapolating from the Yale research cited earlier in this book, we can conclude that if an orchestra performs in the top 10 percent of all orchestral fund-raising departments nationwide, it will only close about 45 percent of the growing cost gap each year moving forward.

As a result, there has been much talk about *sustainability*. But what is it that arts organizations are trying to sustain?

According to Alan Brown, Joseph Kluger, et al. of WolfBrown, sustainability requires a balancing act with three interdependent but sometimes competing priorities: *community relevance*, *artistic vibrancy*, and *capitalization*. Together, these elements give organizations the ability to excel in a permanent state of flux, uncertainty, and creative tension.

Community relevance, says WolfBrown, should be the first and foremost element of sustainability. Achieving relevance in the eyes of the community means that it is seen as a community asset rather than an isolated, self-interested nonprofit. It means creating valued partnerships, not just outreach programs and reduced-price tickets.

Artistic vibrancy is the fuel of sustainability. Artistic vibrancy goes far beyond artistic excellence to include a willingness to experiment with programming through an inclusive and consultative program planning process and the smart use of technologies for engaging audiences and communities in the artistic work. New doorways should be opened before old ones are shut.

Capitalization serves to support the other two elements. There are three purposes for capital: liquidity (enough cash to meet operating needs), adaptability (funds that offer flexibility in adjusting to changing circumstances), and durability (funds to address the range of needs in future years). Good fiscal policy and capitalization are tools of sustainability; financial distress is a symptom, not a cause of the problems arts groups face. Organizations must take care not to lose their sense of vitality and purpose in trying to achieve financial sustainability, or durability will become an empty goal.[11]

Healthy arts organizations require constant regeneration and renewal, which may be in conflict with institutional designs on permanence. Cultural leaders, therefore, have the responsibility of consolidating, reimagining, or phasing out programs that do not have sufficient public value. In that permanence, stability, and durability may be counterproductive ideals in today's rapidly changing society, sustainability may take on the new meaning of a policy of creative invention that is perpetually responsive to the communities in which the organization resides. Says conductor Harvey Felder, "Our challenge is the changing society in which we live. We can fight for the expectations and norms of the past, and in doing so continue to exclude people with different experiential needs. Or we can see the new and evolving expectations as ours."[12]

Some of these broad issues are outside the marketer's realm, but what is within the marketer's purview is a primary factor for sustaining the arts organization: the relationship with the audience. This is an enduring aspect of the organization that managers can influence through the ways they communicate with and treat their patrons at every touch point. Remember that, as Chad Bauman says,

> We need new audiences more than they need us.... We have to make our organizations inviting, accessible and fun. And understand that providing a fun experience doesn't equate to sacrificing artistic credibility. We don't have to sacrifice the core of who we are to attract new audiences.... New audiences need to be cultivated carefully. Create a path for them. Give them an easy entry point. Provide an amazing experience. Steward them so they return soon after their first experience. Build their confidence with multiple experiences, and then provide an opportunity to sample something a little more challenging. Introduce them to new experiences. At some point, if you don't provide them with a challenge, they will grow bored. We are responsible for cultivating our audiences' artistic growth. If we lack audiences for classical, challenging, or new work, perhaps it is because we try to short circuit the system, and ask that new audiences sample what they would at first perceive as vegetables before getting to the dessert.[13]

COLLABORATION

The marketers' tasks are becoming broader as our society evolves and changes. They must work closely with others within and outside of the organization, in addition to meeting or exceeding ticket sales goals.

Within the organization, it is crucial to integrate all functions so that the "whole customer" is served. For the customer, arts attendance, contributions, educational opportunities, and social events are all part of one total experience with the organization, and are not siloed in patrons' minds and hearts as they are siloed by organizational function. The person who makes a donation is typically the person who is deeply moved by the performances and is committed to help the organization thrive. Wouldn't it make sense if marketing and development would work together to realize the same goals, rather than having completely separate budgets and goals? In the climate of separation between departments, huge opportunities are potentially lost. The organization should create a collaborative environment that rewards employees for interdepartmental initiatives.

Educating its patrons is a crucial function of the organization. Central to a love and appreciation of the arts is a deep understanding of the art forms themselves. A study of the effects of arts education on participation in the arts clearly indicated that the richer one's arts education, the greater one's participation in the arts. Arts education was found to be the strongest predictor of arts creation and consumption, stronger even than socioeconomic status, race, ethnicity, and gender.[14] Education is crucial for making art meaningful, important, and necessary.

Because the live performance experience is integral to the educational process, it may well be that arts organizations are the best-suited educational institutions for creating an appreciation and love for the arts and for building future audiences. For their current audiences, arts organizations can make their art more meaningful by providing rich information, music, photos, and video on their websites, and by provoking thought and discussion in their program notes, in special articles in their programs, and at pre- and postconcert discussions (and isn't "discussions" a better word than "lectures"—isn't anything a better word than "lectures"?). They can also offer a variety of enrichment programs in neighborhood venues.

By bringing their art to the people, arts organizations are not only enriching them and exposing them to what they offer, but also possibly encouraging people to come to their venues. For the next generations of arts attenders, more and more arts organizations have taken up the mantle of providing child-centered arts programming to expose, educate, and entertain kids, who hopefully will become appreciative and avid arts attenders later in life. Some organizations also offer young people their own opportunities to act, to sing, to dance, or to play music, under the leadership of seasoned professionals who inspire the children and give them new ways of thinking about the world.

Some Final Insights

Throughout this book I have presented some of the best practices currently in use. But what works well for one organization may not suit another organization's situation. And because environments change and people's needs, interests, and desires change over time, the best practices of today may be obsolete tomorrow. Therefore, it is crucial that arts managers and marketers fully understand and internalize the following universal and enduring principles that underlie each and every successful strategy.

Manage for the Mission

Managing for the mission means that an organization takes no decision or action that is not informed and inspired by its mission statement. The mission is the organization's purpose or reason for being. Every mission statement reflects two things: the competencies and attainment levels sought by the organization, and its purposeful commitments to various stakeholders. Each member of the organization must understand, support, and be able to articulate that mission.[15] If an organization devotes a disproportionate amount of its resources to activities that do not support the mission, then the organization is supporting only the institution and not its reason for being.

Second only to its mission, an arts organization should focus on its core competencies. It is important to concentrate on only a few things, to do them well, and to communicate effectively what those strengths are.

Focus on Quality

The most important factor responsible for the success or failure of an artistic product is its quality. The product that has a quality advantage is destined to capture the minds and hearts of the marketplace. Cutting corners in a way that adversely affects artistic quality is a sure road to failure. Marketing strategies may attract customers, but if the experience itself is not highly satisfying, promotional techniques will not bring them back. Spending advertising and promotion dollars on behalf of an inferior product is counterproductive, if not futile.

Commit Resources to the Marketing Function

Marketing success requires a wholehearted commitment of funds, well-trained personnel, and trust in the marketing function itself. An arts organization cannot muster or maintain a strong audience base without full-fledged campaigns designed to attract, retain, engage, and educate patrons.

Says Michael Kaiser, CEO of the Kennedy Center in Washington, DC,

> Troubled arts organizations invariably get themselves into a vicious spiral. They reduce spending on art and marketing, lessening the interest of audiences and patrons, thereby reducing box office revenue and private contributions. In response, they cut back further, receive less, and so on and so on. Most problems facing arts organizations have to do with inadequate flow of income, not spendthrift artistic directors.

Kaiser firmly believes the solution to these problems is to search relentlessly for new sources of revenue, not to cut costs. This does not mean that reducing expenditures is not important; money must be saved when the organization faces financial shortfalls. But savings should be made on staff costs and administrative costs, *not* on artistic initiatives or marketing. The money Kaiser saves through cost-cutting efforts is devoted to creating vibrant artistic projects and to marketing these projects and the institution in exciting and contemporary ways. Concludes Kaiser, "Box office revenue and contributed funds result from good art, well marketed. Quite simply, effective marketing is an absolute prerequisite for consistent success in the arts."[16]

Know Your Customers

Customer insight—knowing as much as possible about the values, needs, and motivations of existing and potential customers—is the lifeblood of the organization.

Central to effective marketing is a commitment to marketing research. Market research is an investment the organization must make to develop, execute, and monitor intelligent marketing plans and thereby sustain and enlarge its audience. Sizeable arts organizations usually have the resources to employ sophisticated market research techniques to measure and describe their current and potential audiences. But even the smallest organizations with minimal resources can conduct in-house surveys and focus groups, obtain grants for further research, and study published reports on audience characteristics and preferences.

The most successful arts organizations are those whose products and services are directed at separate, distinct, and reachable segments of the population and whose offerings are positioned from the customer's viewpoint, not the organization's.

Present Art That Attracts a Growing Audience

Says Eugene Carr, organizations will thrive if they make art that attracts a growing audience. If managers find ways to continually fill their house, their economic engine will grow. They'll identify new donors, board members, and they'll relate much better to prospective funders. Nothing breeds success like success. This does not mean that organizations must ignore their artistic mission; it means that they must be relevant to their communities.[17]

Attracting Art Lovers to the New York City Ballet

The New York City Ballet (NYCB) is offering a new series of performances designed to attract art lovers to the ballet. The NYCB Art Series commissions contemporary artists to create original works of art inspired by the ballet's unique energy, their spectacular dancers, and the repertory of dances.[18]

For the 2014 winter season, the ballet commissioned works from French street artist J R, who is internationally known for mounting large-scale public photography projects around the world. For this project, J R created two installations. The first one viewers see is outside the David H. Koch Theater at Lincoln Center where J R wheat-pasted enormous ballerinas' legs and toes shoes.

The second installation, which has created the most interest and excitement, is in the promenade of the Koch Theater, where J R covered the inlaid travertine marble floor with a 6,500-square-foot vinyl photograph of more than 80 City Ballet dancers, roughly life size, who are arranged on a sea of crumpled white paper.

From above, it becomes clear that the dancers form a gigantic eye, so many ticket holders in the orchestra section climbed to the Fourth Ring balcony during intermission to get a better view. Said one 24-year-old visitor, "It's like nothing I've ever seen. This is specifically incorporating ballet—the beauty, the power, the vulnerability. It's pure and beautiful." Some patrons got down on the floor and posed for pictures themselves. One man posed looking as if he were lifting a ballerina; some women posed lying on the hands of supine dancers appearing as if being lifted; one young girl mimicked the splits and poses of the City Ballet dancers; another leapt over the dancers' heads. Many such photos have been posted on Instagram.

In addition to commissioning the work, NYCB put on three special performances aimed at art fans new to ballet; every seat cost $29. Tickets were offered to the thousands of people who follow J R on social media, and ad campaigns were mounted at subway stations on the Lower East Side and in some Brooklyn neighborhoods. Standing room tickets were to be released on performance days as the seats were expected to sell out. The installation was also open to the public free of charge for several hours each day for several weeks.

At the 2013 installation, by the Brooklyn artist collective Faile, 70 percent of the people who went to the two special art-themed performances were new to City Ballet. About 7 percent of those first-timers came back to the ballet, which is an unusually high number. Said one patron, a 25-year-old film and theater producer, the Arts Series performances are "one of the hottest tickets in the city. Models, people from film, fashion, art—everyone was there."[19]

Create Accessibility

An important marketing task is creating access for the current or potential customer to the product. Remember that reality is what the consumer perceives it to be, that the solution is not in the product or in the marketer's own mind, but inside the prospect's mind. Success may well lie in the details: offering a performance at a convenient time, emphasizing casual attire, or surrounding the performance with familiar faces and comfortable settings. Modern technology allows organizations to bring alive the arts experience and to provide an abundance of information at little to no cost.

INTEGRATE THE ARTS INTO EVERYDAY LIVES

Many arts organizations have found effective ways to integrate their offerings into the lives of their publics. The New York Philharmonic invites families to integrate music into their weekend outings with its casual, inexpensive, one-hour, Saturday afternoon concerts. The American Symphony Orchestra ties in its musical themes with literature, politics, history, and other interests common to its audience members. The "home base" for Dancing in the Streets is wherever a target audience can be found.

In the fall of 2010, Heart of Los Angeles (HOLA) forged a partnership with the Los Angeles Philharmonic to bring a youth orchestra to the Rampart community. Youth Orchestra Los Angeles (YOLA) at HOLA (YOLA at HOLA) is inspired by El Sistema, the Venezuelan music education system that nurtured Gustavo Dudamel, the Los Angeles Philharmonic's music director. El Sistema uses music education to help young people from impoverished circumstances achieve their full potential and learn values that favor their growth. At its core, El Sistema is about togetherness, a place where children learn to listen to each other and to respect one another. The founder of El Sistema, Jose Antonio Abreu, states it simply: "Music has to be recognized as an . . . agent of social development in the highest sense, because it transmits the highest values—solidarity, harmony, mutual compassion. And it has the ability to unite an entire community and to express sublime feelings."

The impact of the program can be felt well beyond the orchestra room. In a community with limited resources, YOLA at HOLA has brought classical music into neighborhood centers and the homes of hundreds of families. When asked what they like most about the program, many of the students respond that they enjoy being a part of the orchestra and performing at such world-renowned venues as the Walt Disney Concert Hall and the Hollywood Bowl. Many dream of becoming professional musicians; one even has his sights set on becoming a conductor. But more often than not, students express their understanding that there is something much greater than just learning to play music. In the words of one student, "It's not so much about being a great musician, but being an inspiring musician so I can improve my community with my music."[20]

TAKE THE LONG-TERM VIEW

Successful marketing requires a long-term view. Arts organizations commonly recognize the need to attract young audiences. But educating children is costly, and targeting younger audiences is likely to imply smaller packages, lower ticket prices, and lower donation levels—in the short run. However, in the process of absorbing these costs, the organization is building a strong and broad foundation for its future—for a time when these people will have more leisure time and more discretionary income to pay higher ticket prices and make substantial contributions.

View Changes as Opportunities—Not Threats

Embrace change. When it comes to the future, resistance is ultimately futile. Pay attention to societal trends. Gordon Sullivan and Michael Harper, authors of *Hope Is Not a Method*, say that "doing the same thing you have always done—no matter how much you improve it—will get you only what you had before."[21]

Arts managers need strategic *agility*. They need to do long-range strategic planning but be attuned to opportunities and develop adaptive responses and strategies. Says Steve McMillan, former CEO of Sara Lee Corporation, "The perfect offering or cost remains perfect for a fleeting instant. The only advantage that remains sustainable in the long term is innovation."[22]

As customers' needs and preferences change, so must every organization's marketing strategy. Each current and potential arts patron—including single ticket buyers—should be respected, listened to, appreciated, and then, when possible, nurtured into higher levels of commitment. If arts organizations do not respect and meet the public's changing preferences and needs with respect to how the product is offered, there will be an ever-diminishing audience to share in the artistic experiment. So in the long run, such change can only serve to strengthen the organization, not to weaken it.

Revitalization of the Detroit Symphony Orchestra

The Detroit Symphony Orchestra (DSO) has come a long way since 2010, when it faced a deficit of $8.8 million and since its 6-month musician strike in 2011, which resulted in musicians' salaries being cut by 23 percent, as reported in chapter 1. After emerging from the strike, the orchestra faced an urgent need to replace lost support, repair interrupted concert-going habits, and send positive messages to a community benumbed by hardship appeals.

The orchestra developed a goal to become "the most accessible orchestra on the planet," according to Scott Harrison, executive director of board engagement and strategy and executive producer of digital media. To realize the digital component of this goal, the DSO offered free, live transmissions of most of its concert programs, delivered worldwide to computers, mobile phones, and other web-enabled devices. These programs were initiated with little in-house equipment or technical expertise, a staggering financial burden from previous years, and no precedent for determining risk to the subscriber base. Production costs were paid for by a John S. and James L. Knight Foundation grant and support from the Ford Motor Company Fund; Detroit Public Televison partnered with the DSO for cameras, production crew, and broadcast expertise; and the post-strike contract provided for streaming video without additional upfront musicians' fees.[23]

As of June 2012, the first full season of "Live From Orchestra Hall," which included 22 live classical web broadcasts, garnered 120,000 views in more than 75 countries. The DSO's YouTube channel, which offers clips and some complete works, garnered 68,000 views by March of that year. A year later, after two full seasons of webcasting, total audience exceeded 300,000 viewers, the Facebook audience had grown to 25,000 (compared to 4,500 in 2010), the Twitter audience was approaching 15,000 (compared to less than 3,000 in 2010), and lifetime YouTube views increased six-fold to more than 500,000. Importantly, classical ticket sales were up 22 percent compared with the last full season before the strike, showing that the streaming performances did not cannibalize attendance at the concerts. Anecdotally, many subscribers, 74 percent of whom are aware of "Live from Orchestra Hall" webcasts and 67 percent of whom have watched a webcast, report that they tune into webcasts to enjoy a second performance of a favorite program that they attended at Orchestra Hall.[24]

Furthermore, although the webcasts are free, the DSO invites viewers to make a donation when tuning in. During the first full season, "Live from Orchestra Hall" viewers who made donations gave an average of $78—equivalent to three main floor "B" seats in Orchestra Hall. Gifts came from across the United States and from enthusiasts overseas. Partnerships with international distribution partners like Medici.tv and Paraclassics.com extended the DSO's reach to new audiences that would otherwise have cost hundreds of thousands of dollars to reach through marketing efforts. Once the regular season is over, the summer "Encore" series begins, making the concerts available again at dso.org/live.[25] As of January 2014, plans were underway to launch an archive player to allow patrons and supporters 24/7 access to a library of past content.

In fiscal year 2013, the DSO achieved its first balanced budget since 2007. This success was bolstered by a 43 percent increase in fundraising (totaling $18.9 million), a 100 percent increase in the donor base over 2011, and the second consecutive year of subscription sales growth.[26] Also in fiscal year 2013, the Knight Foundation awarded the DSO $2.25 million to start an endownment to support digital operating costs into perpetuity. In addition to direct support for webcast efforts from institutional and individual sources, Harrison reports that a number of major donors list the webcasts and the DSO's commitment to digital accessibility as one of the top reasons they have reaffirmed and increased their support of the orchestra in recent years. The webcasts have become central to the organization's identity and are now viewed as a core activity driving artistic excellence by funders, board members, volunteers, musicians, managers, the community, and local and global audiences.

Focus on the Art and the Audience

The focus of this book has been on the arts organization, for the organization brings the art to the public and the public to the art. Yet the organization is a closed system. It is controlled, systematized, and resistant to change. That resistance must be broken down. Arts organizations must continually change to retain their effectiveness. They must change their internal structures, their ways of doing business, sometimes even their missions. Above all, they must listen to their constituents.

Conductor and composer Pierre Boulez says, "It is a sign of weakness in a civilization that it cannot destroy things." Similarly, conductor and pianist Daniel Barenboim claims, "This mania for keeping everything shows a lack of courage, courage which is needed in order to use the experiences of the past as stepping stones to the vision of the future."[27]

However, it must be remembered that it is the art itself that has the power—the power to move the soul, lift the spirit, expand the mind. Art is an open system; it constantly creates, cajoles, undermines, confronts, challenges. And great art endures, transcending time and space.

If the essence of art is the relationship between the artist and the audience, the arts organization must be vigilant in pursuing both the artists' and the audiences' best interests. Artistic programs should be developed not with the purpose of keeping the institution alive, but of making it viable. In the end, as managers, marketers, board members, and others who work to sustain and develop our arts organizations, our accountability is to the artists and their publics—not to the organization. By attempting to alter public perceptions of our institutions, we only scratch the surface of the challenges we face and, at best, create short-term solutions. By educating our publics about the art we treasure, by being relevant to our communities, and by being sensitive to the continually changing behavior, interests, and needs of various audience segments, we can build enthusiastic and loyal audiences for the future. And by doing that, we can guarantee that art will thrive and prosper.

So many possibilities.

Stephen Sondheim, *Sunday in the Park with George*

Notes

Introduction

1. Margaret Mehl, Blog, "Music after the Tōhoku Disaster, Sendai Philharmonic Orchestra," March 5, 2012, http://www.violinist.com/blog/Ku92me/20123/ (accessed February 11, 2013).
2. Allan Kozinn, "James DePreist, Pioneering Conductor Whose Legs Were Paralyzed, Dies at 76," *New York Times*, February 10, 2013, 25.
3. Alan Brown and Rebecca Ratzkin, *New Beans: Intrinsic Impact and the Value of Aart*, ed. Clayton Lord (San Francisco: Theatre Bay Area, 2012), 71–72.

1 The Performing Arts: History and Issues—An Ongoing Crisis? A Growing Crisis?

1. Patricia Cohen, "A New Survey Finds a Drop in Arts Attendance," *New York Times*, September 26, 2013, http://www.nytimes.com/2013/09/26/arts/a-new-survey-finds-a-drop-in-arts-attendance.html?_r=0 (accessed September 26,2013) .
2. Heloisa Fischer, personal email to the author, March 17, 2005.
3. Clayton Lord and Christopher Shuff, "Taking Your Fiscal Pulse—Fall 2011: A Report on the Fiscal Health of the National Not-for-Profit Theatre," Theatre Communications Group, 10, http://www.tcg.org/pdfs/tools/fiscal/TakingYourFiscalPulse_Fall2011.pdf.
4. Larry Rohter, "In Europe, Where Art Is Life, Ax Falls on Public Financing," *New York Times*, March 24, 2012.
5. Nick Clark, "British Orchestras are in Danger of Losing Top Billing Despite Rising Ticket Sales," *The Independent*, January 26, 2014, http://www.independent.co.uk/arts-entertainment/classical/news/british-orchestras-are-in-danger-of-losing-top-billing-despite-rising-ticket-sales-9085502.html?printService=print.
6. Milton Cummings Jr., "Government and the Arts: An Overview," in *Public Money and the Muse: Essays on Government Funding for the Arts*, ed. Stephen Benedict (New York: Norton, 1991), 31–32.
7. Paul DiMaggio, "Nonprofit Organizations in the Production and Distribution of Culture," in *The Nonprofit Sector: A Research Handbook*, ed. Walter W. Powell (New Haven, CT: Yale University Press, 1987), 204–205.

8. Ibid.,205.
9. Lawrence W. Levine, *Highbrow/Lowbrow: The Emergence of Cultural Hierarchy in America* (Cambridge, MA: Harvard University Press, 1988), 127.
10. Robert Flanagan, *The Perilous Life of Symphony Orchestras: Artistic Triumphs and Economic Challenges* (New Haven & London: Yale University Press, 2012),146–147.
11. Ibid., 2.
12. Kevin Berger, "Classical Music Waltzes with Digital Media," *Los Angeles Times*, August 7, 2011, http://articles.latimes.com/2011/aug/07/entertainment/la-ca-classical-technology-20110807.
13. Zannie Giraud Voss and Glenn B. Voss, "Theatre Facts 2012," Theatre Communications Group, sent to the author September 9, 2013.
14. Miriam Kreinin Souccar, "New York Musicians' Woes Reach New Lows as Work Disappears," *Crain's New York Business*, April 1, 2012.
15. Daniel Wakin, "New York City Opera Announces Deal with Unions," *New York Times*, January 18, 2012, http://artsbeat.blogs.nytimes.com/2012/01/18/new-york-city-opera-announces-deal-with-unions/.
16. Sam Roberts, "Foundation Files Reveal Insights on Culture, *New York Times*, April 9, 2012, The Arts, 1.
17. "The Arts in America," report prepared by the National Endowment for the Arts,1988.
18. Craig Lambert, "The Future of Theater: In a Digital Era, Is the Play Still the Thing?" *Harvard Magazine*, January–February 2012, 35.
19. Flanagan, *The Perilous Life of Symphony Orchestras*,60.
20. NEAst udy ona rtsp articipation,i ni bid.,4 1.
21. League of American Orchestras, *Audience Demographic Research Review*, December 10, 2009, http://www.americanorchestras.org/images/stories/knowledge_pdf/Audience_Demographic_Review.pdf.
22. Flanagan, *The Perilous Life of Symphony Orchestras*,42.
23. Berger," ClassicalM usic Waltzesw ithD igitalM edia."
24. Sudeep Reddy, "Latinos Fuel Growth in Decade," *Wall Street Journal*, March 25, 2011, http://online.wsj.com/article/SB10001424052748704604704576220603247344790.html.
25. Lester M. Salamon, "Arts, Culture and Recreation," in *America's Nonprofit Sector: A Primer* (New York: Foundation Center, 1992), 94.
26. William J. Baumol and William G. Bowen, "Anatomy of the Income Gap," in *Performing Arts: The Economic Dilemma* (New York: Twentieth Century Fund,196 6).
27. Flanagan, *The Perilous Life of Symphony Orchestras*,90.
28. Baumola ndB owen, *Performing Arts*.
29. Craig Lambert, "The Future of Theater," *Harvard Magazine*, January–February 2012. Harvardmagazine.com/2012/01/the-future-of-theater.
30. RobertB rustein, *Dumbocracy in America* (Chicago: Ivan R. Dee, 1994), 26.
31. Ibid.,52–53.
32. Kathleen Watt, "Charles Wuorinen: New Wine, Old Skins," *USOPERAWEB*, http://www.usoperaw eb.com/2004/autumn/wuorinen.htm/.

33. Herbert Gans, *Popular Culture and High Culture* (New York: Basic Books, 1974),75–94.
34. Robert F. Kelly, “Elitism in the Arts,” presented at the Second International Conference on Arts Management, Jouy-en-Josas, France, June 1993.
35. Peter Goodman, *An American Salute* (NJ: Hal Leonard Performing Arts Publishing Group, 2000), 285.
36. Elizabeth Zimmer, “The World Is at Your Door,” *Inside Arts* (Spring 1993): 18.
37. Leopold Segedin, “Interdisciplinary Studies: Evolution or Revolution?” paper prepared for presentation at the conference of the Association of Integrative Studies, Washington, DC, 1980.
38. TheodoreL evitt, *The Marketing Imagination* (New York: Free Press, 1986).
38. Rockefeller Panel Report, “The Performing Arts: Problems and Prospects” (New York: McGraw-Hill, 1965), 4–5.
40. Bradley Morison and Julie Gordon Dalgleish, *Waiting in the Wings* (New York: American Council for the Arts, 1987), 66.
41. Arian Campo-Flores, “Remixing Classical-Music Concerts for the iPod Generation,” *The Wall Street Journal,* May 4, 2012, http://online.wsj.com/news/articles/SB10001424052702304050304577378253248886874
42. From a talk given on November 19, 2005, at Next Theater, Evanston, IL. Used with permission.
43. Ronald Berman, “Art vs. the Arts,” *Commentary* 68, no. 5 (November 1979), 47.
44. Morisona ndD algleish, *Waiting in the Wings*,139.
45. Robert Brustein, “Culture by Coercion,” *New York Times*, November 29, 1994, Op-Ed section.
46. Ibid.
47. Richard Dyer, “What Role Should a Symphony Play?” *Boston Sunday Globe*, July 11, 1993, sec. B, 28.
48. Richard Christiansen, “If Art Is Not Enough, What Is It Worth?” *Chicago Tribune*, December 11, 1994, sec. 13, 2.
49. Fran Liebowitz, “Liebowitz On Wit, Humor, and ‘Public Speaking,’”*NPR: Talk of the Nation,* November 18, 2010, http://www.npr.org/2010/11/18/131420458/lebowitz-on-wit-humor-and-public-speaking
50. SamuelL ipman,“ Designsf orA rtsi nE ducation,”M arch/April1990.
51. *Three NEA Monographs on Arts Participation: A Research Digest*, NEA Research Report #52, *Arts Education in America: What the Declines Mean for Arts Participation*, by Nick Rabkin and E. C. Hedberg, NORC at the University ofC hicago, http://www.arts.gov/research/2008-S PPA-ArtsLearning.pdf.
52. Jeffrey L. Brudney, “Art, Evolution, and Arts Education,” *Society*, September/October 1990, 17–19.
53. “New Harris Poll Reveals That 93 Percent of Americans Believe That the Arts Are Vital to Providing a Well-Rounded Education,” Americans for the Arts Press Release, June 13, 2005.
54. Ichak Adizes, “The Cost of Being an Artist,” *California Management Review* (Summer 1975), 80–84.

55. Jolie Jensen, *Is Art Good for Us?* (Lanham, MD: Rowman & Littlefield, 2002),2–4.
56. Alan Brown, "The Shifting Sands of Demand: Trends in Arts Participation," *Arts Reach* 13, no. 1 (2004), 10; excerpted from *Magic of Music*, Issues Brief No. 5, *Smart Concerts: Orchestras in the Age of Edutainment* (Miami, FL: John S. and James L. Knight Foundation), December 2004, http://www .knightfdn.org.
57. Nello McDaniel and George Thorn, "The Workpapers: *A Special Report, The Quiet Crisis in the Arts,*" FEDAPTS's 1989/1990 Annual Report (New York, 1991).
58. *Back Stage*, 31 (June 8, 1990), 1A.
59. McDaniela nd Thorn," The Workpapers,"1 8–19.
60. Richard A. Peterson, *From Impresario to Arts Administrator: Formal Accountability in Nonprofit Cultural Organizations* (New York: Oxford University Press,1987),169.
61. Winston Churchill, BrainyQuote.com, Xplore Inc, 2013, http:// www.brainyquote.com/quotes/quotes/w/winstonchu131188.html (accessed December 4, 2013).
62. Robert F. Kelly, "The Enemy Within . . . Marketing in the Arts," paper presented at the First International Conference on Arts Management, Montreal, Candada, August 1991.
63. Robert F. Kelly, "Elitism in the Arts," presented at the Second International Conference on Arts Management, Jouy-en-Josas, France, June 1993.
64. *Sunday in the Park with George*, music and lyrics by Stephen Sondheim, first performed at Playwrights Horizons, New York, 1983.

2 The Evolution and Principles of Marketing

1. Philip Kotler, *Marketing 3.0: From Products to Customers to the Human Spirit* (Hoboken, NJ: John Wiley & Sons, 2010), 45.
2. "The American Marketing Association Releases New Definition for Marketing," January 14, 2008, http://www.marketingpower.com/AboutAMA /Documents/American%20Marketing%20Association%20Releases%20 New%20Definition%20for%20Marketing.pdf.
3. Kotler, *Marketing 3.0*,4.
4. Ibid.,x i–xii.
5. Philip Kotler, Dipak C. Jain, and Suvit Maesincee, *Marketing Moves* (Boston: Harvard Business School Press, 2002), x, 21.
6. StephenS ondheim, *You Gotta Get a Gimmick,*" Gypsy,"1962.
7. Arthur Miller, *Death of a Salesman* (New York: Dramatists Play Service, 1998; first published in Great Britain: Crescent Press, 1949), 36.
8. Peter F. Drucker, *Management: Tasks, Responsibilities, Practices* (New York: Harper & Row, 1973), 64–65. Emphasis mine.
9. Miller, *Death of a Salesman*,60.
10. Jim McCarthy, "Bourne with a Natural Talent for Being Audience-Oriented," *Selling Out* (blog), November 26, 2013, http://sellingout.com

/bourne-shows-the-difference-between-audience-oriented-and-pandering/ (accessed December 4, 2013).

11. Alan R. Andreasen and Russell W. Belk, "Predictors of Attendance at the Performing Arts," *Journal of Consumer Research* (September 1980), 112–120.
12. E. Jerome McCarthy, *Basic Marketing: A Managerial Approach* (Homewood, IL: Richard D. Irwin, 1981).
13. Kotleret a l., *Marketing Moves*,x ,21.
14. Steve McMillan, presentation at the Kellogg School of Management Marketing Conference, Evanston, IL, January 29, 2003.
15. Marya Mannes, "Girl Gadfly: A Critic Fights the Fat Life," *Life*, June 12, 1964,64.
16. B. Joseph Pine II and James H. Gilmore, *The Experience Economy: Work Is Theatre & Every Business a Stage* (Boston: Harvard Business School Press, 1999), front matter.

3 UNDERSTANDING THE PERFORMING ARTS MARKET: HOW CONSUMERS THINK

1. Daniel Kahneman, *Ted Talks*, March 2010, http://www.ted.com/talks/daniel_kahneman_the_riddle_of_experience_vs_memory.html.
2. Diane Paulus, *Harvard Magazine* (January–February 2012), 38. Emphases in the original.
3. Ibid.E mphasesi nt heo riginal.
4. Jeremy Eccles, "Focusing on the Audience," *International Arts Manager* (January 1993), 21–23.
5. Philadelphia Arts Market Study, commissioned by the Pew Charitable Trusts, prepared by Ziff Marketing, 1989.
6. Felicia R. Lee, "Breaking the Motherhood Drama with a Drama on Motherhood," *New York Times*, November 24, 2004, http://www.nytimes.com.
7. Heidi Waleson, "Marketing: What Can We Do for You?" *International Arts Manager*, January 1993, 17–18.
8. Alex Ross, "Listen to This," *New Yorker*, February 16 and 23, 2004, http://www.therestisnoise.com/2004/05/more_to_come_6.html.
9. Sheena Iyengar, *The Art of Choosing* (New York: Twelve, Hachette Book Group, 2010), 7.
10. Barry Schwartz, *The Paradox of Choice: Why More Is Less* (New York: HarperCollins Publishers Inc., 2004), 5–6.
11. Ibid; Barry Schwartz, *Ted Talks*, July 29, 2012, http://www.ted.com/talks/barry_schwartz_on_the_paradox_of_choice.html.
12. Kent Greenfield, *The Myth of Choice* (New Haven: Yale University Press, 2011),63.
13. Howard Sherman, "The Broadway Scorecard: Two Decades of Drama," August 21, 2012, http://www.hesherman.com/2012/08/21/the-broadway-scorecard-two-decades-of-drama/.

14. Daniel Kahneman, presentation of Nancy Hanks Memorial Lecture, Kellogg Graduate School of Management, May 19, 2004.
15. Daniel Kahneman, prize lecture, "Maps of Bounded Rationality," December 8, 2002, Stockholm University, http://nobelprize.org/economics/laureates/2002/kahneman-lecture.html.
16. Anne Midgette, "Decline in Listeners Worries Orchestras," *New York Times*, June 25, 2005, A15.
17. Daniel Kahneman, *Thinking Fast and Slow* (New York: Farrar, Straus and Giroux, 2011), 348–349.
18. Janelle Gelfand, "Orchestra League Leader Takes Industry to Task," *Enquirer*, Cincinnati.com, June 6, 2005.
19. Tim Baker and Heather Maitland, *Profile of Dance Attenders in Scotland, Section 3: Qualitative Research Report*, Scottish Arts Council, October 9, 2002,6.
20. Gordon E. Butte, Decision Partners, "Increasing Opera Attendance: The 2002 American Express Audience Research Project," *Opera America Newsline*, June2002.
21. Chris Jones, "Fall Theater Guide 2012: Take a Risk This Fall," *Chicago Tribune*, August 31, 2012, On the Town, 1.
22. Daniel Ariely, *Predictably Irrational* (New York: HarperCollins Publishers, 2010),x x.
23. Gerald Zaltman, *How Customers Think: Essential Insights into the Mind of the Market* (Boston: Harvard Business School Press, 2003), 8–9.
24. Gary Klein, *Sources of Power: How People Make Decisions* (Cambridge: MIT Press,1999) .
25. Chris Blamires, "What Price Entertainment?" *Journal of the Market Research Society* 34, no. 4 (October 1992), 377.
26. Artlink, TheatreR oyalS tratfordE ast Audienceq ualitativer esearch,1987.
27. MOMIX, http://www.mosespendleton.com/.
28. Marilyn Jackson, "Review: MOMIX Soars and Enchants Again," *Philadelphia Inquirer*, May 20, 2012, http://articles.philly.com/2012–05–20/news/31778738_1_momix-dancer-moses-pendleton.
29. John E. Swan and Linda Jones Combs, "Product Performance and Consumer Satisfaction: A New Concept," *Journal of Marketing Research* (April 1976), 25–33.
30. Leon Festinger, *A Theory of Cognitive Dissonance* (Stanford, CA: Stanford University Press, 1957), 260.
31. J. Walker Smith, Ann Clurman, and Craig Wood, *Coming to Concurrence: Addressable Attitudes and the New Model for Marketing Productivity* (Evanston: Racom Communications, 2005), 95.
32. Alan Brown, "Classical Music Consumer Segmentation Study: How Americans Relate to Classical Music and Their Local Orchestras," *Arts Reach* 11, no. 3 (January 2003), 18.
33. Decision Partners, *Increasing Opera Attendance: The 2002 American Express National Audience Research Project* (New York: Opera America, 2003).
34. Smithet a l., *Coming to Concurrence*,211.

35. Alan Brown, "Initiators and Responders: A New Way to View Orchestra Audiences," John S. and James L. Knight Foundation, Issues Brief 4, July 2004,7, www.knightfdn.org.
36. Ibid.,4.
37. "Marketing the Arts in Cleveland: An In-Depth Survey," conducted by Ziff Marketing, Inc., and Clark, Martire & Bartolomeo, Inc., commissioned by the Cleveland Foundation, 1985.

4 Exploring Characteristics of Current and Potential Performing Arts Attenders

1. Jeffrey Passel and D'Vera Cohn, "U.S. Population Projections: 2005–2050," Pew Research Center, February 11, 2008.
2. Ken Dychtwald, "Inspired, Not Retired," *New York Times Magazine*, May 22,2005.
3. David Wolfe, *Marketing to Boomers and Beyond* (New York: McGraw-Hill, 1993),117.
4. Ibid.,164–165.
5. Sally Kane, "Generation Y," about.com, http://legalcareers.about.com/od/practicetips/a/GenerationY.htm.
6. This information was provided to the author in personal emails by Claire Coveney, communications manager, ArtsConnection/High 5 Tickets to the Arts, October 4, 2012.
7. Chris Jones, "Child's Play," *Chicago Tribune*, May 18, 2003, Arts and Entertainment,1.
8. MarkS wed," Who'sK eepingS core?," *Los Angeles Times*, April 7, 2007.
9. Jack McAuliffe, "Voyage of Discovery," *Symphony*, November/December 2004,18–1 9.
10. Rachel L. Swarns, "More Americans Rejecting Marriage in 50s and Beyond," *New York Times*, March 2, 2012, A12.
11. Faith Popcorn's Brain Reserve, July 29, 2012, http://faithpopcorn.com/ContentFiles/PDF/FPBR%20Predictions%202012.pdf.
12. Stephanie Holland, "Marketing to Women Quick Facts," She-Conomy blog, http://www.she-conomy.com/report/marketing-to-women-quick-facts.
13. Brian Sternthal and Alice Tybout, "Segmentation and Targeting," in *Kellogg on Marketing*, ed. Dawn Iacobucci (Hoboken, NJ: Wiley, 2001), 7–18.
14. "Facts on Women," August 26, 2012, http://www.she-conomy.com/facts-on-women.
15. Martha Barletta, *Marketing to Women* (Chicago: Dearborn Trade, 2002), 175.
16. Rachel Lamb, "The Difference in Marketing towards Males and Females via Email," *Luxury Daily*, October 9, 2012, http://www.luxurydaily.com/the-difference-in-marketing-towards-males-and-females-via-email/.
17. Joel Henning, "Joffrey Ballet: Back on Its Feet," *Arts Reach* 14, no. 3 (2006), 5. This article originally appeared in the *Wall Street Journal*, January 11, 2006.

18. Barletta, *Marketing to Women*,4–8.
19. "Marketing to Men," Branding Strategy Insider.com, April 19, 2010, http://www.brandingstrategyinsider.com/2010/04/marketing-to-men.html.
20. Forrester Research, *Gays Are the Technology Early Adopters You Want*, 2003, http://www.forrester.com/ER/Research /Brief/Excerpt/0,1317,17004,00.html.
21. "Marketing to Gay and Lesbian Consumers: Reaching a $600 Billion Market Requires Sensitivity and Support," Business.com, Inc., http://www.business.com/guides/marketing-to-gay-and-lesbian-consumers-1224/ (accessed August 26, 2012).
22. Wilde Marketing, *Pink Dollars*, 2002, http://www.wildemarketing.com/pink_dollars.html.
23. "Marketing to Reach the Gay and Lesbian Community," n.d., www.gaymarketexpress.com.
24. Passela ndC ohn," U.S.P opulationP rojections."
25. David Brooks, "Questions of Culture," *New York Times*, February 19, 2006, WE12.
26. *American Demographics*, April 2003, Packaged Facts, *Report on the U.S. Hispanic Market*, 4th ed., October 2003, http://www.marketresearch.com.
27. Ibid.
28. Securities Industry Association, "The African-American Market," n.d., http://www.sia.com/hrdiversity/html/african-american.html.
29. Ibid.
30. Packaged Facts, *The U.S. Asian-American Market,* March 2002, http://www.marketresearch.com.
31. Emanuel Rosen, *The Anatomy of Buzz: How to Create Word of Mouth Marketing* (New York: Currency, 2002), 62.
32. Jesse McKinley, "Lincoln Center Goes A-Courting," *New York Times*, July 12, 2005, B1.
33. David Meer, "Marketing Trends in the '90s for the Performing Arts," *Dance/USA Journal* (Summer 1991), 16–23.
34. FaithP opcorn, *The Popcorn Report* (New York: Harper Business, 1992), 22.
35. "Americans Back Arts, but Don't Feel Touched," *New York Times*, March 1, 1993, sec. B, 2. Article based on an opinion poll conducted in October 1992 by Research and Forecasts Inc. for the National Cultural Alliance.
36. Philip Kotler, *Marketing Management*, 7th ed. (Englewood Cliffs, NJ: Prentice Hall, 1991), 166.
37. EverettM .R ogers, *Diffusion of Innovations* (New York: Free Press, 1962).
38. Michael L. Ray, "Psychological Theories and Interpretations of Learning," in *Consumer Behavior: Theoretical Sources*, ed. S. Ward and T. S. Robertson (Englewood Cliffs, NJ: Prentice-Hall, 1973), 45–117.
39. Abraham Maslow, *Motivation and Personality* (New York: Harper & Row, 1954),80–106.
40. David B. Wolfe, *Marketing to Boomers and Beyond* (New York: McGraw-Hill, 1993), 135–140.

41. Danah Zohar and Ian Marshall, *Spiritual Capital: Wealth We Can Live By* (San Francisco: Berrett-Koehler Publishers, 2004).
42. Frederick Herzberg, *Work and the Nature of Man* (Cleveland: William Collins,1966) .
43. Harper W. Boyd Jr., and Sidney J. Levy, *Promotion: A Behavioral View* (Englewood Cliffs, NJ: Prentice-Hall, 1967), 38.
44. Wolfe, *Marketing to Boomers and Beyond*,117.

5 Planning Strategy and Applying the Strategic Marketing Process

1. Adrian J. Slywotzky, *Value Migration* (Boston: Harvard Business School Press, 1996), 39–40.
2. Jay Conrad Levinson, Guerilla Marketing, http://www.gmarketing.com/the-process-of-marketing (accessed January 10, 2014).
3. Peter F. Drucker, *The Five Most Important Questions You Will Ever Ask about Your Nonprofit Organization* (San Francisco: Jossey-Bass, 1993), 9.
4. C. K. Prahalad and Gary Hamel, "The Core Competence of the Corporation," *Harvard Business Review*, May/June 1990, 80–84.
5. Joel Brown, "Boston Theater, All in One Place," *Boston Globe*, September 6, 2013, http://www.bostonglobe.com/arts/theater-art/2013/09/06/boston-theater-all-one-place/W6euIBvrkflUardaFlLgjM/story.html.
6. E. B. Knauft, Renee A. Berger, and Sandra T. Gray, *Profiles of Excellence: Achieving Success in the Nonprofit Sector* (San Francisco: Jossey-Bass, 1991), 19.
7. Drucker, *The Five Most Important Questions*,51,53.
8. Guthrie Theater Announces 2012–2013 season, Guthrie Theater website, http://www.guthrietheater.org/guthrie_opportunities/media_room/press_releases/guthrie_theater_announces_20122013_season.
9. Peter F. Drucker, "What Business Can Learn from Nonprofits," *Harvard Business Review*, July/August 1989, 89.
10. A. Tversky and D. Kahneman, "Availability: A Heuristic for Judging Frequency and Probability," *Cognitive Psychology* 5 (1973), 207–232.
11. Max H. Bazerman, *Judgment in Managerial Decision Making* (New York: Wiley,1 990).
12. Kathleen Eisenhardt, "Speed and Strategic Choice: How Managers Accelerate Decision Making," *California Management Review* 32, no. 3 (1990), 39–54.
13. David E. Gumpert, *How to Really Create a Successful Marketing Plan*, 2nd ed. (Boston: Inc. Publications, 1995), 42–43.
14. Robert Lauterborn, "New Marketing Litany: 4P's Passé; C-Words Take Over," *Advertising Age*, October 1, 1991, 26.
15. Mohanbir Sawhney, "Winning Strategies for E-Commerce," Lecture at the Kellogg Graduate School of Management, Evanston, IL, April 1999.
16. Philip Kotler, *Marketing Insights from A to Z* (Hoboken, NJ: Wiley, 2003), 109.
17. Drucker, *The Five Most Important Questions*,51,53.

18. Knauftet a l., *Profiles of Excellence,*120.
19. Donald C. Hambrick and Albert A. Cannella Jr., "Strategy Implementation as Substance and Selling," *The Academy of Management Executive* 3, no. 4 (1989), 278–285.
20. Thomas Wolf, *Managing a Nonprofit Organization* (New York: Fireside/ Simon & Schuster, 1990), 155.
21. Herrington Bryce, *Financial and Strategic Management for Nonprofit Organizations,* 2nd ed. (Englewood Cliffs, NJ: Prentice Hall, 1992), 453.
22. Wolf, *Managing a Nonprofit Organization,*150–151.

6 Identifying Market Segments, Selecting Target Markets, and Positioning the Offer

1. *Counting New Beans: Intrinsic Impact and the Value of Art* (San Francisco: Theatre Bay Area, 2012), 71–72.
2. "Cultural Segments," Morris Hargreaves McIntyre: Lateral Thinkers, downloaded November 2012, http://www.lateralthinkers.com/culturesegments .html.
3. Philadelphia Arts Market Study, prepared by Ziff Marketing, Inc., commissioned by the Pew Charitable Trusts, 1989.
4. "Marketing the Arts in Cleveland: An In-Depth Survey," conducted by Ziff Marketing, Inc., and Clark, Martire & Bartolomeo, Inc., commissioned by the Cleveland Foundation, 1985.
5. Bradley Morison and Julie Dalgleish, *Waiting in the Wings* (New York: American Council for the Arts, 1987).
6. Sidney J. Levy, "Arts Consumers and Aesthetic Attributes," in *Marketing the Arts,* ed. Michael P. Mokwa, William M. Dawson, and E. Arthur Prieve (New York: Praeger, 1980)
7. Ibid.
8. Sidney J. Levy, "Arts Consumers and Aesthetic Attributes," in *Marketing the Arts,* ed. Michael P. Mokwa, William M. Dawson, and E. Arthur Prieve (New York: Praeger, 1980).
9. David A. Aaker and J. Gary Shansby, "Positioning Your Product," *Business Horizon,* May/June 1982, 56–62.
10. Al Ries and Jack Trout, *Positioning: The Battle for Your Mind* (New York: McGraw-Hill, 1986), 34.
11. William Rudman, *Market the Arts!* (New York: Foundation for the Extension and Development of the American Professional Theatre, 1983), 165.
12. Yoram J. Wind, *Product Policy: Concepts, Methods, and Strategy* (Reading, MA: Addison-Wesley, 1982), 79–81; Ries and Trout, *Positioning.*
13. Riesa nd Trout, *Positioning,*5.
14. Patricia Seybold, "Get Inside the Lives of Your Customers," in *Harvard Business Review on Customer Relationship Management* (Boston: Harvard Business School Press, 2001), 30.
15. Philip Kotler and Fernando Trias de Bes, *Lateral Marketing* (Hoboken, NJ: Wiley, 2003), 98.

7 Conducting and Using Marketing Research

1. Peter S. Goodman, "Emphasis on Growth Is Called Misguided," *New York Times,* September 22, 2009, http://www.nytimes.com/2009/09/23/business/economy/23gdp.html?_r=0
2. "Big Data: The Next Frontier for Innovation, Competition, and Productivity," McKinsey Global Institute, May 2011. http://www.mckinsey.com/Insights/MGI/Research/Technology_and_Innovation/Big_data_The_next_frontier_for_innovation.
3. Ethan Roeder, "I Am Not Big Brother," *New York Times,* December 6, 2012, http://www.nytimes.com/2012/12/06/opinion/i-am-not-big-brother.html?_r=0.
4. David Brooks, "What Data Can't Do," *New York Times,* February 19, 2013, http://www.nytimes.com/2013/02/19/opinion/brooks-what-data-cant-do.html?_r=0.
5. Ibid.
6. Chad Bauman, Arts Marketing blog, "What If You Didn't Have to Guess?" March 3, 2013, http://arts-marketing.blogspot.com/2013/03/what-if-you-didnt-have-to-guess.html?m.
7. Leonard Berry, *On Great Service: A Framework for Action* (New York: Free Press, 1995), 39–40.
8. Gerald Zaltman, *How Customers Think: Essential Insights into the Mind of the Market* (Boston: Harvard Business School Press, 2003), 11. Emphases in the original.
9. Zaltman metaphor elicitation technique, Wikipedia, December 27, 2012, http://en.wikipedia.org/wiki/Zaltman_metaphor_elicitation_technique.
10. Alan R. Andreasen, *Marketing Social Change: Changing Behavior to Promote Health, Social Development, and the Environment* (San Francisco, CA: Jossey-Bass, 1995), 101.
11. Pamela L. Alreck and Robert B. Settle, *The Survey Research Handbook* (Homewood, IL: Irwin, 1985), 67–68.
12. Nancy R. Lee and Philip Kotler, *Social Marketing,* 4th ed. (Los Angeles: Sage, 2011), 98.
13. Philip Kotler and G. Armstrong, *Principles of Marketing,* 9th ed. (Upper Saddle River, NJ: Prentice Hall, 2001), 140.

8 Using Strategic Marketing to Define and Analyze the Product Offering

1. Theodore Levitt, *The Marketing Imagination* (New York: Free Press, 1986), 76–77.
2. Olson Zaltman Associates, "Thoughts and Feelings about Broadway Theatre—And Its Role in Your Life," a ZMET Study for the League of American Theatres and Producers, fall 2006.
3. Dominique Bourgenon, presentation at conference of the International Society for the Management of Arts and Culture (AIMAC), Jouy-en-Josas, France, June 1992.

4. Barbara Hoffman, “Future in Footlights,” *New York Post*, January 16, 2012, http://www.nypost.com/p/entertainment/theater/future_in_footlights_hG1JnCayveBLyEiBoOEIFI.
5. Clayton Lord, ed., *Counting New Beans* (San Francisco: Theatre Bay Area, 2012), 179, 181.
6. Theodore Levitt, “Marketing Intangible Products and Product Intangibles,” *Harvard Business Review*, May/June 1981, 94–102; and Leonard Berry, “Services Marketing Is Different,” *Business*, May/June 1980, 24–30.
7. Philip Kotler, “Atmospherics as a Marketing Tool,” *Journal of Retailing* (Winter 1973/74), 48–64.
8. Leonard L. Berry and A. Parasuraman, *Marketing Services* (New York: Free Press, 2004), 190.
9. Ed Rothstein, “Shakespeare as Muse in Symphony Concert,” *New York Times*, September 28, 1993.
10. Robert Schwartz, “The Crises of Tomorrow Are Here Today,” *New York Times*, October 31, 1993, 31–32.
11. HowardR eich,“ MusicL essons,” *Chicago Tribune*, July 17, 1994.
12. Daniel J. Wakin, “Philharmonic Steals a Page From the Art World With a New-Music Biennial,” *New York Times*, January 24, 2013, C1, 8.

9 MANAGING LOCATION, CAPACITY, AND TICKETING SYSTEMS

1. Alan Brown, “All the World’s a Stage: Venues and Settings, and the Role They Play in Shaping Patterns of Arts Participation,” WolfBrown, Cambridge, MA, 2012, 1.
2. Chris Waddington, “UNO Study Explains Why New Orleans Patrons Attend Performing Arts Events,” *Times Picayune*, December 20, 2012, http://www.nola.com/arts/index.ssf/2012/12/uno_study_explains_why_new_orl.html.
3. Brown,“ Allt he World’sa S tage,”2.
4. Alex Ross, “The Met Edges into the Future,” *New York Times*, May 15, 1994, Arts and Leisure.
5. Brown,“ Allt he World’sa S tage,”5.
6. Charles Dickens, *Nicholas Nickleby* (Hertfordshire, Great Britain: Wordsworth Classics, 1995), 362.
7. Waddington,“ UNOS tudyE xplains.”
8. Roger Tomlinson, “Ticketing Is Getting Personal,” *Filling the Seats, Ticketing: The New Age*, Musical America Worldwide, February 2013, http://www.musicalamerica.com/specialreports/TICKETS_2013.pdf.
9. “Can Digital Media Sell New Subscriptions?” Capacity Interactive blog, August 29, 2012, http://www.capacityinteractive.com/ideas/can-digital-media-sell-new-subscriptions/ (accessed February 8, 2013).
10. Susan Elliott, ed., “Filling the Seats, Ticketing: The New Age,” Musical America Worldwide, February 2013.
11. Mike Ewing, “71% More Likely to Purchase Based on Social Media Referrals,” HubSpot’s Inbound Marketing Blog, January 9, 2012, http://blog.hubspot

.com/blog/tabid/6307/bid/30239/71-More-Likely-to-Purchase-Based-on-Social-Media-Referrals-Infographic.aspx.

12. Brett Ashley Crawford, "How to Get the Most Out of Your Ticketing Investment Dollar," Musical America Worldwide, February 2013.

10 Focusing on Value and Optimizing Revenue through Pricing Strategies

1. Warren Buffett, BrainyQuote.com, Xplore Inc, 2013, http://www.brainyquote.com/quotes/quotes/w/warrenbuff149692.html (accessed December 5, 2013).
2. Seth Godin, Blog, October 25, 2007, http://sethgodin.typepad.com/seths_blog/2007/10/i-cant-afford-i.html.
3. Tim Baker, "What Price Value?" Creative New Zealand 21st Century Arts Conference, Auckland, June 26–27, 2008, http://www.creativenz.govt.nz/assets/paperclip/publication_documents/documents/87/original/tim-baker-what-price-value.pdf?1322079828 (accessed April 14, 2013).
4. Ichak Adizes, "The Cost of Being an Artist," *California Management Review* 17, no. 2 (Summer 1975).
5. Ibid.
6. RSGB Study, "Omnibus Arts Survey for the Arts Council of Great Britain," Research Surveys of Great Britain Limited, Arts Council of Great Britain,1991.
7. Ibid.
8. An extensive analysis of this study was originally published as J. Scheff, "Factors Influencing Subscription and Single-Ticket Purchases at Performing Arts Organizations," *International Journal of Arts Management* 1, no. 2 (Winter 1999), 16–27.
9. Fiona Govan, "Spain Abandons the Theatre," *Telegraph*, March 26, 2013, http://www.telegraph.co.uk/news/worldnews/europe/spain/9955016/Spain-abandons-the-theatre.html.
10. Andrew McIntyre, personal email to the author, August 10, 2013.
11. Personal phone conversation with Tom O'Connor, marketing director of Roundabout Theater, March 27, 2013.
12. Gary Show, "American Conservatory Theatre," case study UVA-M-254, Darden Graduate Business School, University of Virginia, rev. December 1991.
13. Zannie Giraud Voss and Glenn B. Voss, "Theatre Facts 2012," Theatre Communications Group, 6 (emailed to author).
14. John Tierney, "Tickets? Supply Meets Demand on Sidewalk," *New York Times*, December 25, 1993, Living Arts.
15. Adapted from case study published in Joanne Scheff Bernstein, *Arts Marketing Insights: The Dynamics of Building and Retaining Performing Arts Audiences* (San Francisco: Jossey-Bass, 2007), 136–141.
16. William Poundstone, *Priceless: The Myth of Fair Value (and How to Take Advantage of It)* (New York: Hill and Wang, 2011), 156.

17. Jennifer Melick, "Just the Ticket," *Symphony*, November–December 2010, http://www.nxtbook.com/nxtbooks/symphonyonline/nov_dec_2010/#/50 (accessed February 22, 2013), 53–54.
18. Ibid.E mphasisi nt heo riginal.
19. Ibid.
20. Andrew McIntyre, personal email to the author, August 10, 2013.
21. Penn Trevella, comments made in the Fuel4Arts Pricing Forum, November 2002, http://www.fuel4arts.com,A pril2006.
22. Dan Ariely, *Predictably Irrational* (New York: Harper Perennial Edition, HarperCollins Publishers, 2010), 55–56.
23. Ibid., 34.
24. RobertC ross, *Revenue Management* (New York: Broadway Books, 1997), 4.
25. Tim Baker, "Defining Revenue Management and Dynamic Pricing—the Difference Is Important!," http://www.thinkaboutpricing.com/left-navigation/revenue-management.html (accessed April 15, 2013).
26. Steven Roth, "Center Theatre Group—Beyond Revenue Management," The PricingInstitute,Center_Theatre_Group_-_Beyond_Revenue_Management.pdf (accessed April 15, 2013).
27. Baker," DefiningR evenueM anagementa ndD ynamicP ricing."
28. Steven Roth, "Center Theatre Group—Beyond Revenue Management," The PricingInstitute,Center_Theatre_Group_-_Beyond_Revenue_Management.pdf (accessed April 15, 2013).
29. Sumac Research, "Dynamic Pricing: It Worked for the Airlines, but Can It Work for Theatres?" *Management in the Not for Profit Organization*, July 2010, http://www.npmanagement.org/Article_List/Articles/Dynamic_Pricing.htm (accessed April 9, 2013).
30. Andrew McIntyre, personal email to the author, August 10, 2013.
31. Ibid.

11 IDENTIFYING AND CAPITALIZING ON BRAND IDENTITY

1. Tim Baker and Steve Roth, "Value Is All in the Mind," Psycho Pricing: Part 4, The Pricing Institute, September 2012, http://www.thinkaboutpricing.com/uploads/uploads_files/Psycho_Pricing_Value_is_all_in_the_Mind.pdf (accessed April 14, 2013).
2. Howard Schultz, *Pour Your Heart Into It: How Starbucks Built a Company One Cup at a Time*, (New York: Hyperion, 1999), 248.
3. Philip Kotler, *Marketing 3.0: From Products to Customers to the Human Spirit* (New Jersey: John Wiley & Sons, 2010), 37.
4. Ibid.
5. Jerome Kathman, "The Experience Business," http://www.theexperiencebusiness.co.uk/inspiration/, December 5, 2013.
6. Scott Bedbury, *A New Brand World: 8 Principles for Achieving Brand Leadership in the 21st Century* (New York: Viking, 2002), 105. Emphasis in theo riginal.

7. MohanS awhney,bl og,m ohansawhney.com,M arch2011.
8. Paul Valerio, *Raiders of the Lost Overture*, Method 10 X 10, http://10x10.method.com/raiders-of-the-lost-overture/(undated).
9. Lisa Baxter, "On Being the Creative Producers of Your Brand," *The Experience Business*, July 27, 2012, http://www.theexperiencebusiness.co.uk/blog/entry/on-being-the-creative-producers-of-your-brand/ (accessed February 8, 2013).
10. Ibid.E mphasesi nt heo riginal.
11. Bedbury, *A New Brand World*, xiii, 21. Emphasis in the original.
12. Philip Kotler, *Kotler on Marketing: How to Create, Win, and Dominate Markets* (New York: Free Press, 1999), 55.
13. Gerri Morris, "It's a Vision Thing" (Manchester, UK: Morris Hargreaves McIntyre Consultancy and Research, 2004), http://www.lateralthinkers.com.
14. "Meaningful Marketing: The New Heart of Corporate/Nonprofit Partnerships," IEG, 2012, 9, www.sponsorship.com.
15. David Ogilvy, 1955 speech to AAAA, quoted by K. Roman, "David Ogilvy: The Most Famous Advertising Man in the World. A Speech at the University Club—New York," 2004, http://www.gandalf.it/m/ogilvy2.htm.
16. William Rudman, "Essentials of Effective Public Relations," in *Market the Arts!* ed. Joseph Melillo (New York: FEDAPT, 1983), 163.
17. DanielB oorstin, *The Image* (New York: Vintage Books, 1992), 187.
18. Robert Jackal, "The Magic Lantern," in *Moral Mazes: The World of Corporate Managers* (New York: Oxford University Press, 1988), 172–173.
19. Charles Conrad, "Analyzing Organizational Situations: Introduction," in *Strategic Organizational Communication* (New York: Holt, Reinhart & Winston, 1985), 202.
20. Examplesa daptedf romB oorstin, *The Image.*
21. Boorstin, *The Image*,185–194.
22. Rudman," Essentialso fE ffectiveP ublicR elations,"165.
23. Stuart Isacoff, "Evgeny Kissin," in *Musical America Directory* (New York: K-III Directory Corp., 1995).

12 FORMULATING COMMUNICATIONS STRATEGIES

1. Roger A. Strang and Jonathan Gutman, "Promotion Policy Making in the Arts: A Conceptual Framework," in *Marketing the Arts*, ed. Michael P. Mokwa, William M. Dawson, and E. Arthur Prieve (New York: Praeger, 1980),225–238.
2. Andrew Druckenbrod, "Music Preview," *Pittsburgh Post-Gazette*, rom http://www.post-gazette.com/pg/04294/398513.stm.
3. AlexR oss," Listent o This," *New Yorker*, February 16 and 23, 2004.
4. Harvey Felder, "The Quest for Generational Diversity," *Symphony*, Winter 2013,69–7 0.

5. "Storytelling That Moves People: A Conversation with Screenwriting Coach Robert McKee," *Harvard Business Review*,J une2003.
6. Caitlin Johnson, "Cutting through Advertising Clutter," *Sunday Morning, CBS News*, February 11, 2009, http://www.cbsnews.com/8301–3445_162–2015684.html (accessed May 29, 2013).
7. Ben H. Bagdikian, "How Much More Communication Can We Stand?" in *Ethics, Morality, and the Media*, ed. Lee Thayer (New York: Hastings House, 1980), 175–180.
8. Peter F. Drucker, "Management: Tasks, Responsibilities, Practices," in *Managerial Communications* (New York: Harper & Row, 1974), chapter 38, 481–493.
9. Wilbur Schramm, "How Communication Works," in *The Process and Effects of Mass Communication*, ed. Wilbur Schramm and Donald F. Roberts (Urbana: University of Illinois Press, 1971), 4.
10. Brian Sternthal and C. Samuel Craig, *Consumer Behavior: An Information Processing Perspective* (Englewood Cliffs, NJ: Prentice-Hall, 1982), 86–116.
11. Otto Ottensen, "The Response Function," in *Current Theories in Scandinavian Mass Communications Research*, ed. Mie Berg (Grenaa, Denmark: G.M.T., 1977).
12. Jamie James, "Sex and the 'Singles' Symphony," *New York Times*, May 2, 1993, sec. H, 1, 27.
13. Dik Warren Twedt, "How to Plan New Products, Improve Old Ones, and Create Better Advertising," *Journal of Marketing* 33, no. 1 (January 1969), 53–57.
14. James F. Engel, Roger D. Blackwell, and Paul W. Minard, *Consumer Behavior*, 5th ed. (Hinsdale, IL: Dryden Press, 1986), 477.
15. C. I. Hovland, A. A. Lumsdaine, and F. D. Sheffield, *Experiments on Mass Communication*, vol. 3 (Princeton, NJ: Princeton University Press, 1948), chapter8.
16. Edward T.H all, *The Silent Language* (Garden City, NY: Doubleday, 1973).
17. Daniel Wakin, "Serious Music? He Loves It. No, Seriously." *New York Times*, December 11, 2009, http://www.nytimes.com/2009/12/13/arts/music/13baldwin.html?pagewanted=all&_r=0 (accessed May 27, 2013).
18. Herbert C. Kelman and Carl I. Hovland, "Reinstatement of the Communication in Delayed Measurement of Opinion Change," *Journal of Abnormal and Social Psychology* 48 (1953), 327–335.
19. Philip Kotler, *Marketing 3.0: From Products to Customers to the Human Spirit* (Hoboken, NJ: John Wiley & Sons, Inc., 2010), 64–66.
20. Rachel Lamb, "What Packs More Punch, Unknown Models or Celebrity Endorsements?" *Luxury Daily*, April 4, 2012, http://www.luxurydaily.com/newsletter-archive/luxury-daily-april-4–2012-what-packs-more-punch-unknown-models-or-celebrity-endorsements/.
21. Emanuel Rosen, *The Anatomy of Buzz: How to Create Word of Mouth Marketing* (New York: Doubleday: 2002), 6 and 249. Emphasis in the original.
22. Ibid.,249.

23. Ibid.,261.
24. Malcolm Gladwell, *The Tipping Point: How Little Things Can Make a Big Difference* (Boston: Little, Brown, 2000), 35, 54.

13 Delivering the Message: Advertising, Personal Selling, Sales Promotion, Public Relations, and Crisis Management

1. Sidney J. Levy, *Promotional Behavior* (Glenview, IL: Scott, Foresman, 1971), chapter4.
2. Bob Schulberg, *Radio Advertising: The Authoritative Handbook* (Lincolnwood, IL: NTC Business Books, 1989).
3. Ibid.
4. Herbert E. Krugman, "What Makes Advertising Effective?" *Harvard Business Review*, March/April 1975, 98.
5. Peggy J. Kreshel, Kent M. Lancaster, and Margaret A. Toomey, "Advertising Media Planning: How Leading Advertising Agencies Estimate Effective Reach and Frequency" (Urbana: University of Illinois Department of Advertising, paper no. 20, January 1985). Also see Jack Z. Sissors and Lincoln Bumba, *Advertising Media Planning*, 3rd ed. (Lincolnwood, IL: NTC Business Books, 1989), Chapter 9.
6. John Wanamaker, Wikipedia, http://en.wikipedia.org/wiki/John_Wanamaker (accessed December 8, 2013).
7. "Personal Recommendations and Consumer Opinions Posted Online Are the Most Trusted Forms of Advertising Globally," press release (New York: The Nielsen Company, July 7, 2009).
8. Philip Kotler, *Marketing 3.0: From Products to Customers to the Human Spirit* (New Jersey: John Wiley & Sons, 2010), 32.
9. David Meerman Scott, *The New Rules of Marketing and PR* (New Jersey: John Wiley & Sons, 2011), 24.
10. Chris Jones, "Courtroom No Place for the Great American Narrative," *Chicago Tribune*, July 21, 2013, Sec. 4, p. 1.
11. Adapted from E. G. Schreiber, "Promoting the Performing Arts," and "Rhoda Weiss's Public Relations Tips for Nonprofit Organizations," in *Media Resource Guide*, 5th ed. (Los Angeles: Foundation for American Communication, 1987), 39–41; and Dianne Bissell, *Marketing Promotions Guide* (Chicago: League of Chicago Theatres, 1985), 51–55.
12. David R. Yale, *The Publicity Handbook* (Lincolnwood, IL: NTC Business Books, 1991), 5.
13. David Meerman Scott, *The New Rules of Marketing and PR* (New Jersey: John Wiley & Sons, 2011), 293.
14. Chris Jones, theater critic, *Chicago Tribune*, personal conversation with the author, July 26, 2013.
15. Nancy Depke, "Marketing Ballet When New in the Neighborhood," *Dance/USA*, March 1985, 6, 15.

16. Yale, *The Publicity Handbook*,39.
17. Dianne Bissell, *Marketing Promotions Guide* (Chicago: League of Chicago Theatres, 1985), 51–55.
18. Yale, *The Publicity Handbook*,31.
19. StevenF ink," Copingw ith Crises," *Nation's Business*, August 1984, 52R.
20. Claudia Reinhardt, "Workshop: How to Handle a Crisis," *Public Relations Journal*(November1987) ,43–44.
21. Hooshang Kuklan, "Managing Crises: Challenges and Complexities," *SAM Advanced Management Journal* Autumn 1986), 39–44.
22. Ibid.
23. James A. Benson, "Crisis Revisited: An Analysis of Strategies Used by Tylenol in the Second Tampering Episode," *Central States Speech Journal* 39, no. 1 (Spring 1988), 50.
24. Stratford A. Sherman, "Smart Ways to Handle the Press," *Fortune*, June 19, 1989,69–72.
25. Adapted from Anthony R. Katz, "Checklist: 10 Steps to Complete Crisis Planning," *Public Relations Journal*(November1987) ,46–47.

14 Harnessing and Leveraging the Power of Digital Marketing Methods

1. Philip Kotler, *Marketing 3.0: From Products to Customers to the Human Spirit* (New Jersey: John Wiley & Sons), 7.
2. Chuck Martin, "The Mobile Shopping Life Cycle," *HBR Blog Network*, June 11, 2013, http://blogs.hbr.org/cs/2013/06/the_mobile_shopping_life_cycle.html (accessed August 3, 2013). Emphasis in the original.
3. Eugene Carr, "Patron Technology, Arts Patron Information," shared privately with the author, October 12, 2012.
4. Sarah Jones, "Ringtones? MP3s? Beethoven Would Have Been Proud," March26,2006, http://www.Living.Scotsman.com.
5. "Digital Media Marketing in the Arts: *Musical America* Special Reports," April 2012, http://www.musicalamerica.com/specialreports/DMM_2012.pdf.
6. Julie Aldridge and Roger Tomlinson, *Pump Up Your Website: Improving the Effectiveness ofYour E-Marketing*, Arts Marketing Association of Great Britain, seminar presented across the United Kingdom, October and November 2004, http://www.a-m-a.co.uk/pumpupyourWebsite.
7. Laurie Windham, *The Soul of the New Consumer* (New York: Allworth Press, 2000),45.
8. Aldridgea nd Tomlinson, *Pump Up Your Website.*
9. "An Introduction to Business Blogging: How to Use Business Blogging for Marketing Success," an online publication of HubSpot, April 4, 2012, http://offers.hubspot.com/introduction-to-business-blogging-via-blog (accessed June 17, 2013).
10. Seth Godin, *Permission Marketing: Turning Strangers into Friends, and Friends in Customers* (New York: Simon & Schuster, 1999), 130.

11. Julie Aldridge, *Word of Mouse: Practical Online Marketing* (London: Arts Marketing Association, 2002).
12. Eugene Carr, *Wired for Culture: How E-mail Is Revolutionizing Arts Marketing* (New York: Patron Technology, 2003), 68.
13. Ibid.,65.
14. "Your Music Lesson Is in the Mail," *Los Angeles Times*, August 21, 2005.
15. Carr, *Wired for Culture*,68.
16. Larry Freed, "Social Media Marketing: Do Retail Results Justify Investment?" ForSee Results, February 3, 2011, http://www.foreseeresults.com/research-white-papers/_downloads/social-media-marketing-u.s.-2011-foresee.pdf (accessed June 17, 2013).
17. Dina Gerdeman, "Catching Up with Our Mobile Society, Mobile Marketing: The Arts in Motion," *Musical America*: Special Report, May 2013, 5, http://www.musicalamerica.com/specialreports/MOBILE_2013.pdf.
18. Luke Wroblewski, "Mobile Has Changed When and Where People Read E-Mail," *Mobile First*, July 1, 2013, http://www.lukew.com/ff/entry.asp?933 (accessed August 3, 2013).
19. James Orsini, "Marketers Must Keep Mobile Simple," LuxuryDaily.com, June 11, 2013, http://www.luxurydaily.com/marketers-must-keep-mobile-simple/.
20. Ibid.
21. Jeff Bullas, *Stunning Social Media Statistics*, September 2011, blog post, http://www.jeffbullas.com/2011/09/02/20-stunning-social-media-statistics/.
22. "E-mail + Twitter + Facebook: 22 Tips to Cross-Channel Success," Lyris™, 2013, http://www.lyrislabs.com/media/pdf/E-mail-Facebook-Twitter-22-Cross-Channel-Tips-and-Takeaways.pdf.
23. Mohan Sawhney, lecture to alumni, Kellogg Graduate School of Management, March 31, 2011.
24. Steve Nicholls, "Why Most Companies Fail at Social Media," *Chief Executive Magazine*, September 19, 2012, http://chiefexecutive.net/why-most-companies-fail-at-social-media (accessed June 13, 2013).
25. Roxanne Divol, David Edelman, and Hugo Sarrazin, "Demystifying Social Media," *McKinsey Quarterly*, April 2012, http://www.mckinseyquarterly.com/Marketing/Digital_Marketing/Demystifying_social_media_2958.
26. Scott Monty, "The Best Marketing Advice: Tweet This," Silverpop, 2012, http://www.silverpop.com/downloads/white-papers/Silverpop-WP-Best-Marketing-Advice-unwhite.pdf.
27. Geri Jeter, "How Social Media Has Energized Small Arts Organizations—An Interview with Diablo Ballet's Dan Meagher," *California Literary Review*, September20,2012, http://calitreview.com/30459.
28. Ibid.
29. Ibid.
30. Nicola Clark, "Making the Skies Friendlier," *New York Times*, February 24, 2012,B 1.

31. Kyle MacMillan, "Arts Organizations Take to Social Media to Share Collections, Shows," *Chicago Sun Times*, February 11, 2013, http://www.suntimes.com/entertainment/18160927–421/chicago-arts-organizations-take-to-social-media-to-share-collections-shows.html (accessed February 12, 2013).
32. Kate Taylor, "Quiet in the Audience, Please," *Globe and Mail*, April 21, 2012, http://www.theglobeandmail.com/arts/theatre-and-performance/quiet-in-the-audience-please/article4101599/ (accessed June 18, 2013).
33. Ibid.
34. Mohan Sawhney, lecture to alumni, Kellogg Graduate School of Management, March 31, 2011.
35. Scott Forshay, "Transitional Marketing and the Connected Screen," Luxury Daily, October 1, 2012, http://www.luxurydaily.com/transitional-marketing-and-the-connected-screen/

15 Building Audience Frequency and Loyalty

1. Danny Newman, *Subscribe Now!* (New York: Theatre Communications Group, 1983), 15–16.
2. Alan Brown, "Classical Music Consumer Segmentation Study: How Americans Relate to Classical Music and Their Local Orchestras," *Arts Reach* 11, no. 3 (2003), 16.
3. Newman, *Subscribe Now!*17–24.
4. Zannie Giraud Voss and Glenn B. Voss, *Theatre Facts 2012* (New York: Theatre Communications Group, 2012), http://www.tcg.org/pdfs/tools/TheatreFacts_2012.pdf.
5. Philip Kotler, *Marketing Insights from A to Z* (Hoboken, NJ: John Wiley & Sons, 2003), xiii.
6. DavidM amet, *Glengarry Glen Ross*" (New York: Grove Press, 1982), 31.
7. Brown," ClassicalM usicS egmentationS tudy,"16.
8. Femke Colborne, "Subscriptions under Scrutiny as Opera Europa Meets in Latvia," *International Arts Manager*, December 2004/January 2005, 13.
9. Bil Schroeder, marketing director of South Coast Repertory, in personal conversation with the author, March 5, 2012.
10. Jim Royce, personal email to the author, June 24, 2004.
11. TRG, "How Valuable Is Loyalty? A Performing Arts Example," National Arts Marketing, Development & Ticketing Conference, New York, March 16–18,201 2.
12. Chad Bauman, Arts Marketing Blog, "The Subscription Equation (and Other Tactics)," February 3, 2013, http://arts-marketing.blogspot.com (accessed February 8, 2013).
13. Ibid.
14. Kotler, *Marketing Insights from A to Z*,98.
15. W. Chan Kim and Renée Mauborgne, "Blue Ocean Strategy," *Harvard Business Review*, October 2004, 81.
16. Vossa nd Voss, *Theatre Facts 2012.*

17. Andrew McIntyre, personal email to the author, August 10, 2013.
18. Eugene Carr, "E-Marketing Essentials: Marketing Lessons from Obama and Romney," September 11, 2012, http://patrontechnology.com/newsletter-marketing-lessons-segmentation/.
19. Andrew McIntyre and Helen Dunnett, *Practical Guide: Move on Up: How Test Drive and TelePrompt Develop Audiences*, 2003, http://www.newaudiences2.org.uk/downloads/move_on_up.pdf(accessed2006.

16 Focusing on the Customer Experience and Delivering Great Customer Service

1. "2013: The Year of the Patron," TRG Arts, http://trgarts.blogspot.com/2013/01/2013-year-of-patron.html (accessed February 19, 2013).
2. Mohanbir Sawhney, *A Manifesto for Marketing: What Ails the Profession and How to Fix It*,S ummer 2004, http://www.mohansawhney.com.
3. C. K. Prahalad and Venkatram Ramaswamy, "Co-Opting Customer Competence," in *Harvard Business Review on Customer Relationship Management* (Boston: Harvard Business School Press, 2001), 2–5.
4. J. Walker Smith, Ann Clurman, and Craig Wood, *Coming to Concurrence: Addressable Attitudes and the New Model for Marketing Productivity* (Evanston: Racom Communications, 2005), 98.
5. Frederick Newell, *Why CRM Doesn't Work: How to Win by Letting Customers Manage the Relationship* (Princeton, NJ: Bloomberg Press, 2003).
6. Susan Fournier, Susan Dobscha, and David Glen Mich, "Preventing the Premature Death of Relationship Marketing," in *Harvard Business Review on Customer Relationship Management* (Boston: Harvard Business School Press, 2001),134–137.
7. Ibid.
8. Newell, *Why CRM Doesn't Work*,7–8.
9. Judith Allen, personal conversation with the author, July 8, 1999.
10. Jim Royce, personal conversation with the author, March 7, 2012.
11. B. Joseph Pine II and James H. Gilmore, *The Experience Economy: Work Is Theatre & Every Business a Stage* (Boston: Harvard Business School Press, 1999),77–78.
12. Quoted in Clayton Collins, "Five Minutes with J. D. Power III," *Profiles*, October 1996, 23. Emphases in the original.
13. Baldridge.com, "Be Careful How You Measure Customer Satisfaction," March 24, 2010, baldridge.com (accessed June 2, 2013).
14. Pine II and Gilmore, *The Experience Economy*, 77–78. Emphasis in the original.
15. Baldridge.com, "9 Ways to Get Closer to Customers," September 16, 2009, baldridge.com (accessedJune 2, 2013).
16. Baldridge.com, "Kano Satisfaction Model," August 30, 2010, http://www.baldrige.com/criteria_customerfocus/kano-satisfaction-model/ (accessed June 2, 2013).

17 Audiences for Now; Audiences for the Future

1. Mohanbir Sawhney, Alumni Lecture for Kellogg Graduate School of Management, March 21, 2011.
2. Ivor Royston, Wikipedia, http://en.wikipedia.org/wiki/Ivor_Royston (accessed January 10, 2014).
3. Doug Borwick, "Engaging Matters," May 1, 2013, http://www.artsjournal.com/engage/2013/05/engagement-vocabulary/.
4. Rick Lester, "Audience Engagement's No Man's Land," *Analysis from TRG Arts,* June 27, 2013, http://www.trgarts.com/Blog/BlogPost/tabid/136/ArticleId/175/Audience-Engagement-s-No-Man-s-land.aspx?utm_source=enews&utm_medium=patronmail&utm_campaign=summer2013 (accessed July 11, 2013).
5. Morris Hargreaves McIntyre, "This Time It's Personal: Interactivity, Personalization and the 21st Century Arts Organization," n.d., http://www.lateralthinkers.com/resources/comment/conference_reader.pdf.
6. Karen Freeman, Patrick Spenner, and Anna Bird, "Three Ways You're Wrong about What Your Customers Want," BRW, August 29, 2012, http://brw.com.au/p/sections/the_business_end/ways_you_re_wrong_about_what_your_2fBruIPi1oMoB8Cxj7WU2K.
7. Tim Baker, "What Price Value?" Creative New Zealand 21st Century Arts Conference, Auckland, June 26–27, 2008, 4, http://www.creativenz.govt.nz/assets/paperclip/publication_documents/documents/87/original/tim-baker-what-price-value.pdf?1322079828 (accessed April 14, 2013).
8. Imogen Tilden, "Damon Albarn Kickstarts ENO's 'Undress for Opera' Scheme," guardian.co.uk, Wednesday, October 3, 2012, http://www.guardian.co.uk/music/2012/oct/03/damon-albarn-kickstarts-eno-opera-scheme.
9. Holly Hickman, "The 10 Commandments of Classical Music Audience Building," on *Matt Lehrman on Audience Building, Audience Wanted,* July 31, 2013, http://www.artsjournal.com/audience/2013/07/the-10-commandments-of-classical-music-audience-building/ (accessed August 1, 2013).
10. PhilipK otler, *Marketing 3.0* (New Jersey: John Wiley & Sons, 2010), 45.
11. Alan Brown, Joseph Kluger, et al. (principal consultants of WolfBrown with Joanna Woronkowicz), "Is Sustainability Sustainable?" *Sounding Board,* publication of WolfBrown, http://wolfbrown.com/images/soundingboard/WB_SoundingBoardv30_1114_d.pdf (accessed September 14, 2013).
12. Harvey Felder, "The Quest for Generational Diversity," *Symphony,* Winter 2013,71–7 2.
13. Chad Bauman, Arts Marketing blog, "Is Your Organization Fun?" November 17, 2012, http://arts-marketing.blogspot.com (accessed February 8, 2013).
14. Louis Bergonzi and Julia Smith, "Effects of Arts Education on Participation in the Arts," Research Division Report #36, National Endowment for the Arts, Seven Locks Press, Santa Ana, California, 1996, 32–47.
15. Peter F. Drucker, *The Five Most Important Questions You Will Ever Ask about Your Nonprofit Organization* (San Francisco: Jossey-Bass, 1993), 12.

16. Michael Kaiser, speech to the Arts Marketing Association of Great Britain, Cardiff, Wales, July 1999.
17. Gene Carr, "A Springtime of Discontent: Two Paradigms for Thriving Arts Organizations," Patron Technology Newsletter, June 2011, http://pm.patrontechnology.com/newsletters/June2011.htm (accessed July 29, 2013).
18. NYCB Art Series 2014, New York City Ballet, http://www.nycballet.com/artseries/ (accessed January 27, 2014).
19. Michael Cooper, "A Giant Photo Connects Fans to Ballet Stars," *New York Times,* January 26, 2014, http://www.nytimes.com/2014/01/27/arts/dance/a-giant-photo-connects-fans-to-ballet-stars.html?_r=1.
20. "Creating Social Change through Music," posted by HOLA on September 17,2012, http://heartofla.org/2012/09/news/creating-social-change-music.
21. Gordon R. Sullivan and Michael V. Harper, *Hope Is Not a Method* (New York: Random House, 1996), 152.
22. Steve McMillan, presentation at the Kellogg School of Management Marketing Conference, Evanston, IL, January 29, 2003.
23. Nancy Malitz, "Special Reports, Case Study No. 1: Detroit Symphony Orchestra," Musical America Worldwide, April 2, 2012, http://www.musicalamerica.com/news/newsstory.cfm?archived=0&storyid=26952&categoryid=7 (accessed January 10, 2014).
24. Dso.org digital media consumption survey, July 2013 to January 2014. Provided to the author by Scott Harrison, executive producer of digital media at the Detroit Symphony Orchestra, January 22, 2014.
25. Detroit Symphony Orchestra, "'Live from Orchestra Hall' Webcasts Make Detroit Symphony Orchestra Among the 'Most Attended' Orchestras in America," blog, dso.org, June 12, 2012, http://blog.dso.org/2012/06/'live-from-orchestra-hall'-webcasts-make-detroit-symphony-orchestra-among-the-most-attended-orchestras-in-america/ (accessed January 10, 2014).
26. Detroit Symphony Orchestra, "2013 Annual Meeting: DSO Achieves First Balanced Budget Since 2007," December 12, 2013, blg.dso.org/category/breaking-news/ (accessed January 3, 2014).
27. DanielB arenboim, *A Life in Music* (New York: Scribner's, 1991), 171.

Name Index

Organization Index

Subject Index

Printed by Printforce, the Netherlands